世界建筑大师名作图析

（原著第四版）

［美］ 罗杰·H·克拉克
迈克尔·波斯 著

卢健松　包志禹　译

中国建筑工业出版社

著作权合同登记图字：01-2013-7608号

图书在版编目（CIP）数据

世界建筑大师名作图析（原著第四版）/（美）克拉克，波斯著；
卢健松，包志禹译.—北京：中国建筑工业出版社，2015.6（2022.10重印）
ISBN 978-7-112-17746-2

Ⅰ.①世…　Ⅱ.①克…②波…③卢…④包…　Ⅲ.①建筑设计-
世界-图集　Ⅳ.①TU206

中国版本图书馆CIP数据核字（2015）第027137号

Precedents in Architecture: Analytic Diagrams, Formative Ideas, and
Partis, Fourth Edition/Roger H. Clark and Michael Pause, IBAN 13
978-0470946749/0470946741

责任编辑：董苏华　戚琳琳
责任校对：陈晶晶　姜小莲

本书专注于一种思考建筑的方法，强调不同大师作品中的共同属性，探寻建筑形态生成的原型理论。本书以特有的图解方式，重新解析了勒·柯布西耶、密斯、阿尔托、迈耶等38余位世界著名建筑大师的118余个著名建筑实例。本书通过对这些建筑总平面图以及平、立、剖面图的分析与对比，发掘出设计构思中的共性，揭示建筑的形体构思是如何超越文化和时代界限的。

本书所选的建筑师，既包括历史上重要的建筑师，也包括当代极富原创性、设计独到、设计兴趣和建筑品质卓尔不凡的建筑师。本书所选建筑实例均具有重要的历史意义，代表了一个时代、一类功能或一种风格；设计富有创意，功力深厚，品位独到。

本书是研究设计范例的经典之作，是解剖建筑方案的必备工具，提供了理解建筑的一种途径、一种对建筑学进行分析思考的方法，有助于建立一种形式语汇，使读者在对建筑历史演进过程的分析中，找到有价值的指引，可以帮助建筑系的学生们和建筑师们去理解设计大师的作品，同时创造自己的设计。

第四版新增了7位前卫建筑大师的分析：史蒂芬·比尔、戴维·奇普菲尔德、伊东丰雄、汤姆·库迪格、布赖恩·迈克-里昂、托马斯·菲弗、阿尔瓦罗·西扎；对原第三版的建筑大师（如矶崎新、黑川纪章、麦金，米德和怀特、皮亚诺和罗杰斯、科尔托纳、SOM等）的分析进行了增补，以使本书更加与时俱进。

本书内容丰富，资料翔实，系统性强，是讲解世界建筑设计的经典著作，是建筑师及建筑学专业学生具有极好参考借鉴价值的教学参考书。

世界建筑大师名作图析（原著第四版）

［美］　罗杰·H·克拉克　　著
　　　　迈克尔·波斯

　　　　卢健松　包志禹　译

*

中国建筑工业出版社出版、发行（北京海淀三里河路9号）
各地新华书店、建筑书店经销
北京嘉泰利德公司制版
北京云浩印刷有限责任公司印刷

*

开本：880×1230毫米　横1/16　印张：24　字数：684千字
2016年6月第一版　2022年10月第七次印刷
定价：**98.00**元
ISBN 978-7-112-17746-2
（31347）

To Judy and Kathy

献给　朱迪和凯西

目　录

CONTENTS

前　　言

第四版前言

本书的第一、二、三版评论仍具重要意义，谨向读者推荐。分析方法及概念形成方法，始终是解析和创作建筑作品的语汇表。这种方法超越时间及起源，联结历代建筑作品。因此，它将是超越建筑风格、文化、形式的一次契机。它使我们意识到，建筑不仅是一张美丽的照片或是一组精美的图集那么简单。

与先前的版本相比，我们增加了本书的"分析"部分，以期通过对这些建筑的分析，呈现这些建筑的真实信息和图样。新增信息包括了7位建筑大师，每人2件作品。通过先前已经应用的技术和形式，这些新的作品将无缝穿插在本书的"分析"部分。书中的分析图剔除了建筑的平面图、立面图以及剖面图中的某些信息，既是我们的解析，也是一种抽象。这种抽象的目的在于，突显那些被研究过的特定主题。而通过把工程信息放在与分析图相邻的前一页，是希望能有助于读者将分析与工程信息关联起来。将所有的分析图汇聚在一页上的做法，则有助于为读者提供一个集中理解该建筑信息的机会。当然，读者也可以通过逐页的阅读，比较各张分析图，发现不同的建筑师如何解决特定问题。除此，读者还可以涉猎书中构思形成部分，

阅读不同建筑的图样集锦，研究不同建筑师如何处理同一设计原型。

我们意识到，任何建筑师所创造出的建筑形式，都是社会、技术、经济、文化、法律以及政治因素的综合考虑，其中，最重要的是方案的标新立异以及业主的利益与需求。众所周知，我们在这版中增加的建筑师都在某个领域举足轻重，比如布赖恩·迈克－里昂（Brian MacKay-Lyons）。他的作品汲取了当地建造技艺的优点，同时回应了建造场地特殊的地理与气候特征。甚至有人称他为"场所诗人"。然而，场所的重要性并未改变他在其他作品形式中体现出的明显的旨趣与特征，那些自始至终都会呈现在他作品中的几何形态、比例关系、空间运用以及平面与剖面关系。

汤姆·库迪格（Tom kundig）多次公开表示，他灵感的来源总是"大景观"，很明显，他的建筑作品中，景观处理无处不在。他还撰文坦陈，早年与雕塑家的互动对他影响至深，至今犹存。这些影响表现在建筑师对建筑材料的精巧使用上；表现在他在作品中大量使用被称为"小宇宙"的灵巧装置；表现在他可以在合适的时间、合适的地点使用合适的工匠。然而，这些也显示了，在其大尺度景观营造与细部语言迷恋之间，他

仍然执迷于某些建筑原型的思想。

如果称布赖恩·迈克－里昂（Brian Mackay–Lyons）为场所诗人，那么托马斯·菲弗（Thomas Phifer）一定就是场馆诗人。采用比 20 世纪现代主义更为通俗的现代建筑语言，菲弗（Phifer）创造出精确的极简主义雕塑，这些作品有时是实体的，但通常是透明的。这些场馆均为形态几何，质感细腻，视线通透，可通过一系列格栅和面板改变光线品质，同时保持自身的形态。这些场馆的存在转瞬即逝，通常位于如建筑一般可控的景观之中，在不断改变气候条件的过程中，实现内外空间的持续转换。

在史蒂芬·比尔（Stephane Beel）的两所住宅中，显然，他试图通过入口来处理该住宅与基地的关系。梅森住宅（Villa Maesen）是一栋线性建筑，接近 200 英尺长，位于一座墙面连续突起大宅的花园之中。事实上，梅森住宅成为了与既有的长墙平行，与现有房屋等宽的，一座可居住的墙壁。鉴于该住宅与墙面基本等高，因此需要穿过它的消极空间方能进入住宅。P 住宅（Villa P）同样是线性设计方案，但在这个案例中线条被折叠成四条边而成为一个院子。入口通过折线上的缝隙，经由一座小桥到达庭园对面的入口。虽然该住宅的形式看起来像是位于没有高差的基地上，但穿越的小桥揭示了它实际位于一个坡地上。比尔因此称这个院子为"没有楼板的中庭"。

从 2009 年 10 月到 2010 年 1 月末，戴维·奇普菲尔德（David Chipperfield）将他的作品集中在伦敦设计博物馆展出，名为"形式思想"，意在区分"形状"与"形式"。在他的理解里，"形状"是有机的，无非是结果的呈现。然而"形式"意味着可以被构建的规律。因此，毋庸置疑，我们设计的是形式。而奇普菲尔德本人是一位保守的"形式生成者"。他所创造出的建筑，既不屑于展现建筑师的聪明，也不为表达而表达。他静谧的建筑作品非常特别。

本书第四版之所以收录这些建筑作品，是因为我们相信它们足够有力，能让我们对议题的分析更加深入。无论建筑如何复杂，无论建筑师关注多少问题，抑或他们的动机与兴趣如何，建筑师们所创造出的，总是可以被分析的形式。综上所述，这些建筑形式将会比建筑师的兴趣、时尚与思考更加源远流长。我们清楚，建筑师并未意识到我们所分析的内容，但这些分析图能如我们所分析的那样，展示出建筑的形式法则。因此，我们可以用对所建成之"形式"予以图解，尽管我们明白这不一定是它被建造的原因。通过分析，我们可以创造出与建筑相关的故事，当然并不是所有的故事都存在。

如第三版的前言所述，《世界建筑大师名作图析》（Precedents in Architecture）一书在 1985 年由范·诺斯特兰德·赖因霍尔德（Van Nostrand Reinhold）公司首度出版，随后由约翰·威利出版公司接手。该书至少有 4 种语言的译版，在 2006 年，中国建筑工业出版社将其第三版翻译为中文出版。本书的成功与长久屹立，可见读者对本书所包含信息的需求，同时也强化了本书的初衷，探索设计思想的需求必将超越时间，也将超越建筑形式生成所依托的场所。

除了之前版本的前言所表述的感谢外，我们同样希望再次感谢所有和此版书有关的人。我们非常感激布赖恩·迈克－里昂以及托马斯·菲弗慷慨地提供其两栋住宅尚未出版的宝贵信息，允许我们将其收录在本版之中。在布赖恩·迈克－里昂的工作室里，丽萨·莫里森（Lisa Morrison）和沙瓦·诺斯特卡斯卡（Sawa Rostkowska）给予了我们特别的帮助。在菲弗的办公室里，史蒂芬·瓦拉迪（Stephen Varady）同样对我们提供了帮助。

本书的出版当然离不开约翰·威利出版公司的玛格丽特·卡敏斯（Margaret Cummins）的支持。与第三版书一样，她和我们接洽第四版书的工作。如前所述，她的说服、建议和鼓励对此版书的发展有着关键的作用。我们由衷地感谢她对本书的出版以及我们的权益所付出的关怀。我们同样感谢约翰·威利出版公司其他的编辑、美工和制作组对我们的帮助。我们不可能无一遗漏地感谢所有对本书有所贡献或者鼓励我们继续进行案例分析的个人。就像多年前刚开始时，美国建筑学会荣誉会员乔治·E·哈特曼（George E. Hartman, Jr., FAIA Emeritus）那样。

同之前的版本一样，本书内所有页面均取自原图。我们以手工制图方式在描图纸上绘制分析图，因此我们将对其负责。与第三版相同，杰森·米勒（Jason Miller）将我们的草图和分析图精确地转变为本版中的 28 个新页面。在此，我们将特别感谢他在制图时的精准、奉献、勤勉、耐心和技巧，同时也感谢他与我们相处时的幽默。

最后，多年以来，我们的学生和其他学校的学生，都以本

书所提供的经典建筑作品为研习设计的重要工具。他们的提问和发现，不断地挑战我们，也使得日常教学充满乐趣。

第三版前言

我们建议读者阅读本书第一版和第二版的前言，其中大部分内容对我们，以及就我们对本书的感觉而言，依然中肯。本书提供的理解建筑学的方法仍然有用，第三版让我们有机会增加 8 位建筑师每人两件作品的实例分析，来充实"分析"部分。

与前两版一样，我们继续挑出一系列的图解来剖析建筑原型的构思，意在挖掘设计构思的共性以供比较。当然我们也意识到，处理这些问题时，他们也许不会认同这些图解。所以，这些图解是我们自己的解释，其中一些比别人的解释更加详尽。显然，这些图解是从我们所关注并限定的主题中提取出来的。就一个特定的建筑师或建筑而言，单张的图解可能更加清晰或更富有启迪，它也许暗示了这个建筑师当时的关注点。通过对建筑某些同类问题的探讨，也许能看见建筑师与建筑之间不断互动的微妙关系。我们也知道建筑有多样的表达——社会的、技术的、经济的、文化的、法律的以及政治的，其一或是全部，就如同建筑师或业主的个人喜好或一时冲动那样，会决定建筑的最终形式。

第三版增收的建筑师中，例如西古德·莱韦伦茨，我们知道他并不著述也不教学。幸好近年里，有几家出版社整理出版了他的生平和作品。我们发现有意思的是，他的早期作品是用一种精致的，却是原创的、古典的语言（例如复兴小教堂）；

后期作品，例如本书中位于瑞曲克利潘的圣彼得教堂，拒绝了这种语言。然而，正如分析图解所揭示的那样，早期作品和晚期作品之间却又有几分相似。他的作品证明的是一种有节制的和收敛的想象力，这种想象力带来了并不拘谨的神秘的建筑。

斯蒂文·霍尔似乎从生物学和地质学借用了一些概念，创造出雕塑般的流动空间。他的建筑面向文脉，而他显然对在建筑室内引入自然光和控制自然光感兴趣。如何用草图和水彩来把握他对建筑的感觉，非常重要，所述甚详，然而他早期对于几何关系的兴趣依然体现在他最近的作品里。

这一版收录的拉斐尔·莫内奥作品，显示出他对场地的充分利用，其结果是紧凑的建筑与基本占满的场地。通过这种紧凑，莫内奥在作出自治的和生动的内部空间的同时，回应了场地文脉。另一方面，赫尔佐格和皮埃尔·德梅隆把建筑的外表面，即把表皮置于作品的优先地位。也许他们想要创造一种视觉的和有质感的表面，以此产生建成形式中已失去的感觉。

抛开各自的兴趣点和切入点，这些建筑师的共同思路是，建造出建筑学范畴内物质和空间所构成的形式。建筑不是没有形式的。最终的建成形式也许比时下关注的热点和思考更加持久。本书剖析的话题也许不在这些思考之列。我们的图解提供了理解建筑的一种途径，其中的一些实例也许有助于建立一种形式语汇。书中解析的实例也许是把一种思路整理或组织起来的手段，也许是产生一款设计的途径。在任何一个实例中，我们可以图解已经存在的形式，但不能解析它们何以存在。

第三版的工作和前两版的形式一样。新增的版面按照字母顺序融入在"分析"部分。这个部分现在有 31 位建筑师。他们一起代表了历史上的重要建筑师，以及近年来创作出有意义的建筑作品的建筑师。他们的入选不仅是因为作品的质量和实力，而且因为通过比较，他们对得起建筑探索的机会和他们的构思。

20 世纪 70 年代，我们开始探索分析建筑的名作，第一次发表在北卡罗来纳州立大学设计学校（现学院）的学生出版物上。1987 年本书面世，书名为《名作分析》（Analysis of Precedent）。1985 年，Van Nostrand Reinhole 出版了本书的第一版，1996 年出版了第二版，分别重印多次，并被译成西班牙文和日文。我们还知道它们竟被出人意料地译成了中文和朝鲜文。第二版获得了美国建筑师学会的国际建筑学书籍奖。该奖的评审委员，在世界各地出版的书籍基础上，评价道，"本书提供了一种建筑学的分析语汇，有助于建筑师理解他人的作品并产生原创理念。无论是初学者还是职业建筑师，该书都能充实读者的设计语汇。"

本书的成功和持久生命力表明了有一种对建筑学知识的需求。当我们开始着手收集第三版的素材之际，我们敏锐地意识到这项研究的最初前提是设计理念的共性和意义超越时空。随着工作的进展，这些假设得到了支持。建筑构思是建筑学的基础，扎根在社会的、技术的、经济的、文化的、法律的以及政治的种种之上。

除了第一版和第二版序言中提到的所有应当致谢的人以外，我们希望感谢与第三版直接相关的人；当然，向所有对本书产生影响或作出贡献的人——致谢是殊为困难的。我们还要感谢那些他们自己也不一定知道是否影响过我们的人们，但其中有几位值得特别提及。约翰威立出版公司的玛格丽特·卡敏斯，她和我们进行了第三版的接洽，并从约翰威立公司为我们的工作争取到了赞助；她的说服、建议和鼓励极其关键。威立出版公司的其他编辑、美编和制作团队也让我们受益。彼得·Q·博林、詹姆士·L·内格尔以及维克托·赖格纳则以鼓励和推荐的方式支持我们。我们也要感谢设计学院全体员工的真诚援助。

本书所有的图从原作而来。同时，我们对这些图的内容负责，詹森·米勒以其勤奋、耐心和杰出技巧诠释了我们的草图，并作出了新增的 32 页内容，特此致谢。

最后，和以前一样，我们要向我们的学生致谢。他们充实、挑战，且不断地质疑，同时证明了分析过程作为设计的一种工具是有价值的。他们让每天都妙不可言。

第二版前言

初版的成功表明了对有关建筑学的概念生成和分析方法的需求。我们初版以后的十年里的经验说明，这种素材，作为教授建筑学的一种工具颇为实用。它提供了一种分析语汇，它有助于建筑师和建筑学子理解别人的作品，并创造自己的设计。这种方法至今仍然有用，看来似乎无需修改第一版的内容。然而，再版还是给了我们在"分析"部分增加 7 位建筑师的作品，

进一步丰富本书内涵的机会。起初，这些作品的选定，是为了与先前 16 位建筑师的内容争鸣。增选的案例，一些具有重要的历史价值；一些是文献中鲜见的；还有一些是因为第一版发行后建筑师及其作品的声名鹊起。所有选定的作品兼具力度、品质以及内涵。本书的目的则旨在进一步展示其超越时间和文化的设计构思。本书仍按照原有的版式，7 位新建筑师，每人选定 2–4 个建筑实例，记述工程信息，展开构思分析。

本书或许对特定的建筑师或作品分析详尽，但我们的主旨却并非要殚精竭虑地（利用照片、笔记、文案等）研究单栋建筑或单个建筑师。事实上，我们意图通过比较来深入探索设计构思中的共同之处。为了达此目的，我们仍运用初版的图解方法。尽管有些建筑师或建筑作家也曾运用图解方式解释或介绍过本书中的建筑，但本书中的图解方法是我们的原创。

除了在初版序言中提到的所有应当感谢的人以外，以下各位都为本书再版给予帮助。

格雷姆美术高级研修基金会又一次支持了我们的工作，对此我们深表感谢。VNR 出版公司为实现再版此书提供了赞助金。这两项资金赞助了我们的研究，且包括插图的制作。

虽然难以向所有对我们的计划作出贡献或产生影响的人——致谢，但这几位应予特别介绍。首先，感谢温迪·洛克纳，是她说服我们再版，她的支持和鼓励尤为关键。VNR 出版公司的编辑人员给予了我们积极与可贵的帮助。詹姆斯·L·内格尔、维克多·赖格纳和马克·西蒙则以鼓励、建议和推荐作品等方式支持我们的工作。彼得·博林和卡罗尔·鲁舍慷慨地

贡献了其中一些建筑师作品的珍贵资料。设计学院的工作人员自愿给予的帮助，也在此一并致谢。

还应特别推介的是玛拉·默多克，她独自一人，技能非凡，饱含热情与耐心，绘制了所有的新插图。

最后，还要向我们的所有学生表示感谢，是他们使我们看到研究范例是学习设计的极有价值的手段；也是他们，不断地挑战与激励我们。

v 第一版前言

本书研究建筑学。

其特色在于，它专注于一种思考建筑的方法，强调其本质上的共同属性，而非其差异之处。本书关注那绵延不绝的传统，它们是蕴于当下的过往。但这种关注并非全然或部分地对其照搬，而是希望通过自觉领悟彰显不同风格和主题的建筑案例，探寻建筑形态生成的原型理论。

建筑学研究所涉甚广，但本书仅关注建筑形态问题。多言无益，我们将不涉足建筑学中的社会、政治、经济或技术领域。设计构思主要囿于建筑学的形态与空间领域，它也将成为本书研讨的重点。

显然，合理的建筑构思，并不能作为设计手段，导出优秀的设计作品。可以想见，很多让人大跌眼镜的建筑也是精心构思的产物。在设计中，精于形式原型的应用，并不会削弱建筑自身以及其他问题的重要性。无论怎样，那些当下以及过往伟大建筑的共通之处，正是理解建筑基本法则，那些可被认知的形式法则的明证。

本书的分析和阐述都将关乎建筑形式，因而未必与建筑师的初衷以及他人的阐释相符。分析也将局限于可以图解的那些特质，未必面面俱到。

本研究有助于促进对建筑历史的理解；审视建筑设计中超越时间的基本相似点；识别出漫漫时光中设计问题的普遍解决之道；使分析成为设计的手段。发展一套以案例分析来研讨建筑构思的方法至关重要。与那种只关注名字和日期的研究方法相比，这种调研方法要付出更为艰辛的劳动才会有所收获。这种努力所获得的回报，是一份历经时间洗礼与验证的设计语汇。我们深信，建筑师将从其形态构思、概念组织以及建筑组成局部的综合理解中获益匪浅。

本书内容丰富，共包含64个实例的图解资料，案例由不同的建筑师创作，各个案例均配有详细分析，是一本设计形态生成与构思的作品集，不同建筑形象的汇编，技术分析的参考书。其中相当多的资讯，仅在本书中可得。

衷心感谢格雷姆美术高级研修基金会对本研究的支持。

本书的任何成就都是许多人和思想不期而遇所形成的火花，但其间有一位的贡献特别突出。数年之前，和小乔治·E·哈特曼的一系列长谈，使我们对建筑及其历史的构思与观念方得以逐渐成形。自那以后，他还不断地、热情地给予我们支持和鼓励。除此，詹姆斯·L·内格尔、路德维希·格拉泽、威廉·N·摩根和已故的威廉·考迪尔等人都慷慨地予以帮助，使我们获得格雷姆基金会的资助。罗杰·坎农、罗伯特·休曼和戴比·布法林在查找资料和信息方面提供了可贵的帮助。此外，我们还要感谢建筑学院的克劳德·E·麦金尼主任、威妮弗雷德·霍奇、学院的秘书与图书馆员们的支持与帮助。我们班的学生们，丰富、充实、挑战我们的思想，并鼓励我们将这些内容纳入书中，在此也一并衷心感谢。

还应特别感谢丽贝卡·H·门茨和迈克尔·A·涅米宁，他们以卓越的才能绘制了本书中的那些插图。没有他们的技能、耐心、勤勉和奉献精神，本书是难以成书的。

接下来，还要感谢家人，他们的牺牲、奉献和理解支持了我们的工作。

其他所有对本书给予鼓励或作出贡献的人们，也在此一并致谢。

我们希望能借此书中所展示的资料，拓宽对建筑学的理解；以实例阐明一种对学生、教师以及设计师都有用的学习方法；并展示一种确定建筑形态与空间形式的分析技巧。

罗杰·H·克拉克（Roger H.Clark）

迈克尔·波斯（Michael Pause）

2011 年 6 月

---

* 正文两边的页码为英文版原书的页码，便于检索。——编者注

PREFACES

PREFACE TO THE FIRST EDITION

This book is about architecture.

In particular, it focuses on a way of thinking about architecture that emphasizes what is in essence the same, rather than different. Our concern is for a continuous tradition that makes the past part of the present. We do not wish to aid the repetition or revival of style whether in whole or part. Rather, by a conscious sense of precedent that identifies patterns and themes, we hope to pursue archetypal ideas that might aid in the generation of architectural form.

While architecture embodies many realms, we concentrate on built form. Without apology, we make no attempt to discuss the social, political, economic, or technical aspects of architecture. The domain of design ideas lies within the formal and spatial realm of architecture, and thus it is this arena that is explored in this book.

Obviously, a sound architectural idea will not, as a tool for design, inevitably lead to a good design. One can imagine many undesirable buildings which might originate with formative ideas. To be sensitive to the potential of archetypal pattern in design does not lessen the importance of concern for other issues or for the building itself. However, one commonality shared by the great buildings of this era with those of the past is a demonstrated understanding of basic architectural ideas which are recognizable as formative patterns.

Our analysis and interpretations are of built form and, therefore, may not necessarily coincide with the architect's intentions or the interpretations of others. The analysis is not all-inclusive in that it is limited to characteristics which can be diagrammed.

The intentions of this study are to assist the understanding of architectural history, to examine basic similarities of architects' designs over time, to identify generic solutions to design problems which transcend time, and to develop analysis as a tool for design. Of importance is the development of a vehicle for the discussion of ideas through the use of example. The understanding of history derived from this kind of investigation can only be obtained by far greater labor than that involved in acquiring a knowledge of history that focuses on names and dates. The reward for this effort is a design vocabulary that has evolved and been tested over time. We believe designers benefit from a comprehensive understanding of formative ideas, organizational concepts, and partis.

As a resource, this book offers factual graphic information on 64 buildings, a detailed analysis of each of these buildings, a range of designs by individual architects, a compilation of formative ideas for design generation, a collection of architectural images, and a reference for a technique of analysis. Some of this information is not readily available in other sources.

We are indebted to the Graham Foundation for Advanced Studies in the Fine Arts for support to make this study possible.

Any effort of this nature is the fruit of many encounters with individuals and ideas, but one debt in particular stands out as significant. Through a series of conversations with George E. Hartman, Jr., several years ago, some of our thoughts and ideas about architecture and history were focused. Since that time, he has continuously and enthusiastically offered support and encouragement. James L. Nagel, Ludwig Glaser, William N. Morgan, and the late William

Caudill each generously sponsored our efforts to secure assistance from the Graham Foundation. Roger Cannon, Robert Humenn, and Debbie Buffalin provided valuable help in locating material and information. For their assistance and support we thank several persons in the School of Design: Dean Claude E. McKinney, Winifred Hodge, the secretaries, and the librarians. The students in our classes have enriched, stimulated, and challenged our ideas, and encouraged us to record them in this volume. We fully acknowledge our debt to them.

A special acknowledgment is reserved for Rebecca H. Mentz and Michael A. Nieminen, whose considerable talents were used to draw the sheets reproduced in this volume. Without their skill, patience, diligence, and dedication this volume would not have been possible.

Our gratitude is extended to our families who have aided our efforts through sacrifice, devotion, and understanding.

To all other persons who have encouraged or in some way contributed to this study we collectively give thanks.

By making available the information that is presented in this volume, we hope to expand the understanding of precedents in architecture; to illustrate an educational technique that is useful to students, educators, and practitioners; and to demonstrate an analytic technique that can have impact on architectural form and space decisions.

PREFACE TO THE SECOND EDITION

The success of the first edition indicated that there was a need for conceptual and analytic information about architecture. Our experience with the first edition over the past decade demonstrated that the material has been useful as a tool for teaching architecture. It has provided a vocabulary for analysis that helps students and architects understand the works of others and aids them in creating their own designs. This approach continues to be useful and there was no apparent need to revise the information. Instead, the second edition gave us the opportunity to enrich the content of the analysis section by adding the works of seven architects. They were chosen initially to augment the content of the original sixteen architects. Some were selected for historical significance, some for lack of widespread documentation of their work. Others were picked because of emerging reputations and the production of a meaningful body of work since the publication of the first edition. All were selected because of the strength, quality, and interest of their designs. It is our intent to continue to show that design ideas transcend culture and time. Keeping the same format, we have added factual and analytic information on two or four buildings by each of the seven new architects.

While some may find this book useful for information about a particular architect or building, it is not our primary purpose to present any one building or architect exhaustively (e.g., photographs, written descriptions, or contract documents). Rather, our intention is to continue to explore the commonality of design ideas through comparison. To achieve this we have used the diagrammatic technique that was developed in the original study. While some of the architects and architectural authors have used diagrams to explain or inform others about the buildings included in this volume, the diagrams in this book are our own creation.

In addition to the acknowledgments cited in the preface of the first edition the following have helped make this edition a reality. The Graham Foundation for Advanced Studies in the Fine Arts supported our work for a second time; for this we are grateful. Van Nostrand Reinhold also contributed grant money to make this edition possible. Both of these sources aided our research and allowed for the production of the drawings.

While difficult to acknowledge all individuals who have contributed to or influenced our ideas, certain people's efforts deserve recognition. We are indebted to Wendy Lochner for persuading us to attempt a second edition. Her support and encouragement were critical. The editorial staff at Van Nostrand Reinhold provided us with willing and valuable assistance. James L. Nagle, Victor Reigner, and Mark Simon supported our efforts through encouragement, suggestions, and recommendations. Peter Bohlin and Carole Rusche generously contributed valuable information on the works of some of the architects. Collectively, we thank the staff of the School of Design for their willing assistance.

Special recognition goes to Mara Murdoch who single-handedly, with great skill, dedication, and patience, drew all of the new pages.

Finally, we wish to acknowledge all of our students, who have shown us that the study of precedents is a valuable tool for learning to design, and who continue to challenge us.

PREFACE TO THE THIRD EDITION

We commend to the reader the Prefaces to the first and second editions of this volume. Much of what is included in those Prefaces remains pertinent to us and our feelings about this work. The approach to understanding architecture presented herein continues to be useful and this edition again gave us the opportunity to enrich the Analysis section by adding factual and analytic information on two buildings by each of eight architects.

As with the previous editions, we have chosen to continue to present the buildings as a series of analytical diagrams that examine archetypal ideas. Our intention is to continue to explore the commonality of design ideas for comparison. We, of course, are aware that the architects examined herein may not have embraced the subjects of the diagrams nor, if they did consider the issues, approached them in the same way we have interpreted them. Thus, the diagrams are our own interpretations and some are more interpretive than others. Obviously these diagrams are then abstractions that focus on an issue that we have identified. For a particular architect or building a single diagram may be clearer or more revealing, which might suggest the identification of an issue of interest to the architect involved. By examining the buildings through the same issues it is possible to see relationships and nuances of development between architects and their buildings. We also understand that architecture has many manifestations—social, technical, economical, cultural, legal, and political. Any or all of these areas can impact the final form of the building, as can an individual architect's or client's personal predilection or whim.

Of those architects, for instance, that have been added for this edition, we know of Sigurd Lewerentz's interest in not doing things the conventional way. He is perhaps not as well known as some of the other architects in this volume, probably because he did not write about his work and did not teach. Fortunately, some publications have appeared in recent years that have chronicled his life and his work. We found it interesting that while he began with a refined, yet original, Classical language (at the Chapel of the Resurrection, for instance), his later work, represented here by the St. John's Church in Klippan, rejected that language. Yet there are similarities between the earlier and later work, as revealed by the analytical diagrams. His work demonstrates a subdued and restrained imagination that resulted in uncompromising and mysterious buildings.

Steven Holl seems to borrow from concepts of biology and geology in making sculpturally fluid spaces. While his buildings gesture toward their context, he has an obvious interest in the introduction and manipulation of natural light for the interior spaces of his buildings. Much has been written about the importance of his sketches and watercolors in capturing the feelings he desires for a building, yet his early interest in geometries is still demonstrated in his recent buildings.

Rafael Moneo's work included in this edition shows his intense use of the site, resulting in a building that is compact and basically fills the site. Through this compactness, Moneo reacts to the urban context while providing an autonomous and animated inner world. Herzog and de Meuron, on the other hand, give obvious priority in their work to the skin, the surface, of their buildings. Perhaps their desire is to create a visual and tactile surface that will create the perception that the built form has disappeared.

The common thread is that each of these architects has, regardless of their interest or considerations, produced built forms that include the physical and spatial realms of architecture. Architecture is not formless. In the end the built form may outlast the current fascinations and considerations. The issues we examine here may not be part of those considerations. Our analytical diagrams afford a way to understand buildings. In some cases they may help build a formal vocabulary. The issues examined could be the means for ordering or organizing an idea, or they may possibly be a way to generate a design. In any case, we can diagram what has been done, but not necessarily why it has been done.

The work that has been used for this third edition is in the same format as the previous editions. The new pages have been seamlessly inserted into the Analysis section in alphabetical order. This section now includes the work of thirty-one architects. Collectively they represent architects of historic importance and those who have produced meaningful work recently. All were selected not only because of the quality and strength of their work, but also because they afford the opportunity to explore buildings, their organizations, and ordering ideas, through comparison.

We began exploring the analysis of architectural precedents in the 1970s and first published such work in a student publication of the School (now College) of Design at North Carolina State University. That volume, titled *Analysis of Precedent*, appeared in 1978. Van Nostrand Reinhold published the original edition of *Precedents in Architecture* in 1985 and the second edition followed in 1996. Both editions have been through several printings, and each has been translated into Spanish and Japanese. We are also aware that these editions have been translated on an ad-hoc basis into Korean and Chinese. The second edition received an International Architecture Book Award from the American Institute of Architects. The jury for this awards program, which included books from publishers worldwide, commented that "*Precedents in Architecture* provides a vocabulary for architectural analysis that helps architects understand the works of others and aids in creating original ideas. Whether a novice or professional, this work enriches the reader's design vocabulary."

The success and longevity of this work suggests there is a need for this information about architecture. As we started to produce the material for this third edition, we were keenly aware of the initial premise for the study—the commonality and significance of design ideas that transcend time and place. As the work progressed, these assumptions have been reinforced. Architectural ideas are the underpinnings of architecture upon which other concerns—social, technical, economical, cultural, legal, and political—are layered.

In addition to the acknowledgments cited in the prefaces to the first and second editions, we wish to recognize some people directly related to this edition. It is always difficult to thank adequately all of the individuals who have had an influence on this work or have contributed to its development. We are indebted to each of them whether they knew they had an influence or not. Certain people, however, deserve to be mentioned specifically. This edition would not have existed at all without the efforts of Margaret Cummins of John Wiley and Sons. She approached us about considering a third edition, and she made it all possible by securing for us a grant from John Wiley to support our work. Her powers of persuasion, suggestions, and encouragement were critical. The other members of the editorial, art, and production staff at Wiley were also helpful. Peter Q. Bohlin, James L. Nagle, and Victor Reignier encouraged us through suggestions and recommendations. We also thank the College of Design, its administration and staff, for their willing assistance.

As with previous editions all of the pages in this edition are from original drawings. While we are responsible for the content of the drawings, Jason Miller has with diligence, patience, and great skill interpreted our sketches to create these thirty-two new pages. We owe him a special thank you.

Finally, as we have done previously, we wish to thank our students, who reinforce, challenge, and question constantly while demonstrating that analytical processes are valuable as a tool for design. They make each day an interesting pleasure.

PREFACE TO THE FOURTH EDITION

Our commentary in the original, second, and third editions remains relevant and we commend the reader to them. The methodology of analysis and the formative ideas presented continue to be a useful means for providing a vocabulary for understanding the architectural work of others and for creating architecture. It provides a tool for connecting architectural works regardless of time or origin. Thus, it affords the opportunity to transcend style, culture, and type. It reminds us that there is more to architecture than a picture or a well composed photograph.

As with the previous editions we have added to the Analysis section of the book with the desire to present factual drawings and information about the buildings, along with our analysis of these buildings. The new information presents the work of seven architects with two buildings by each of them. This new work has been seamlessly inserted into the Analysis section using the techniques and format that were developed previously. The analytic diagrams are our interpretations and are thus abstractions that purposely eliminate some information found in the plans, elevations, and sections of the buildings. The desire through these abstractions is to highlight the particular issue being examined. By presenting the factual information on a page adjacent to the analytic diagrams our intention is to aid the reader in connecting the factual information with our interpretation. Placing all of the analytic diagrams on one page affords the reader the opportunity to accumulate information about the building. One can also read from page to page to compare any one analytic diagram to see how different architects addressed that particular issue. Alternatively, one can refer to the Formative Idea section of the book to see collections of diagrams of buildings by various architects about one archetypal idea.

We are aware that the built form from any architect is the result of multiple considerations – social, technical, economic, cultural, legal, and political – not the least of which are the programmatic peculiarities and the client's interests and concerns. Of the architects we have added to this edition we know, for instance, of the importance that the region has had on Brian MacKay-Lyons. His architecture takes advantage of local building skills while responding to the particular geography and climate of the site where he builds. Others have even referred to him as "the poet of place." However, the importance of place does not change his apparent interest and abilities in other issues of form like geometry, proportion, spatial manipulation, and the relationship between the plan and section that consistently appear in his buildings.

Tom Kundig has indicated on many occasions that his source of inspiration has always been "the large landscape" and clearly he makes gestures in his work to that landscape. He has also written about the seminal influence of a sculptor through interactions early in his life that still impact his thinking and his work. It seems that this early influence is manifest in his sophisticated use of materials and the importance of craftsmen in creating his custom-made mechanical devices or contraptions, most often referred to as "gizmos," that are found in his work. But it appears that between his interest in the language of details and the larger landscape he is also interested in other archetypal ideas.

If Brian MacKay-Lyons is the poet of place, Thomas Phifer could be considered the poet of the pavilion. Using a more universal language of twentieth-century modernism, Phifer creates precise minimalist sculptures that are sometimes solid, but more often transparent. These pavilions are geometrically derived and, when transparent, visually delicate with a series of layers of scrims and mesh panels that alter the light quality while maintaining views. The ephemeral quality of these pavilions, often setting within a landscape that is equally as controlled as the architecture, is constantly altered both internally and externally by the changing climatic conditions.

In the two houses by Stephane Beel it is obvious he reveals much about the house and its site through the process of entry. Villa Maesen is a linear building of close to two hundred feet in length located in a former kitchen garden of a nearby chateau that features a series of prominent walls. In essence, the villa becomes a new inhabited wall sited parallel to the longest existing wall and is located the same distance from that existing wall as is the width of the house. With the villa approximately the same height as the wall, one then enters into the house through a space that is the negative of the house. Villa P is also a linear scheme, but in this instance the line is bent to form four sides of a court. Entry is through a gap in the bent linear form, across a bridge, through the court, to the door on the opposite side of the court. While the form of the house might be expected on a flat site, in this case there is a sloped site that is revealed as one crosses the bridge. Beel refers to this court as a "floorless patio."

From October 2009 until the end of January 2010, David Chipperfield had a comprehensive exhibit of his work at the Design Museum in London titled "Form Matters" that distinguishes between shape and form. In his terms shape is organic, more the result of consequence; while form implies discipline and is something that could be constructed. Therefore, whatever its provocation, it is form that we design. Chipperfield himself is a conservative form maker. He is not interested in creating buildings that constantly tell us how clever the architect is, nor is he interested in expression for the sake of expression. His quiet architecture nonetheless is special.

We have chosen each of the architects to add to this edition of the book because we believe their work is strong and that their buildings add depth to the issues we have analyzed. However complex the architecture may be, or however many concerns the architect grappled with, or whatever their motives or interest may be, each of the architects has produced built forms that can be analyzed. As stated previously, these built forms may very well outlast the architect's interests and their current fascinations and considerations. We understand that the architect may not have considered what we have diagrammed, but the diagrams can describe the formal aspects of the building as we interpret them. So we can diagram what has been done – the form – while we also understand that may not be why it was done. Through the analysis we have created one story about the building that can be related, not all of the stories that are possible.

As indicated in the Preface to the Third Edition, *Precedents in Architecture* has been continuously published since 1985 first by Van Nostrand Reinhold and subsequently by John Wiley & Sons. It has been translated into at least four languages and in 2006 the China Architecture and Building Press published the Third Edition in Chinese. The success and longevity of this work suggests the desire for this information and reinforces the initial premise of the need for exposing the design ideas that transcend time and place and that underpin the making of architectural form.

In addition to the acknowledgments cited in the previous prefaces, we wish to recognize some individuals directly related to this edition. We are indebted to Brian MacKay-Lyons and to Thomas Phifer for each generously agreeing to provide us with information about two of their houses that had previously not been published so that we could include their work in this edition. At Brian Mackay-Lyons' office, Lisa Morrison and Sawa Rostkowska were especially helpful. At Phifer's office, Stephen Varady was similarly helpful.

This edition would not have existed at all without the efforts of Margaret Cummins of John Wiley & Sons. As with the third edition, she approached us about considering a fourth edition. As previously stated, her powers of persuasions, her suggestions, and her encouragement were each critical to the development of this edition. We are profoundly grateful that she cares about this book and is willing to act on its, and thus our, behalf. The other members of the editorial, art, and production staffs at Wiley were also helpful and deserve our thanks. It is impossible to adequately thank all the individuals who had an influence on this work, contributed to its development, or encouraged us to continue pursuing the analysis of precedent and this book in particular. Look what you started, George E. Hartman, Jr., FAIA Emeritus, those many years ago.

As with previous editions all of the pages in this edition are from original drawings. We have produced the analytic diagrams in freehand on tracing paper, thus we are responsible for their content. As was the case with the third edition, Jason Miller has interpreted our sketches and diagrams to precisely draw the twenty-eight new pages in this edition. We owe him a special thank you for his precision, dedication, diligence, patience, and great skill in producing the drawings; and for his sense of humor in dealing with us.

Finally, over many years now our students, as well as those from other schools, have demonstrated that the study of precedents as presented herein is a valuable tool for design. They have challenged us and made each day we teach interesting through their questioning and discovery.

Roger H. Clark
Michael Pause
June 2011

导　言

建筑历史与经典历史建筑的研究兴趣已然焕发并不断高涨，它源于澄清建筑历史与设计关联的必须。学究式的历史研究将我们的场所置于连续的、严格的学术历史认知之中，这使得我们的建筑知识仅局限在人名、日期与风格之中。只有在各个历史风格的图层之间或之外，做个大体的建筑分类与呈现，历史方能成为建筑设计的丰富源泉。

本研究旨在揭示无拘于时代的建筑构思方法，所用的技巧是仔细地审查与分析建筑物本身，以期总结出相应的建筑设计方法。

本书分两部分。第一部分着重以传统的总图、平面以及立面、经济指标等，分析书中的 118 个案例。第二部分图解和分析了建筑得以生发的形态模式原型和设计构思法则。可以看到，一些特定的模式超越了时代，与场地特性没有明显关联。

本书所选择的建筑均体现了一定时代、功能与风格的特征；建筑师则代表了不同的建筑创作方法。但受资料来源的局限，一些建筑与建筑师由于缺乏素材而未能入选。

实例与方案，我们优先前者。设计方案仅作为特定构思方法示范才会收录于本书第二部分。尽管本书的分析技巧同样适用于建筑组群，但本研究仅就建筑单体展开。

在不同领域，入选建筑的资讯并不一致。差异存在时，我们竭尽所能地核准资料。如果不能完全明辨，亦做合理推理。例如，罗伯特·文丘里（Robert Venturi）从未绘制塔克住宅（Tucker House）总图，因此，本书的总图是从其他的资料推断而来的。

除此，一些案例在文献中有多个名称。例如，安德烈亚·帕拉第奥（Andrea Palladio）的 La Rotonda（圆厅别墅）通常被称为 Villa Capra（卡普拉别墅），偶尔又会根据最初所有者的家族姓氏被称为 Villa Almerico（阿尔梅里科别墅）。当有多个名称的情况下，本书使用最常用的名称，且在索引中标明其他名称。

由于完成一栋建筑耗时良久，亦或不精确的历史记载，某些建筑的年代也众说纷纭。确定建筑的一个或一组准确日期很难。日期可以将作品置于编年的文脉之中，非常重要。当资料来源冲突时，本书采信较常用的年代。

毋庸置疑，建筑的复杂性使其难以归功于某位个人的名下。显然，无论何时建造的房屋，都是同仁、伙伴们共同的作品，是好些人的集体功劳。然而，为了简明起见，本研究所引建筑都归在共识的设计者名下。例如，查尔斯·穆尔（Charles

Moore）名字后并未列出那些或许参与了所有建筑的合作者。同样，罗马尔多·朱尔戈拉（Romaldo Giurgola）而非其合伙的事务所，被公认为设计人。

本书的分析部分，所有建筑单体平面、立面以及剖面均以同样比例绘制。但不同建筑的比例尺则根据建筑的尺度和表达的需要予以调整。总图与建筑平面的朝向统一，并标明朝北方向。

相同的图解方式贯穿始终，确保本书"分析"与"形体构思"两个部分相互关联。本书的图解是一种抽象图示，旨在传达建筑中的核心要素及其本质关联。因此，图解着眼于建筑的物质特征，确保它们之间可以不拘于风格、类型、功能与时代地进行比较。这些图解源自构成建筑的三维形式与空间。它们所考虑的信息要较一般的平面、立面、剖面为多。为简化建筑至其精髓，这些图解刻意简化，除却最重要的因素，其他均被忽略，突出剩余部分，使人印象深刻。

为了分析，有必要建立一个图示标准，使图解之间得以比较。一般而言，粗线用于强调；在"形态构思"部分，建筑的平面、立面抑或剖面都以细线绘制，而分析、对比的内容则以粗线、阴影表达。第 xxiv 页的图例，说明了这些分析图解所用的图示标准。

本书远非面面俱到，相反，所引案例仅用于图解构思的微差。很难找到一栋绝对纯粹的、单一形式主题的建筑。更常见的是大量模式相互层叠，使得建筑含义丰富，内容多解。本研究中，主导模式得以彰显，但其他模式也不容忽略。

INTRODUCTION

The renewed and growing interest in architectural history and historic architectural example has focused the need to clarify the link between history and design. History studied in the academic sense of seeing our place within a continuum, or in the strictly scholarly sense of knowing the past, can limit our knowledge as architects to little more than names, dates, and style recognition. Seeing between and beyond the layers of historical styles, within which architecture is generally categorized and presented, can make history a source of enrichment for architectural design.

The search, in this study, is for theory which transcends the moment and reveals an architectural idea. The technique for this search is the careful examination and analysis of buildings. The desired result is the development of theory to generate ideas with which to design architecture.

This volume is organized into two parts. The first concentrates on the analysis of 118 buildings which are presented in both conventional drawings—site plan, plan, and elevation—and diagrams. The second identifies and delineates formal archetypal patterns or formative ideas from which architecture might evolve. It can be observed that certain patterns persist through time, with no apparent relationship to place.

Buildings that represent a range of time, function, and style, and architects who exemplify seemingly different approaches to architecture were selected. This selection was tempered by availability of information; some architects and some buildings were not included because the material available did not permit thorough analysis.

Preference was given to built buildings in lieu of projects, which are included in the second part only when they represent pertinent examples of an idea. While the analytic technique utilized in this volume is applicable to groups of buildings, this study is limited to single works of architecture.

The information available for the selected buildings contained inconsistencies in some areas. When discrepancies did occur, every effort was made to verify the accuracy of the information. If it could not be totally verified, then reasonable assumptions were made. For example, a site plan was never drawn by Robert Venturi for the Tucker House; therefore, the site plan indicated in this volume is inferred from other information.

In some instances, particular buildings are cited in the literature by more than one name. For example, La Rotonda by Andrea Palladio is often referred to as Villa Capra. Less frequently it is called Villa Almerico, after the name of the family for whom it was originally built. In cases where such multiplicity occurs, buildings are identified in the body of this study by the most frequently used name and in the index by the several names used.

Opinion also differs about dates attributed to several buildings. Because of the length of time it takes to complete a building or because of the imprecision of recorded history, it is often difficult to establish an exact date or series of dates for a building. The significance of the date is simply to place the work in a chronological context. When conflict did occur between sources, the date that is ascribed most often is the one used.

Undoubtedly, the complexity of architecture often makes it difficult to attribute a building to a single person. It is clear that buildings, regardless of when executed, are the products of partnerships or collaborations and the result of inputs from several persons. However, for the sake of clarity, the buildings in this study are assigned to the person who is normally recognized as the designer. For instance, Charles Moore is listed rather than the several associations which might be included for each building. Similarly, Romaldo Giurgola is acknowledged instead of the firm in which he is a partner.

In the analysis part of the study, the plan, elevation, and section for any individual building are drawn at the same scale. However, the scale between any two buildings varies depending upon building size and presentation format. Site plans are oriented to correspond generally to the orientation of the floor plan, and north is indicated where known.

To communicate the analysis of the buildings and the formative ideas in this study, a diagram or a set of diagrams is utilized. The diagrams are drawings that, as abstractions, are intended to convey essential characteristics and relationships in a building. As such, the diagrams focus on specific physical attributes which allow for the comparison of that attribute between buildings independent of style, type, function, or time. The diagrams are developed from the three-dimensional form and space configurations of the building. They take into account more information than is normally apparent in a plan, an elevation, or a section. To reduce the building to its essentials, the diagrams have been intentionally simplified. This elimination of all but the most important considerations makes those that remain both dominant and memorable.

For the analysis, it was necessary to establish a graphic standard so that comparison could be made between the diagrams. In general, heavy lines are used in each diagram to accent a particular issue. In the formative idea part of the study, the plan, elevation, or section of the building is drawn lightly for orientation purposes, while the issue being analyzed and compared is indicated by heavy lines or shading. The following legend indicates the specific graphic standard used for the diagrams in the analysis section.

This study is not exhaustive; rather, examples are included to illustrate the nuances of the idea. It is rare to find a building configuration which embodies a single formal theme in absolute purity. More normal is a variety of patterns layered upon one another—the consequence of which is the potential for the richness that can evolve from multiple interpretations. In this study dominant patterns have been identified, but this is not to suggest that others do not exist.

LEGEND 图例

STRUCTURE 结构

WALLS 墙

COLUMNS 柱

MAJOR BEAMS OVERHEAD 主梁

PLAN TO SECTION 平面到剖面

RELATED CONFIGURATION 相关的形状

REMAINDER OF BUILDING 保留建筑

REPETITIVE TO UNIQUE 重复到独特

UNIQUE 独特的

REPETITIVE 重复的

REMAINDER OF BUILDING 保留建筑

SYMMETRY AND BALANCE 对称和均衡

OVERALL SYMMETRY 总体对称

LOCAL SYMMETRY 局部对称

OVERALL BALANCE 总体平衡

LOCAL BALANCE 局部平衡

REFERENCED COMPONENTS 参照体

POINT AND COUNTERPOINT 相对应之部位

NATURAL LIGHT 自然光

DIRECT 直接光

DIFFUSED 漫射光

INDIRECT 非直接光

INTERIOR SPACE 内部空间

CIRCULATION TO USE-SPACE 交通流线到使用空间

MAJOR CIRCULATION 主交通流线

SECONDARY CIRCULATION 次交通流线

USE-SPACES 使用空间

REMAINDER OF BUILDING 保留建筑

VERTICAL CIRCULATION 垂直交通

GEOMETRY 几何关系

SQUARE 正方形

1.4 RECTANGLE 1.4 长方形

1.6 RECTANGLE 1.6 长方形

DIMENSION OR UNIT 尺度或单位

ANGLE 角度

GRID LINES 方格网

RADIUS CENTER 半径圆心

ADDITIVE AND SUBTRACTIVE 加法和减法

ADDITIVE UNITS 加法单元

SUBTRACTION 减法

WHOLE 整体

SUBTRACTIVE UNIT 减法单元

FACTUAL SHEET

NORTH INDICATOR 指北针

ELEVATION 立面

SECTION 剖面

MASSING 体量关系

MAJOR MASSING 主要体量

SECONDARY MASSING 次要体量

UNIT TO WHOLE 单元到整体

UNITS 单元

REMAINDER OF BUILDING 保留建筑

HIERARCHY 等级关系

MOST DOMINANT 主导的

TO 至

LESS DOMINANT 次要的

分　析　目　录

ANALYSIS

分　析

本章共选编 38 位建筑师，118 个作品。通常每位建筑师选择 4 个代表作品。研究素材按建筑师姓名英文字母排序，建筑师个人作品则按年代先后排序。

每栋建筑占用一个对页，左侧页面标明建筑的名称、时间以及总图、平面、立面与剖面；右侧页面是 11 张一组的图解与图示，简明扼要地解析了该栋建筑。图示是建筑的主导性构思，彰显了该栋建筑的主要特征。它是建筑精髓极度精简的提炼，有它，建筑得以生发；无它，建筑不复存在。

图解旨在洞察每个作品的形式与空间特质；借此，建筑图示方可理解。为此，在众多特性中，挑选出 11 个条目，它们是所有建筑的基本要素，且在特性与形式法则上存在关联。首先，各个条目被逐一分析，然后做关联研究。研究的信息用于甄别需要强化的条件，进而确认潜在的主导要素。从各栋建筑的图解及所形成的图示中，建筑设计的异同一目了然。

本章所选的条目分别为：结构、自然光、体量，以及平面与剖面，流线与使用空间、单元到整体、重复与独特，甚至还包括对称与均衡、几何性、加减法、等级关系。

结构

抽象来讲，结构意味着支撑，因此它存在于所有建筑之中；具体而言，结构是梁、柱、板、墙及其组合。究其所有，设计师均可以刻意应用，以强化和实现构思。这一语境下，梁、柱、板、墙可作为频度、模式、简约、规律、随意、复杂等术语进行构思。如此，结构可以用来界定空间，创造单元，塑造流线，引导行为，或进行组合调整。这样，结构与所有建筑创造性要素关联，是其品质与精彩之处不可或缺的因素。结构条目的图解，潜在强化了自然光、单元与整体、几何性等条目，还强化了流线与使用空间的关系，以及对称性、均衡性与等级性的定义。

自然光

自然光关注日光采用何种方式，在何处进入建筑。光线是渲染形态与空间的工具，光线的量、品质、色彩都将影响建筑的体量感。决策一栋建筑的立面、剖面之时，自然光的引入方式已然决定。经过过滤、屏障、反射，日光的品质明显不同。通过屏障的调节，从侧墙进入空间的光线，与直接进入室内的顶光不同。二者又与经过围护结构反射进入空间的光线大异其趣。尺寸、位置、形状、开口频率；材料质感、纹理与色彩；入射室内之前、之间、之后的调节等概念，都是光线的设计构思。自然光可以强化体量、几何性、等级感及单元到整体、重复与独特，以及流线与使用空间等条目。

体量

作为设计问题，体量组成了建筑三维形体中最常见，由知觉主导的命题。体量问题并非建筑的立面或轮廓问题。它是建筑作为整体的知觉形象。尽管体量是外形和立面的具体化，偶尔也与其类似、雷同，但绝非仅此而已。例如，建筑立面的开窗布局不会影响建筑的体量；相似的，建筑的轮廓不会对建筑形态知觉有多少作用。

虽视作设计的结果，但体量并非是仅关乎三维的构造。作为一种设计构思，体量与概念及文脉，单元模式荟集，单个与集合形体，主次元素等相关。体量具有界定外部空间、适应场地条件、界定入口空间、引导交通流线、彰显建筑重要性的潜质。作为图解条目，体量可以与强化单元与整体、重复与独特、平面与剖面、加减法、等级关系等条目相关。

4

平面到剖面及立面

平面、剖面和立面，是描绘建筑垂直与水平方向形态最普遍，最常规的方法。案例解析中所有构思，其平面构形到立面的生成均由其他条目所决定。平面是行为活动的有机组织，因此也被视为形式的生发器。它可以说明诸如何处是通道，何处可停留等很多问题。立面和剖面则与知觉的关联更紧密，因为这些是建筑更易直接感受的一面。然而，平面和剖面的应用基于一个体量可感知的假定，也就是说，平面和剖面中的每个线条都是有三维向度。平面与剖面之间的依存与交互关系，是设计决策与设计策略的工具。平面、剖面或者立面中的思考，将通过平衡、相似、比例以及相异与相对等概念，相互影响构形。

平面与剖面及立面可以在房间、局部或者整体等多个尺度层级上产生关联。作为案例分析条目，平面到剖面的关系强化了体量、均衡、几何性、等级关系，加法、减法以及单元到整体，重复与独特等的构思。

4

流线与使用空间

流线与使用空间代表了所有建筑中最重要的动态与静态组成。使用空间与功能相关，是建筑决策的首要重点；流线是设计意图得以实施的手段。动静相宜，二者共同构成了建筑的精髓。流线既然决定了人对建筑的体验，因此，它也可以是理解结构、自然光、单元界定、要素重复与独特、几何性、均衡、等级等建筑问题的工具。流线由特定的空间活动定义，或由潜在的空间使用暗示。因此，它既可以从使用空间中分离、穿越、

中止，也可以占据入口、中心、尽端等重要部位。

作为自由平面和开放空间的部分或全部，使用空间可以被暗示，甚至可以作为单独的房间分离。本条目图解中，它是主要功能空间相互关联所生成的一种模式。这些模式可以是集中的、线状的或簇状的。流线和使用空间的关联表明了空间的私密性与关联性。应用本条目作为设计方法，其基本问题是如何理解构形，如何通过流线来构形，抑或如何通过使用空间与其他要素之间的相互作用方式直接构形。

4

单元到整体

单元到整体，将建筑作为相互关联的单元予以考量，是一种建筑生成方法。一个单元可被视为建筑整体组成的实体部分。建筑可以就是一个单元，这样单元等同于整体；或者建筑是一组单元体。单元可以是空间或者形式，它与使用空间、结构组件、体量、体积，抑或这些元素的集合相关。单元也可以与此无关，独立存在。

本条目作为设计策略应用时，自然、可识别性、表达方法，以及单元与单元、单元与整体之间的关联均应考虑。在此意义上，单元体被视为连接体、分离体、叠合物，某种小于整体的东西。单元到整体的关联，可以由结构、体量、几何性等强化；它也可以支撑对称、均衡、集合、加法、减法、等级关系以及重复到独特等条目。

5

重复到独特

"重复到独特"的关联，将建筑空间与形态视为多个或单个实体，并着力研究其组成。如果将"独特"理解为与众不同，那么同一范畴中，元素的比较及其所显示的可识别性，造就了独特要素的差异性。通过共同的参照系，同一范畴中差异性使"重复"与"独特"相互关联。本质上，一个属性的内涵由其他内涵的范畴决定。因此，其组成独特还是重复，取决于其是否具备这种特性。尺度、朝向、位置、形状、布局、色彩、材

料以及纹理的概念，均可用来定性重复还是独特。尽管重复与独特可以在不同尺度上以多种方式发生，本研究将着墨于其主要关联。本研究中，本条目与结构、体量、单元与整体，平面与剖面，几何性，对称与均衡等条目相辅相成。

对称和均衡

建筑学产生之初，对称和均衡的概念已被运用。作为形态组成的基本条目，建筑学中的均衡是空间和形体的组合过程。均衡是感性或概念上的平衡；对称是均衡的特殊状态。形体组合的均衡和重量的均衡类似。这里，一定个数的"A"单元可以和一定数量的"B"单元等价。组分之间的平衡使二者之间存在确实关联，甚至可以画出其潜在的平衡线。为确保平衡存在，必须确立二者之间的关联属性，也就是说，建筑中的一些元素必须与另一些元素建立可感知的平衡。这种平衡由建筑可感知特性决定。概念上的平衡，仅当个人或团体赋予建筑元素特定的价值或额外的意义时才能发生。例如，小一点的圣坛可以与大一些的辅助空间或次级空间取得平衡。

均衡可以在不同属性上发生，但对称仅在平衡线两侧单元相同时方可发生。建筑学中，对称可以有三种形式，镜像对称、中心点对称、平移对称。

对称和均衡都可以存在于整栋建筑、局部组成、房间等不同层次上。尺度的变化造就了整体对称性与局部整体对称性的区别。

考虑所涉的尺寸、朝向、位置、连接、形状和价值等多方面问题，对称与均衡可作为塑形的方法使用。均衡和对称对所有其他分析条目都有很多关联。

几何关系

几何关系是建筑形态的构思方法，是决定建成形式平面及实体几何性的共同信条。本条目中，通过倍增、融合、细分、调控等手段对基本几何性予以重复，逐渐形成可辨析的网格。

在建筑历史演化的源起之时，几何关系已然成为一种设计工具。几何关系是建筑最为共通的基本特质与决定性因素。通过简单几何形、多形式语汇、比例体系，以及高难度几何性操作所实现的复杂形式生成手段，几何性得以在空间与形式领域广泛应用。借由可计量性与可度量性，几何关系成为建筑形式的生成方法。作为本图解的关注条目，几何关系着重于尺寸、位置、形状、形式，以及比例的概念；除此，还关注由融合、衍生、基本几何构形手段控制所引发的几何与形式语言持续变化。本图解中，网格将用于观察频度、形构、复杂、持续以及变化。作为建筑的普遍属性，几何关系与本图解部分所有条目密不可分。

加法和减法

作为构形方法的加法和减法，由增加、聚集、削减等建筑塑形手段发展而来。二者均依靠对建筑物的知觉认知。"加法"作为建筑形式生成的方法时，将局部视作为主体。在加法设计中，应用个人知觉，将建筑视为可识别单元或局部形成的集聚体。作为设计工具生成建筑时，"减法"将整体形态作为主体。在人们眼中，减法设计将建筑视为一个局部被拿走了，但整体可认知的形态。两种方法如果同时用于建筑形态构思，那将变化多端含义丰富。例如，可以将各个单元聚集为一个局部缺失，但整体完型的形态；也可以从可认知的整体形态中减去片段，然后在负形的位置稍作增添，以此生成建筑。

建筑整体如何连接，形态如何生成，对于本图解至关重要。这可以通过观察体形、体量、色彩以及材质的变化来实现。作为构思，加法和减法与体量、几何性、平衡、等级以及单元到整体的关联性、重复到独特、平面到剖面都相辅相成。

等级关系

作为形态构思的方法，建筑设计中的等级关系，是一个或多个属性差序关系的物质表现，在本研究中是指对一系列特征的具体评级。它基于一个评级体系，针对某个选定特性中，对某个属性的质量予以评价。等级关系暗示了从一种状态向另一种状态演化的有序变化，其范畴是指对主次、开合、简繁、公私、雅俗、主从以及个体与集体关系的应用。在此范畴之中，其差序变化可以在空间、形态中分别或同时发生。

本图解之中，等级关系在建成形式中举足轻重，可以通过审视模式、尺度、构形、几何性以及关联性予以确定。质量、丰富度、细节、有机性以及特殊材质上的应用，也彰显了其不同的重要性。作为设计构思方法，等级关系与本图解的其他各个条目均相辅相成。

ANALYSIS

In this section, 118 works of architecture are documented. The buildings are the designs of 38 architects. For most architects, four buildings are presented which are representative of that person's work. The material is ordered with the architects arranged alphabetically and the buildings for each architect presented chronologically and successively.

Each building is recorded on two adjacent pages; the left-hand page documents the building with name, date, and location as well as drawings of the site plan, floor plans, elevations, and sections; illustrated on the right-hand page is a series of eleven analysis diagrams and the parti diagram which culminates in and summarizes the analysis for the building. The parti is seen as the dominant idea of a building which embodies the salient characteristics of that building. It encapsulates the essential minimum of the design, without which the scheme would not exist, but from which the architecture can be generated.

A major concern of the analysis is to investigate the formal and spatial characteristics of each work in such a way that the building parti can be understood. To accomplish this, 11 issues were selected from the widest range of characteristics: fundamental elements which are common to all buildings, relationships among attributes, and formative ideas. Each issue is first explored in isolation and then in relationship to the other issues. This information is studied to discern reinforcement and to identify the dominant underlying idea. From the analysis and the resulting parti for each building, similarities and differences among the designs can be identified.

The issues selected for the analysis are: structure; natural light; massing; and the relationships of plan to section, circulation to use-space, unit to whole, and repetitive to unique. Also included are symmetry and balance, geometry, additive and subtractive, and hierarchy.

STRUCTURE

At a basic level, structure is synonymous with support, and therefore exists in all buildings. At a more germane level, structure is columnar, planar, or a combination of these, all of which a designer can intentionally use to reinforce or realize ideas. In this context, columns, walls, and beams can be thought of in terms of the concepts of frequency, pattern, simplicity, regularity, randomness, and complexity. As such, structure can be used to define space, create units, articulate circulation, suggest movement, or develop composition and modulations. In this way, it becomes inextricably linked to the very elements which create architecture, its quality and excitement. This analysis issue has the potential to reinforce the issues of natural light, unit to whole relationships, and geometry. It can also strengthen the relationship of circulation to use-space and the definition of symmetry, balance, and hierarchy.

NATURAL LIGHT

Natural light focuses on the manner in which, and the locations where, daylight enters a building. Light is a vehicle for the rendering of form and space, and the quantity, quality, and color of the light affect the perceptions of mass and volume. The introduction of natural light may be the consequence of design decisions made about the elevation and section of a building. Daylight can be considered in terms of qualitative differences which result from filtering, screening, and reflecting. Light which enters a space from the side, after modification by a screen, is different from light which enters directly overhead. Both examples are quite different from light which is reflected within the envelope of the building before entering the space. The concepts of size, location, shape, and frequency of opening; surface material, texture, and color; and modification before, during, or after entering the building envelope are all relevant to light as a design idea. Natural light can reinforce structure, geometry, hierarchy, and the relationships of unit to whole, repetitive to unique, and circulation to use-space.

4   MASSING

As a design issue, massing constitutes the perceptually dominant or most commonly encountered three-dimensional configuration of a building. Massing is more than the silhouette or elevation of a building. It is the perceptual image of the building as a totality. While massing may embody, approximate, or at times parallel either the outline or the elevation, it is too limiting to view it as only this. For example, on the elevation of a building the fenestration may in no way affect the perception of the volume of the building. Similarly, the silhouette may be too general and not reflect productive distinctions in form.

Massing, seen as a consequence of designing, can result from decisions made about issues other than the three-dimensional configuration. Viewed as a design idea, massing may be considered relative to concepts of context, collections and patterns of units, single and multiple masses, and primary and secondary elements. Massing has the potential to define and articulate exterior spaces, accommodate site, identify entrance, express circulation, and emphasize importance in architecture. As an issue in the analysis, massing can strength-

en the ideas of unit to whole, repetitive to unique, plan to section, geometry, additive and subtractive, and hierarchy.

PLAN TO SECTION OR ELEVATION   4

Plan, section, and elevation are conventions common to the simulation of the horizontal and vertical configurations of all buildings. As with any of the design ideas in this analysis, the relationship of plan configuration to vertical information may result from decisions made about other issues. The plan can be the device to organize activities and can, therefore, be viewed as the generator of form. It may serve to inform about many issues such as the distinction between passage and rest. The elevation and section are often considered to be more closely related to perception since these notations are similar to encountering a building frontally. However, the use of plan or section notations presumes volumetric understanding; that is, a line in either has a third dimension. The reciprocity and the dependence of one on the other can be a vehicle for making design decisions, and can be used as a strategy for design. Considerations in plan, section, or elevation can influence the configuration of the others through the concepts of equality, similarity, proportion, and difference or opposition.

It is possible for the plan to relate to the section or elevation at a number of scales: a room, a part, or the whole building. As an issue for analysis, the plan to section relationship reinforces the ideas of massing, balance, geometry, hierarchy, additive, subtractive, and the relationships of unit to whole and repetitive to unique.

CIRCULATION TO USE-SPACE   4

Fundamentally, circulation and use-space represent the significant dynamic and static components in all buildings.

Use-space is the primary focus of architectural decision making relative to function, and circulation is the means by which that design effort is engaged. Together, the articulation of the conditions of movement and stability form the essence of a building. Since circulation determines how a person experiences a building, it can be the vehicle for understanding issues like structure, natural light, unit definition, repetitive and unique elements, geometry, balance, and hierarchy. Circulation may be defined within a space that is for movement only, or implied within a use-space. Thus, it can be separate from, through, or terminate in the use-spaces; and it may establish locations of entry, center, terminus, and importance.

Use-space can be implied as part or all of a free or open plan. It can also be discrete, as in a room. Implicit in the analysis of this issue is the pattern created by the relationship between the major use-spaces. These patterns might suggest centralized, linear, or clustered organizations. The relationship of circulation and use-space can also indicate the conditions of privacy and connection. Basic to employing this issue as a design tool is the understanding that the configuration given to either circulation or use directly affects the manner in which the relationship to the other takes place.

UNIT TO WHOLE

The relationship of unit to whole examines architecture as units which can be related to create buildings. A unit is an identified entity which is part of a building. Buildings may comprise only one unit, where the unit is equal to the whole, or aggregations of units. Units may be spatial or formal entities which correspond to use-spaces, structural components, massing, volume, or collections of these elements. Units may also be created independently of these issues.

The nature, identity, expression, and relationship of units to other units and to the whole are relevant considerations in the use of this idea as a design strategy. In this context, units are considered as adjoining, separate, overlapping, or less than the whole. The relationship of unit to whole can be reinforced by structure, massing, and geometry. It can support the issues of symmetry, balance, geometry, additive, subtractive, hierarchy, and the relationship of repetitive to unique.

REPETITIVE TO UNIQUE

The relationship of repetitive to unique elements entails the exploration of spatial and formal components for attributes which render these components as multiple or singular entities. If unique is understood to be a difference within a class or a kind, then the comparison of elements within a class can result in the identification of the attributes which make the unique element different. This distinction links the realms of the repetitive and the unique through the common reference frame of the class or kind. Essentially, the definition of one is determined by the realm of the other. In this context, components are determined to be repetitive or unique through the absence or presence of attributes. Concepts of size, orientation, location, shape, configuration, color, material, and texture are useful in making distinctions between repetitive and unique. While repetitive and unique elements occur in numerous ways and at several scales within buildings, the analysis focuses on the dominant relationship. In the analysis, this issue generates information which strengthens or is reinforced by the concepts of structure, massing, units related to whole, plan related to section, geometry, and symmetry or balance.

SYMMETRY AND BALANCE

The concepts of symmetry and balance have been in use since the beginning of architecture. As a fundamental issue of composition, balance in architecture occurs through the use of spatial or formal components. Balance is the state of perceptual or conceptual equilibrium. Symmetry is a specialized form of balance. Compositional balance in terms of equilibrium implies a parallel to the balance of weights, where so many units of "A" are equal to a dissimilar number of units of "B." Balance of components establishes that a relationship between the two exists, and that an implied line of balance can be identified. For balance to exist, the basic nature of the relationship between two elements must be determined; that is, some element of a building must be equivalent in a knowable way to another part of the building. The equivalency is determined by the perception of identifiable attributes within the parts. Conceptual balance can occur when a component is given additional value or meaning by an individual or group. For example, a smaller sacred space can be balanced by a much larger support or secondary space.

Whereas balance is developed through differences in attributes, symmetry exists when the same unit occurs on both sides of the balance line. In architecture this can happen in three precise ways: reflected, rotated about a point, and translated or moved along a line.

Both symmetry and balance can exist at the building, component, or room level. As scales change, a distinction is made between overall and local symmetry or balance. Consideration of size, orientation, location, articulation, configuration, and value is involved in its use as a formative idea. Balance and symmetry may have an impact on all of the other analysis issues.

GEOMETRY

Geometry is a formative idea in architecture that embodies the tenets of both plane and solid geometry to determine built form. Within this issue, grids are identified as being developed from the repetition of the basic geometries through multiplication, combination, subdivision, and manipulation.

Geometry has been used as a design tool since the very beginnings of architectural history. Geometry is the single most common determinant or characteristic in buildings. It can be utilized on a broad range of spatial or formal levels that includes the use of simple geometric shapes, varied form languages, systems of proportions, and complex form generated by intricate manipulations of geometries. The realm of geometry as an architectural form generator is a relative one of measurement and quantification. As a focus for this analysis, it centers on the concepts of size, location, shape, form, and proportion. It also concentrates on the consistent changes in geometries and form languages that result from the combination, derivation, and manipulation of basic geometric configurations. In the analysis, grids are observed for frequency, configuration, complexity, consistency, and variation. As the pervasive attribute of buildings, geometry can reinforce all of the issues used in the analysis.

ADDITIVE AND SUBTRACTIVE

The formative ideas of additive and subtractive are developed from the processes of adding, or aggregating, and subtracting built form to create architecture. Both require the perceptual understanding of the building. Additive, when used to generate built form, renders the parts of the building as dominant. The perception of a person engaging an additive design is that the building is an aggregation of identifiable units or parts. Subtractive, when utilized in designing, results in a building in which the whole is dominant. A person viewing a subtractive scheme understands the building as a recognizable whole from which pieces have been subtracted. Generally, additive and subtractive are formal considerations which can have spatial consequences.

Richness can occur when both ideas are employed simultaneously to develop built form. For example, it is possible to add units together to form a whole from which pieces are subtracted. It is also possible to subtract pieces from an identifiable whole and then to add the subtracted parts back to create the building.

The manner in which the building whole was articulated, and the ways in which the forms were rendered, was important to the analysis. This was achieved by observing massing, volumes, color, and material changes. Additive and subtractive, as ideas, can strengthen or be reinforced by massing, geometry, balance, hierarchy, and the relationships of unit to whole, repetitive to unique, and plan to section.

HIERARCHY

As a formative idea, hierarchy in the design of buildings is the physical manifestation of the rank ordering of an attribute or attributes. Embodied in this concept is the assignment of relative value to a range of characteristics. This entails the understanding that qualitative differences within a progression can be identified for a selected attribute. Hierarchy implies a rank ordered change from one condition to another, where ranges such as major-minor, open-closed, simple-complex, public-private, sacred-profane, served-servant, and individual-group are utilized. With these ranges, the rank ordering can occur in the realm of the formal, spatial, or both.

In the analysis, hierarchy was explored relative to dominance and importance within the built form through examination of patterns, scale, configuration, geometry, and articulation. Quality, richness, detail, ornament, and special materials were used as indicators of importance. Hierarchy, as a design idea, can be related to and support any of the other issues explored in the analysis.

阿尔瓦·阿尔托
ALVAR AALTO

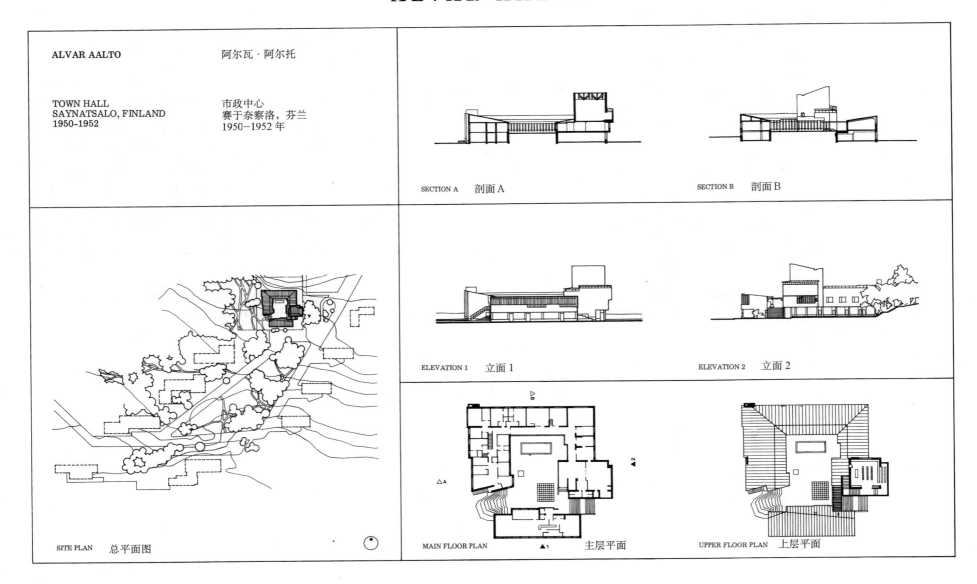

ALVAR AALTO   阿尔瓦·阿尔托

TOWN HALL   市政中心
SAYNATSALO, FINLAND   赛于奈察洛，芬兰
1950-1952   1950-1952 年

SECTION A 剖面 A   SECTION B 剖面 B

ELEVATION 1 立面 1   ELEVATION 2 立面 2

SITE PLAN 总平面图

MAIN FLOOR PLAN ▲1 主层平面   UPPER FLOOR PLAN 上层平面

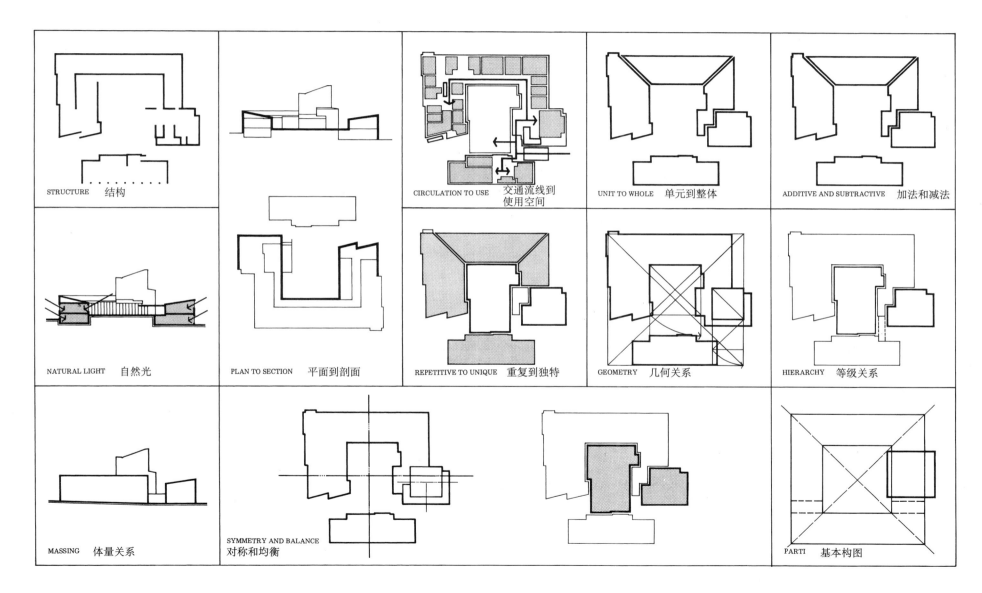

STRUCTURE　结构

CIRCULATION TO USE　交通流线到使用空间

UNIT TO WHOLE　单元到整体

ADDITIVE AND SUBTRACTIVE　加法和减法

NATURAL LIGHT　自然光

PLAN TO SECTION　平面到剖面

REPETITIVE TO UNIQUE　重复到独特

GEOMETRY　几何关系

HIERARCHY　等级关系

MASSING　体量关系

SYMMETRY AND BALANCE　对称和均衡

PARTI　基本构图

15

ALVAR AALTO

阿尔瓦·阿尔托

VOUKSENNISKA CHURCH
IMATRA, FINLAND
1956-1958

伏克塞涅斯卡教堂
伊马特拉，芬兰
1956-1958 年

SECTION A 剖面 A

SECTION B 剖面 B

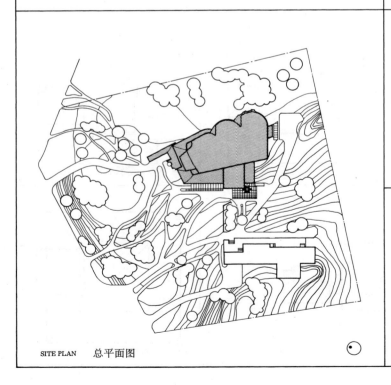

SITE PLAN 总平面图

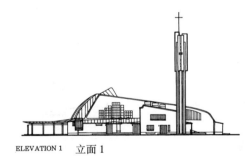

ELEVATION 1 立面 1

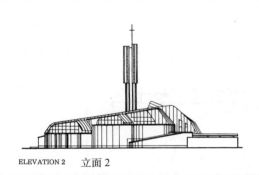

ELEVATION 2 立面 2

FLOOR PLAN 平面

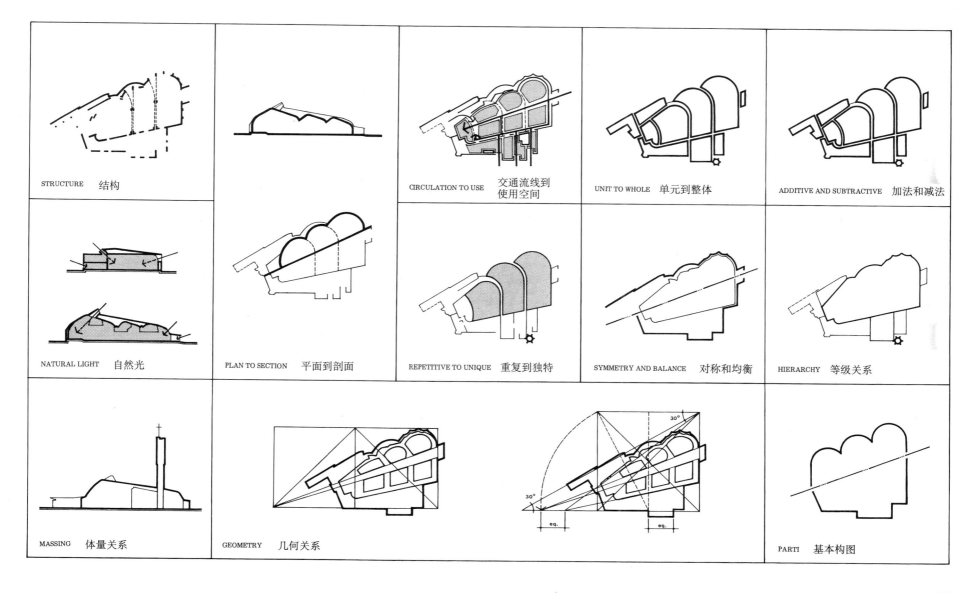

STRUCTURE　结构

CIRCULATION TO USE　交通流线到使用空间

UNIT TO WHOLE　单元到整体

ADDITIVE AND SUBTRACTIVE　加法和减法

NATURAL LIGHT　自然光

PLAN TO SECTION　平面到剖面

REPETITIVE TO UNIQUE　重复到独特

SYMMETRY AND BALANCE　对称和均衡

HIERARCHY　等级关系

MASSING　体量关系

GEOMETRY　几何关系

PARTI　基本构图

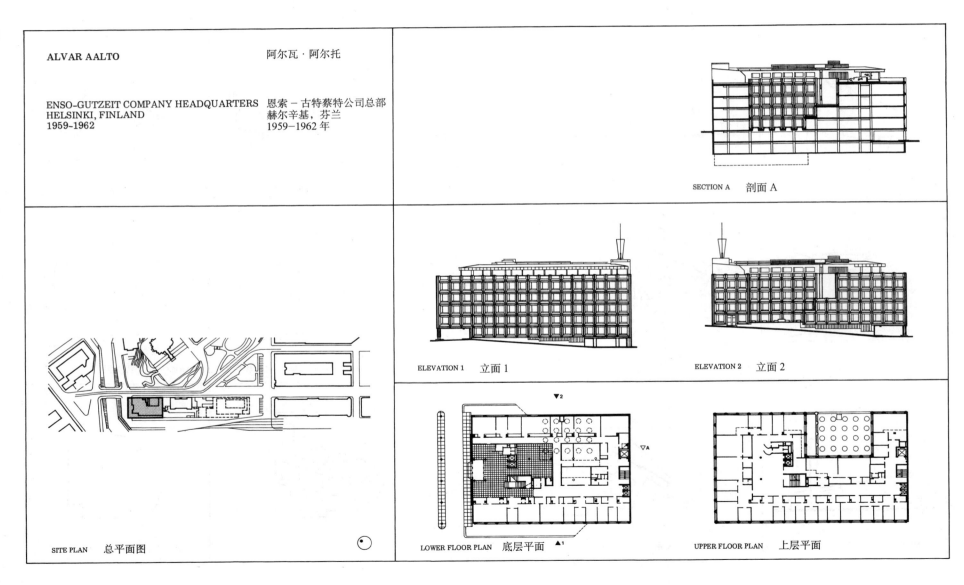

ALVAR AALTO   阿尔瓦·阿尔托

ENSO-GUTZEIT COMPANY HEADQUARTERS   恩索－古特蔡特公司总部
HELSINKI, FINLAND   赫尔辛基，芬兰
1959–1962   1959–1962 年

SECTION A   剖面 A

ELEVATION 1   立面 1

ELEVATION 2   立面 2

SITE PLAN   总平面图

LOWER FLOOR PLAN   底层平面 ▲1

UPPER FLOOR PLAN   上层平面

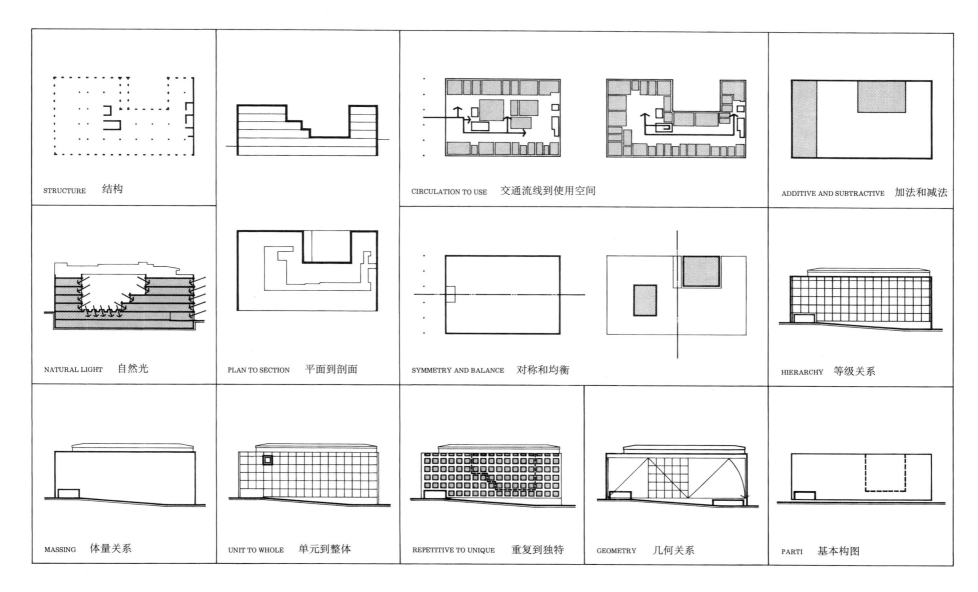

| STRUCTURE 结构 | | CIRCULATION TO USE 交通流线到使用空间 | ADDITIVE AND SUBTRACTIVE 加法和减法 |

NATURAL LIGHT　自然光　　PLAN TO SECTION　平面到剖面　　SYMMETRY AND BALANCE　对称和均衡　　HIERARCHY　等级关系

MASSING　体量关系　　UNIT TO WHOLE　单元到整体　　REPETITIVE TO UNIQUE　重复到独特　　GEOMETRY　几何关系　　PARTI　基本构图

ALVAR AALTO   阿尔瓦·阿尔托

CULTURAL CENTER   文化中心
WOLFSBURG, GERMANY   沃尔夫斯堡，德国
1958-1962   1958—1962 年

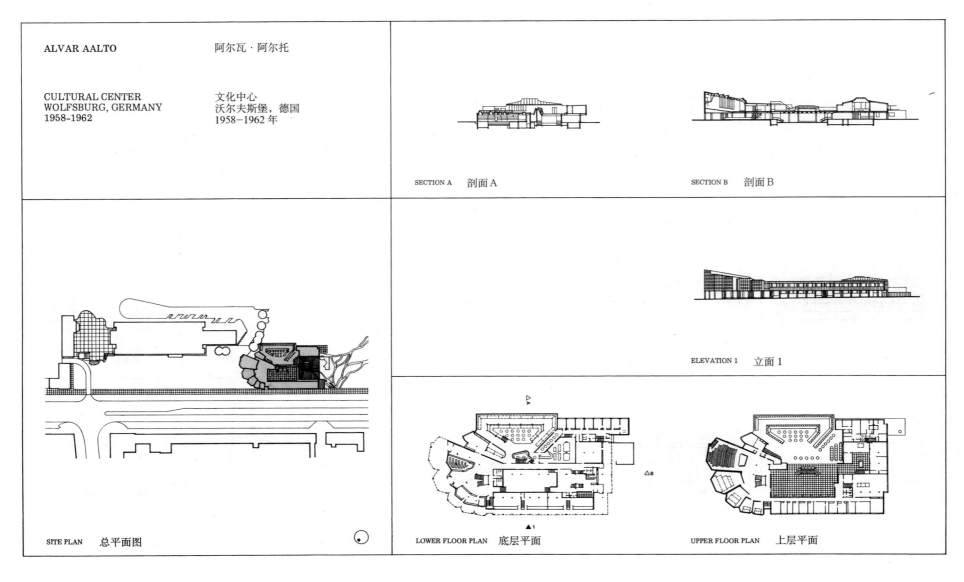

SECTION A   剖面 A

SECTION B   剖面 B

ELEVATION 1   立面 1

SITE PLAN   总平面图

LOWER FLOOR PLAN   底层平面

UPPER FLOOR PLAN   上层平面

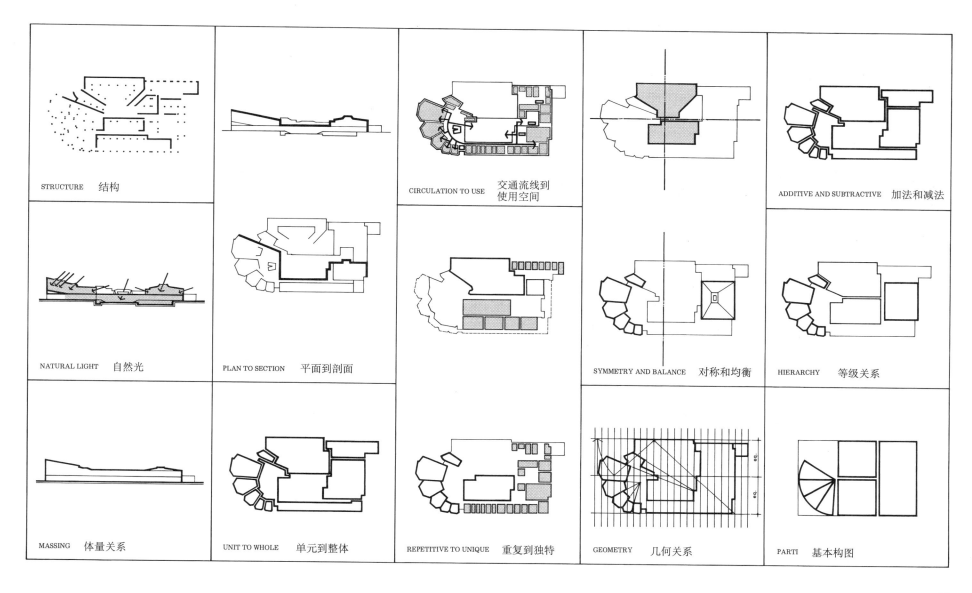

STRUCTURE　结构

CIRCULATION TO USE　交通流线到使用空间

ADDITIVE AND SUBTRACTIVE　加法和减法

NATURAL LIGHT　自然光

PLAN TO SECTION　平面到剖面

SYMMETRY AND BALANCE　对称和均衡

HIERARCHY　等级关系

MASSING　体量关系

UNIT TO WHOLE　单元到整体

REPETITIVE TO UNIQUE　重复到独特

GEOMETRY　几何关系

PARTI　基本构图

安 藤 忠 雄
TADAO ANDO

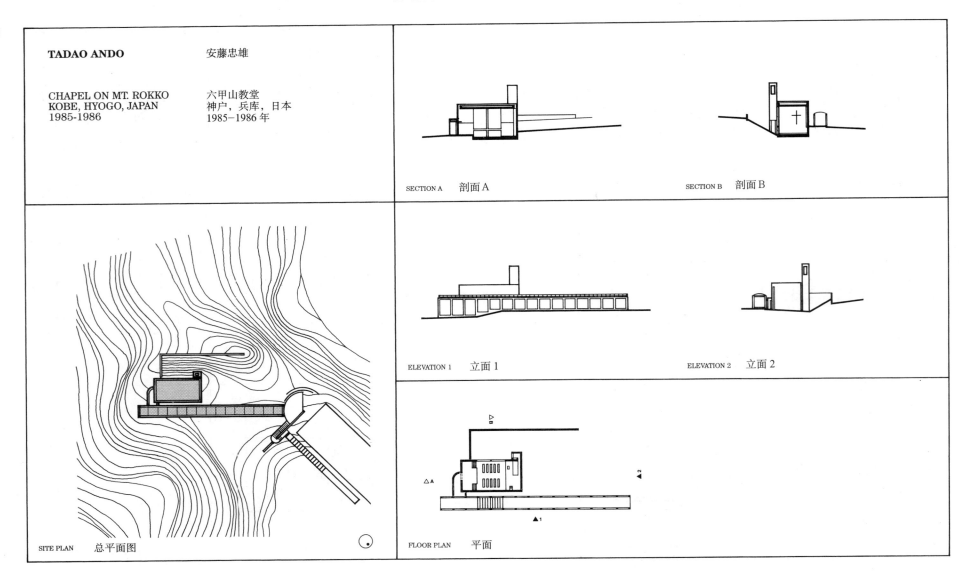

TADAO ANDO　　　　安藤忠雄

CHAPEL ON MT. ROKKO　六甲山教堂
KOBE, HYOGO, JAPAN　神户，兵库，日本
1985-1986　　　　　　1985－1986 年

SECTION A　剖面 A

SECTION B　剖面 B

ELEVATION 1　立面 1

ELEVATION 2　立面 2

SITE PLAN　总平面图

FLOOR PLAN　平面

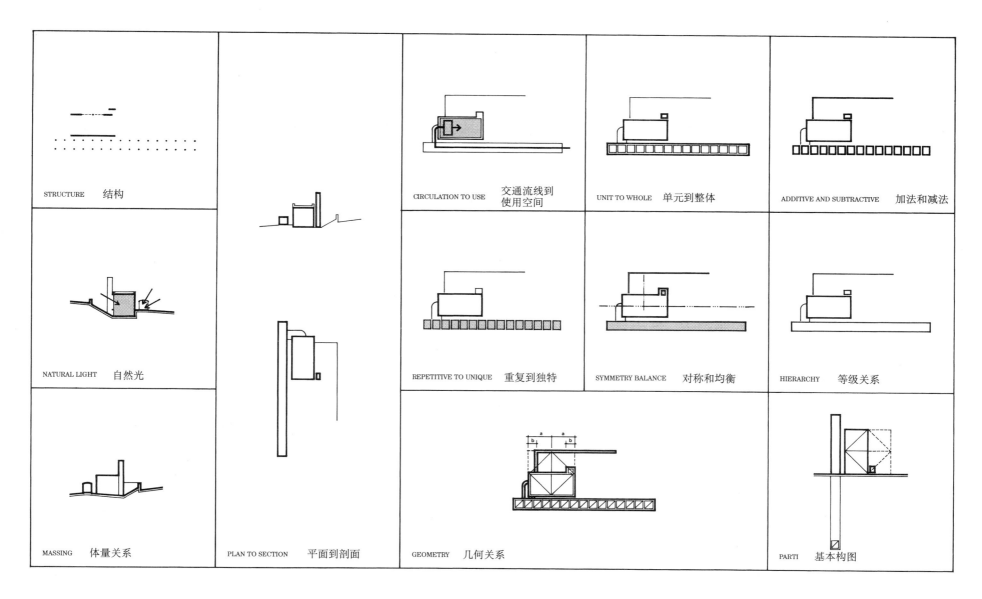

STRUCTURE 结构

NATURAL LIGHT 自然光

MASSING 体量关系

PLAN TO SECTION 平面到剖面

CIRCULATION TO USE 交通流线到使用空间

REPETITIVE TO UNIQUE 重复到独特

GEOMETRY 几何关系

UNIT TO WHOLE 单元到整体

SYMMETRY BALANCE 对称和均衡

ADDITIVE AND SUBTRACTIVE 加法和减法

HIERARCHY 等级关系

PARTI 基本构图

23

TADAO ANDO　　　　安藤忠雄

CHURCH ON THE WATER　　水上教堂
TOMAMU, HOKKAIDO, JAPAN　斗满，北海道，日本
1985-1988　　　　　　　1985-1988 年

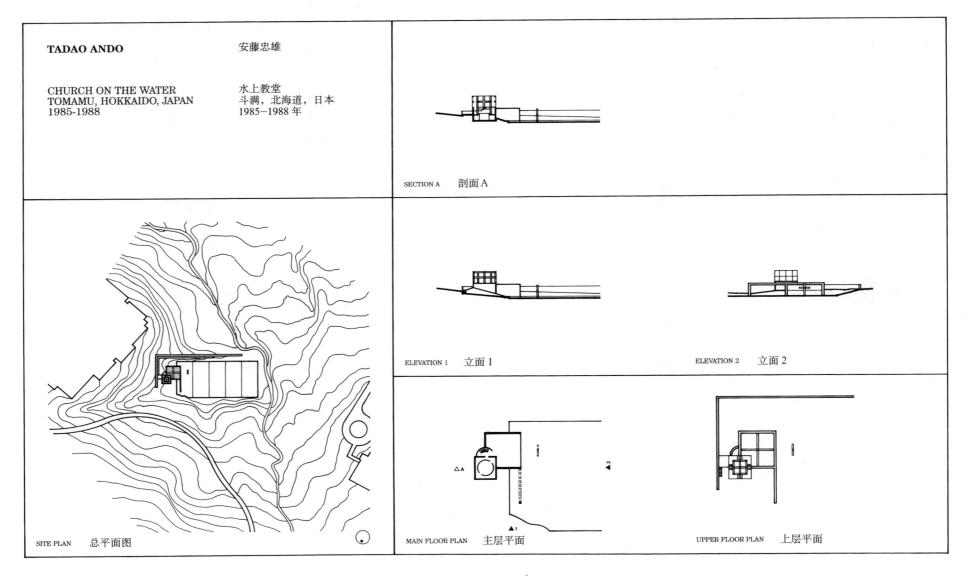

SECTION A　　剖面 A

ELEVATION 1　立面 1

ELEVATION 2　立面 2

SITE PLAN　　总平面图

MAIN FLOOR PLAN　主层平面

UPPER FLOOR PLAN　上层平面

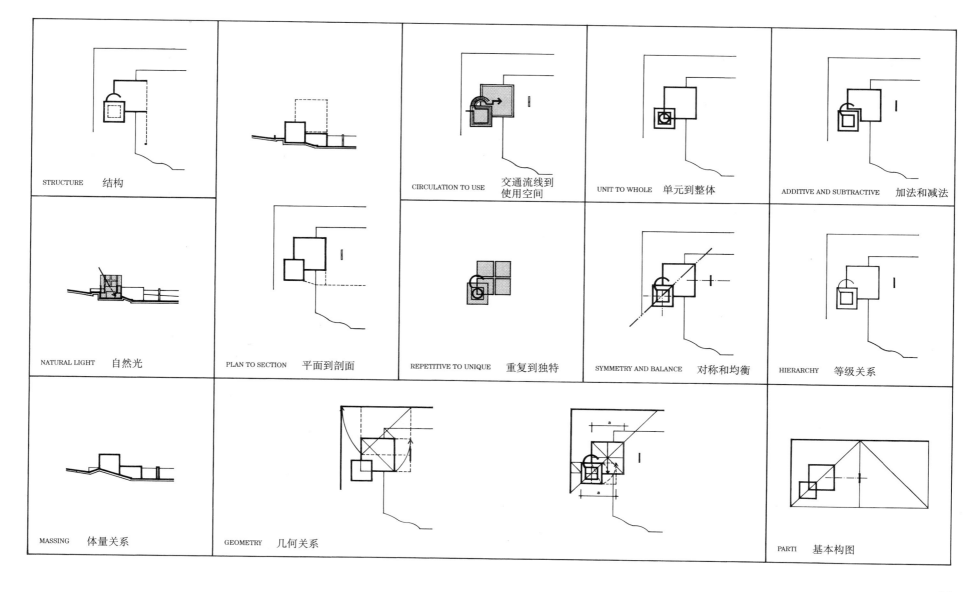

STRUCTURE　结构

CIRCULATION TO USE　交通流线到使用空间

UNIT TO WHOLE　单元到整体

ADDITIVE AND SUBTRACTIVE　加法和减法

NATURAL LIGHT　自然光

PLAN TO SECTION　平面到剖面

REPETITIVE TO UNIQUE　重复到独特

SYMMETRY AND BALANCE　对称和均衡

HIERARCHY　等级关系

MASSING　体量关系

GEOMETRY　几何关系

PARTI　基本构图

埃里克·贡纳尔·阿斯普隆德
ERIK GUNNAR ASPLUND

ERIK GUNNAR ASPLUND　　埃里克·贡纳尔·阿斯普隆德

SNELLMAN HOUSE
DJURSHOLM, SWEDEN
1917-1918

斯内尔曼住宅
于什霍尔姆，瑞典
1917－1918 年

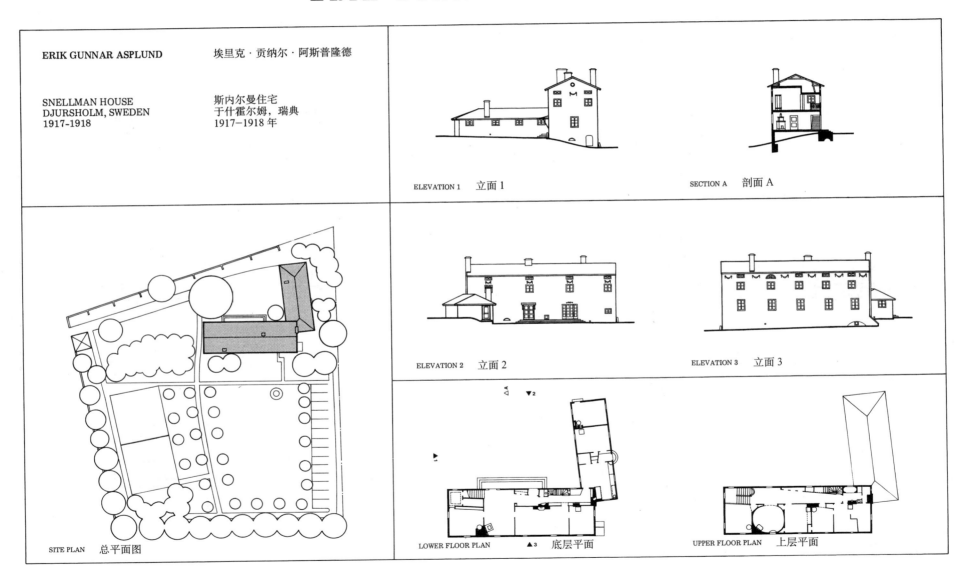

ELEVATION 1　立面 1

SECTION A　剖面 A

ELEVATION 2　立面 2

ELEVATION 3　立面 3

SITE PLAN　总平面图

LOWER FLOOR PLAN　▲3　底层平面

UPPER FLOOR PLAN　上层平面

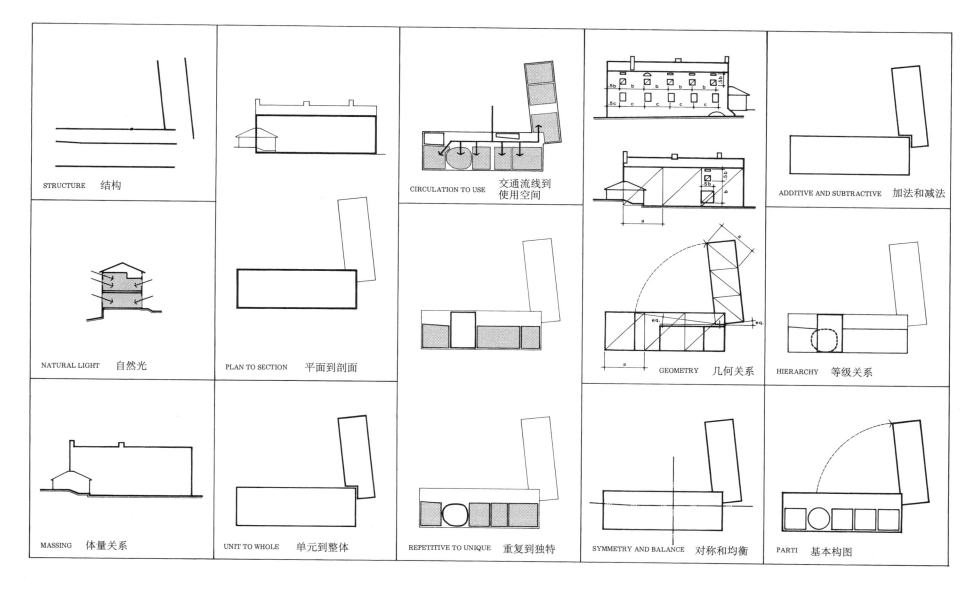

STRUCTURE　结构

CIRCULATION TO USE　交通流线到使用空间

ADDITIVE AND SUBTRACTIVE　加法和减法

NATURAL LIGHT　自然光

PLAN TO SECTION　平面到剖面

GEOMETRY　几何关系

HIERARCHY　等级关系

MASSING　体量关系

UNIT TO WHOLE　单元到整体

REPETITIVE TO UNIQUE　重复到独特

SYMMETRY AND BALANCE　对称和均衡

PARTI　基本构图

ERIK GUNNAR ASPLUND

埃里克·贡纳尔·阿斯普隆德

WOODLAND CHAPEL
STOCKHOLM SOUTH BURIAL GROUND
STOCKHOLM, SWEDEN
1918-1920

伍德兰礼拜堂
斯德哥尔摩南墓地
斯德哥尔摩，瑞典
1918-1920 年

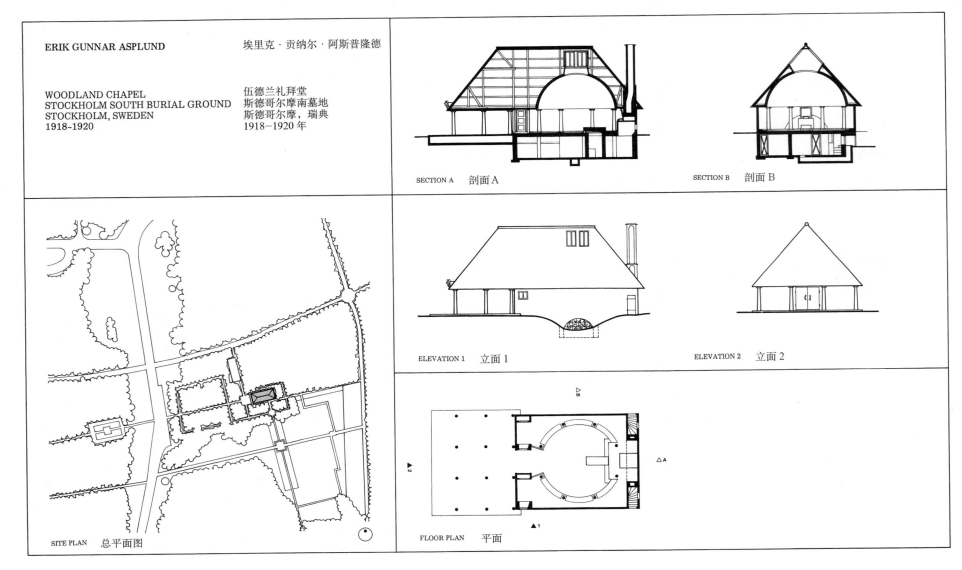

SECTION A　剖面 A

SECTION B　剖面 B

ELEVATION 1　立面 1

ELEVATION 2　立面 2

SITE PLAN　总平面图

FLOOR PLAN　平面

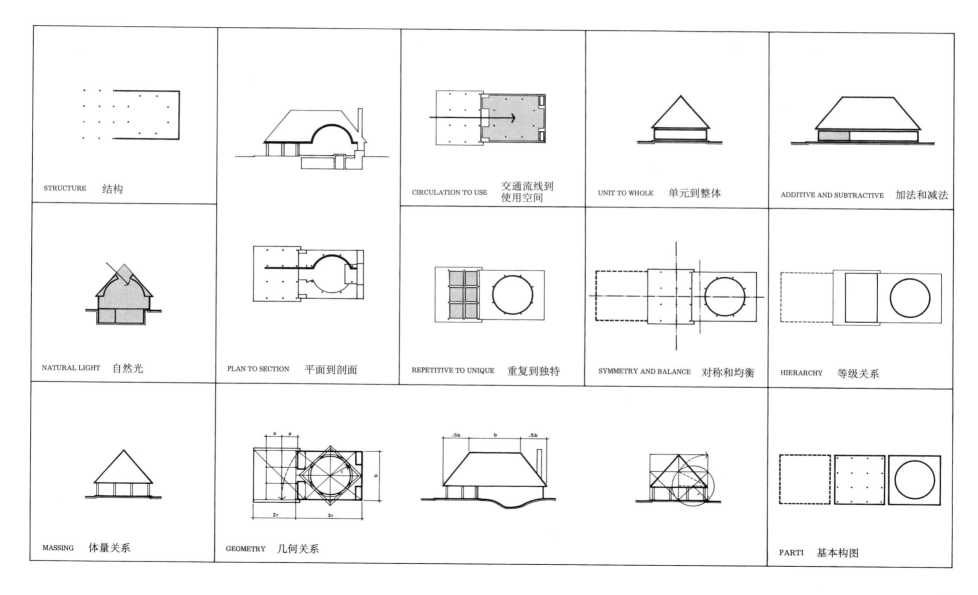

STRUCTURE 结构

CIRCULATION TO USE 交通流线到使用空间

UNIT TO WHOLE 单元到整体

ADDITIVE AND SUBTRACTIVE 加法和减法

NATURAL LIGHT 自然光

PLAN TO SECTION 平面到剖面

REPETITIVE TO UNIQUE 重复到独特

SYMMETRY AND BALANCE 对称和均衡

HIERARCHY 等级关系

MASSING 体量关系

GEOMETRY 几何关系

PARTI 基本构图

ERIK GUNNAR ASPLUND　　　埃里克·贡纳尔·阿斯普隆德

LISTER COUNTY COURTHOUSE　　利斯特县法院
SOLVESBORG, SWEDEN　　　　瑟尔沃斯堡，瑞典
1917-1921　　　　　　　　　1917-1921 年

SECTION A　　剖面 A

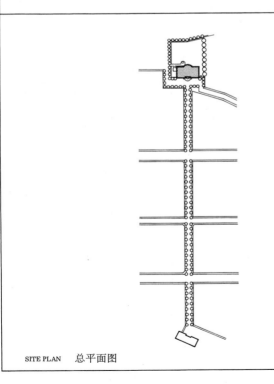

SITE PLAN　　总平面图

ELEVATION 1　立面 1

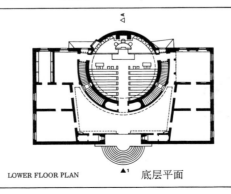

LOWER FLOOR PLAN　　▲¹　底层平面

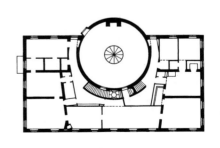

UPPER FLOOR PLAN　　上层平面

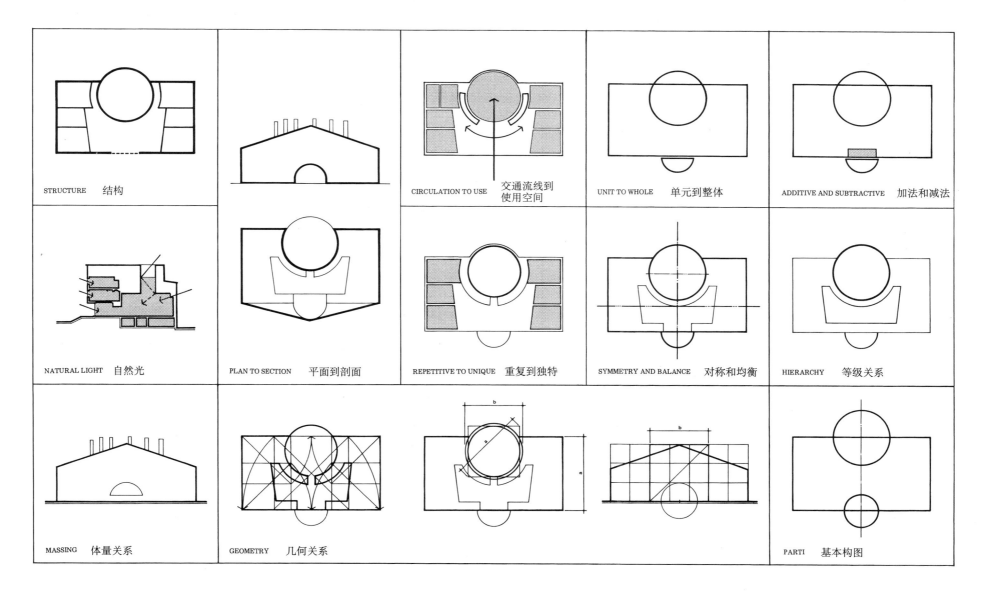

STRUCTURE　结构		CIRCULATION TO USE　交通流线到使用空间	UNIT TO WHOLE　单元到整体	ADDITIVE AND SUBTRACTIVE　加法和减法
NATURAL LIGHT　自然光	PLAN TO SECTION　平面到剖面	REPETITIVE TO UNIQUE　重复到独特	SYMMETRY AND BALANCE　对称和均衡	HIERARCHY　等级关系
MASSING　体量关系	GEOMETRY　几何关系			PARTI　基本构图

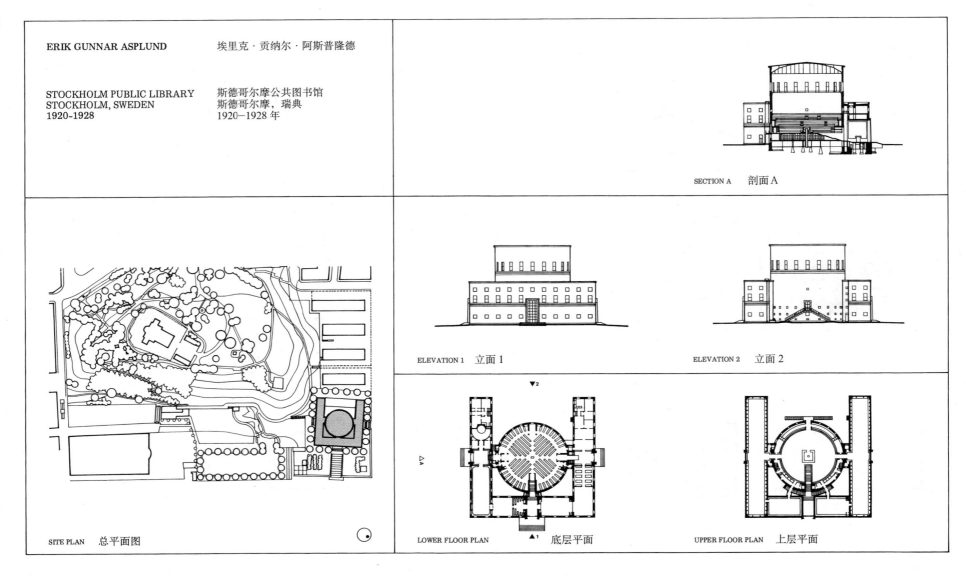

ERIK GUNNAR ASPLUND　　　埃里克·贡纳尔·阿斯普隆德

STOCKHOLM PUBLIC LIBRARY　斯德哥尔摩公共图书馆
STOCKHOLM, SWEDEN　　　　斯德哥尔摩，瑞典
1920-1928　　　　　　　　　1920—1928 年

SECTION A　剖面 A

ELEVATION 1　立面 1

ELEVATION 2　立面 2

SITE PLAN　总平面图

LOWER FLOOR PLAN　底层平面

UPPER FLOOR PLAN　上层平面

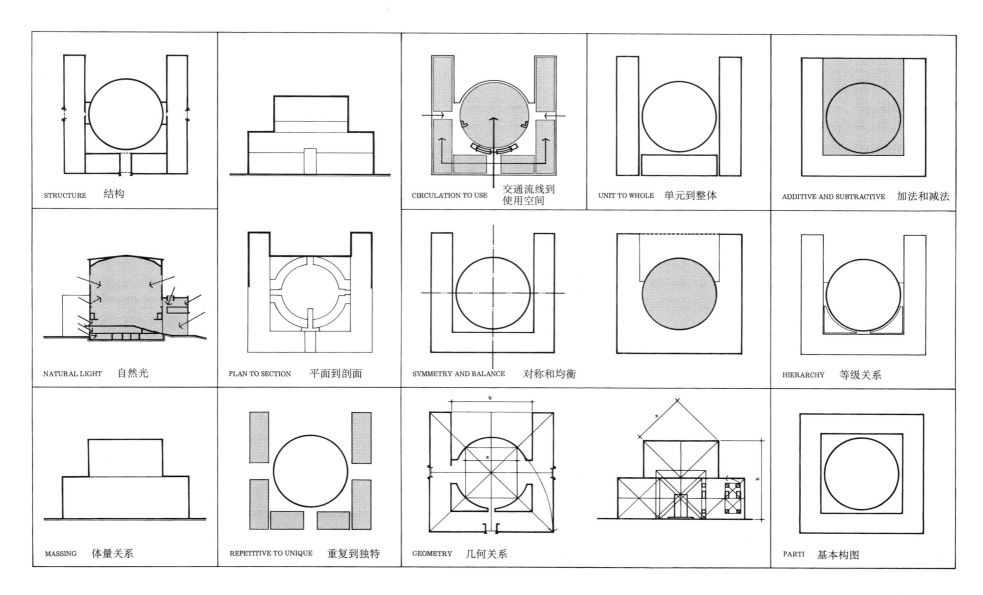

STRUCTURE　　结构

CIRCULATION TO USE　　交通流线到使用空间

UNIT TO WHOLE　　单元到整体

ADDITIVE AND SUBTRACTIVE　　加法和减法

NATURAL LIGHT　　自然光

PLAN TO SECTION　　平面到剖面

SYMMETRY AND BALANCE　　对称和均衡

HIERARCHY　　等级关系

MASSING　　体量关系

REPETITIVE TO UNIQUE　　重复到独特

GEOMETRY　　几何关系

PARTI　　基本构图

史蒂芬·比尔
STEPHANE BEEL

STEPHANE BEEL   史蒂芬·比尔

VILLA MAESEN   梅森住宅
ZEDELGEM, BELGIUM   泽德尔海姆，比利时
1987-1992   1987—1992 年

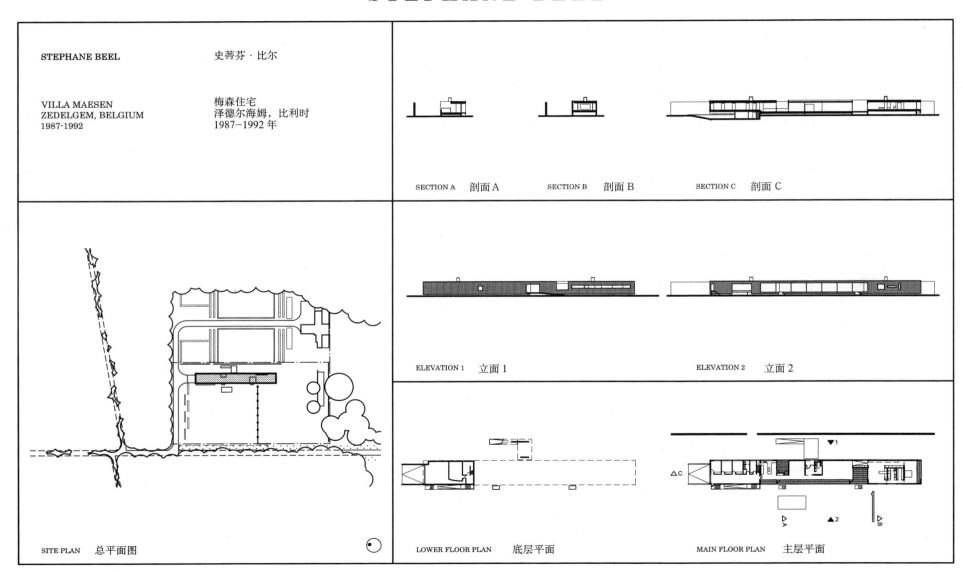

SECTION A 剖面 A   SECTION B 剖面 B   SECTION C 剖面 C

ELEVATION 1 立面 1   ELEVATION 2 立面 2

SITE PLAN 总平面图

LOWER FLOOR PLAN 底层平面   MAIN FLOOR PLAN 主层平面

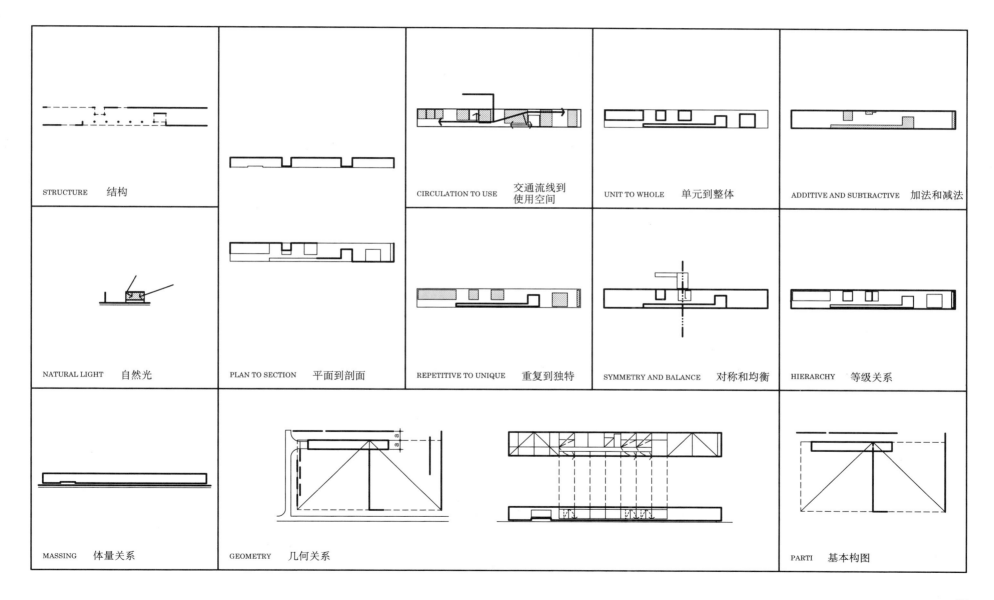

STRUCTURE　　结构

CIRCULATION TO USE　　交通流线到使用空间

UNIT TO WHOLE　　单元到整体

ADDITIVE AND SUBTRACTIVE　　加法和减法

NATURAL LIGHT　　自然光

PLAN TO SECTION　　平面到剖面

REPETITIVE TO UNIQUE　　重复到独特

SYMMETRY AND BALANCE　　对称和均衡

HIERARCHY　　等级关系

MASSING　　体量关系

GEOMETRY　　几何关系

PARTI　　基本构图

STEPHANE BEEL 　　　　史蒂芬·比尔

VILLA P 　　　　　　　P 住宅
ROTSLAAR, BELGIUM 　　Rotslaar，比利时
1991-1993 　　　　　　1991—1993 年

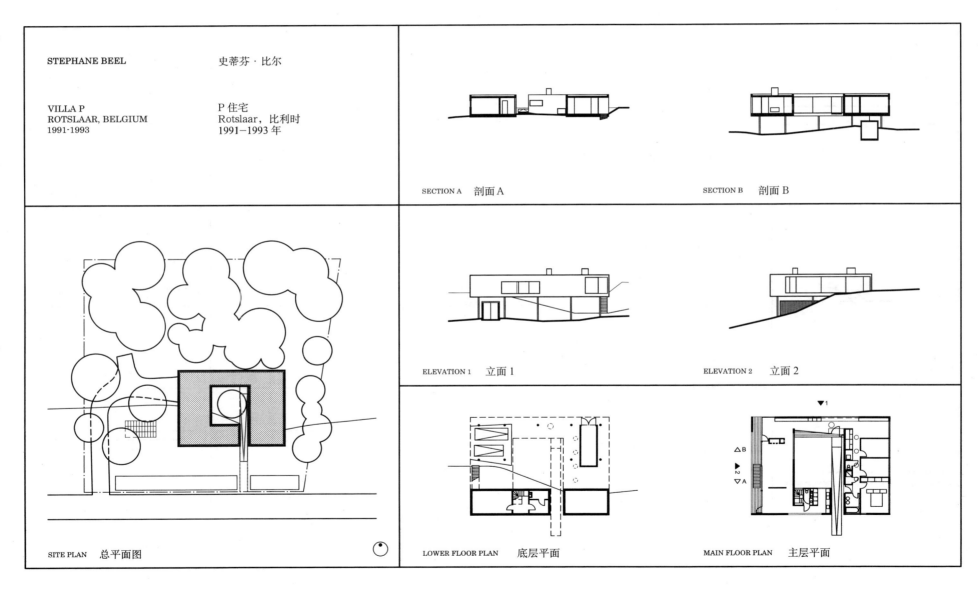

SECTION A 　剖面 A

SECTION B 　剖面 B

ELEVATION 1 　立面 1

ELEVATION 2 　立面 2

SITE PLAN 　总平面图

LOWER FLOOR PLAN 　底层平面

MAIN FLOOR PLAN 　主层平面

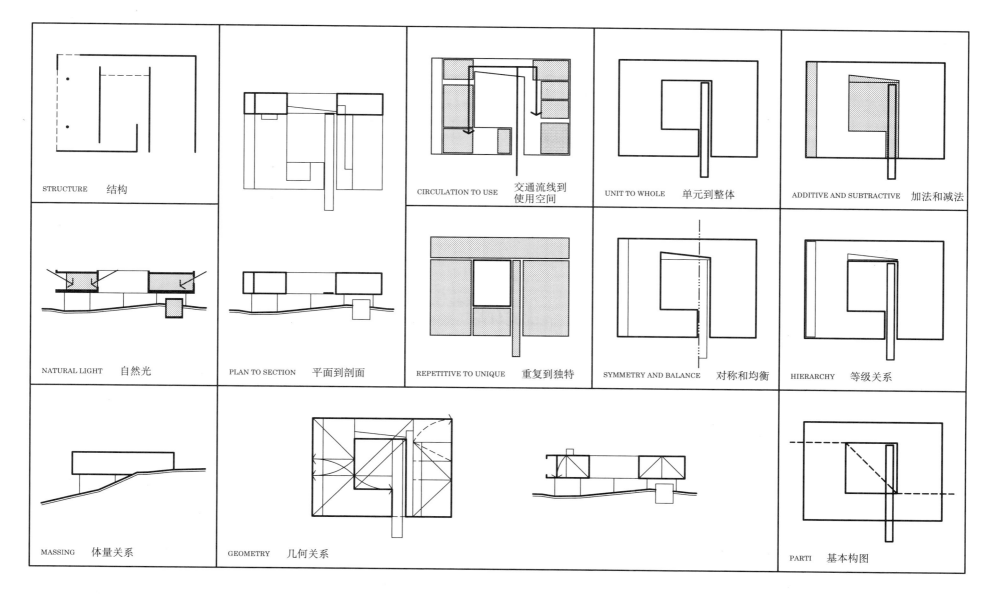

STRUCTURE 结构

CIRCULATION TO USE 交通流线到使用空间

UNIT TO WHOLE 单元到整体

ADDITIVE AND SUBTRACTIVE 加法和减法

NATURAL LIGHT 自然光

PLAN TO SECTION 平面到剖面

REPETITIVE TO UNIQUE 重复到独特

SYMMETRY AND BALANCE 对称和均衡

HIERARCHY 等级关系

MASSING 体量关系

GEOMETRY 几何关系

PARTI 基本构图

彼得·Q·博林
PETER Q.BOHLIN

BOHLIN AND POWELL (PETER BOHLIN)
博林和鲍威尔（彼得·博林）

WEEKEND RESIDENCE FOR MR. AND MRS. ERIC Q. BOHLIN
WEST CORNWALL, CONNECTICUT, USA
1973-1975

埃里克·Q·博林夫妇周末住宅
西康沃尔，康涅狄格州，美国
1973－1975 年

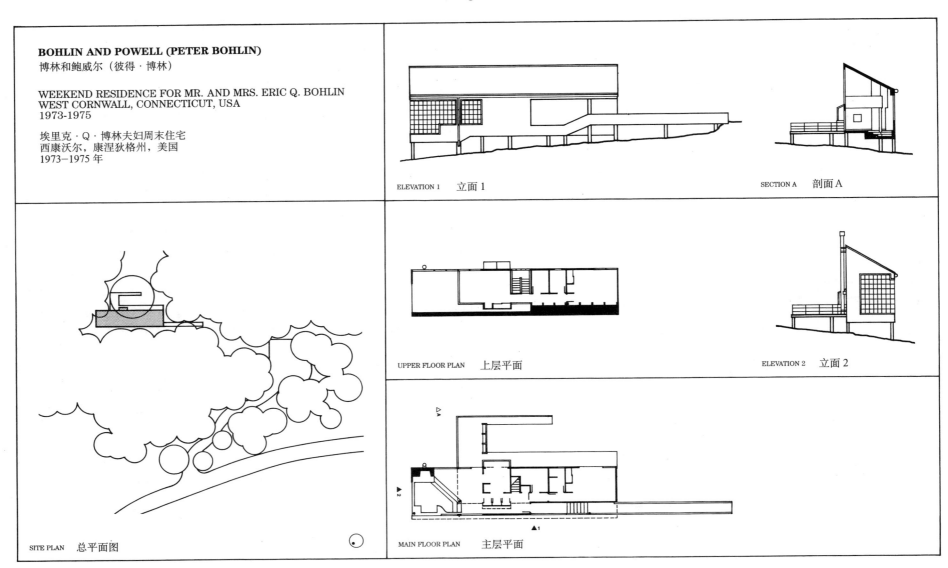

ELEVATION 1　立面 1

SECTION A　剖面 A

UPPER FLOOR PLAN　上层平面

ELEVATION 2　立面 2

SITE PLAN　总平面图

MAIN FLOOR PLAN　主层平面

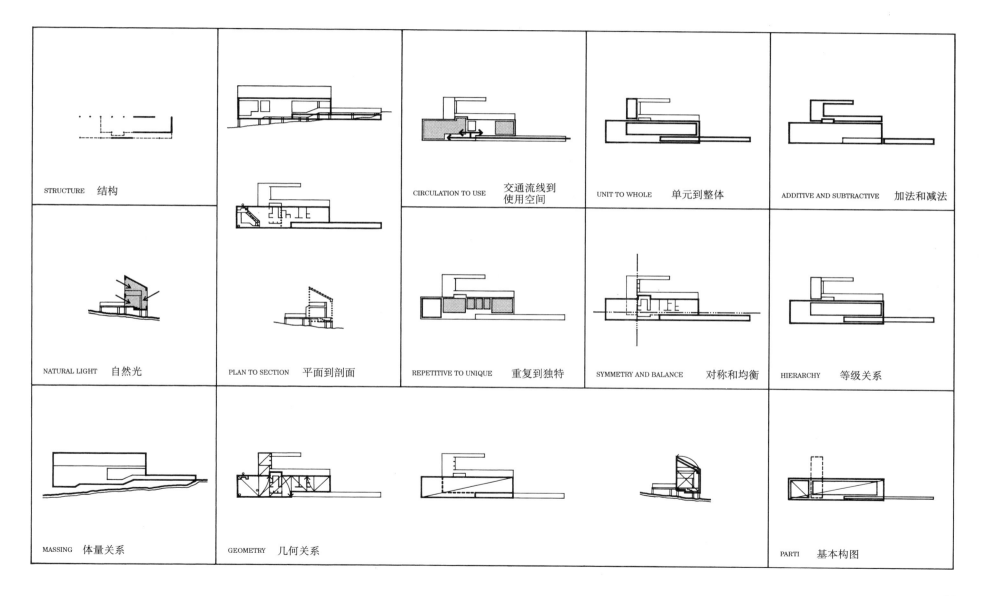

STRUCTURE　结构

CIRCULATION TO USE　交通流线到使用空间

UNIT TO WHOLE　单元到整体

ADDITIVE AND SUBTRACTIVE　加法和减法

NATURAL LIGHT　自然光

PLAN TO SECTION　平面到剖面

REPETITIVE TO UNIQUE　重复到独特

SYMMETRY AND BALANCE　对称和均衡

HIERARCHY　等级关系

MASSING　体量关系

GEOMETRY　几何关系

PARTI　基本构图

BOHLIN CYWINSKI JACKSON (PETER BOHLIN)
博林·西万斯基·雅克松（彼得·博林）

GAFFNEY RESIDENCE
ROMANSVILLE, PENNSYLVANIA, USA
1977-1980

加夫尼住宅
罗芒斯维勒，宾夕法尼亚州，美国
1977-1980 年

SECTION A　剖面 A

SECTION B　剖面 B

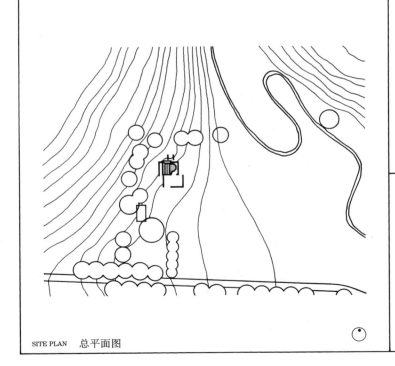

SITE PLAN　总平面图

ELEVATION 1　立面 1

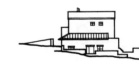

ELEVATION 2　立面 2

ELEVATION 3　立面 3

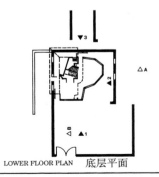

LOWER FLOOR PLAN　底层平面

MIDDLE FLOOR PLAN　中层平面

UPPER FLOOR PLAN　上层平面

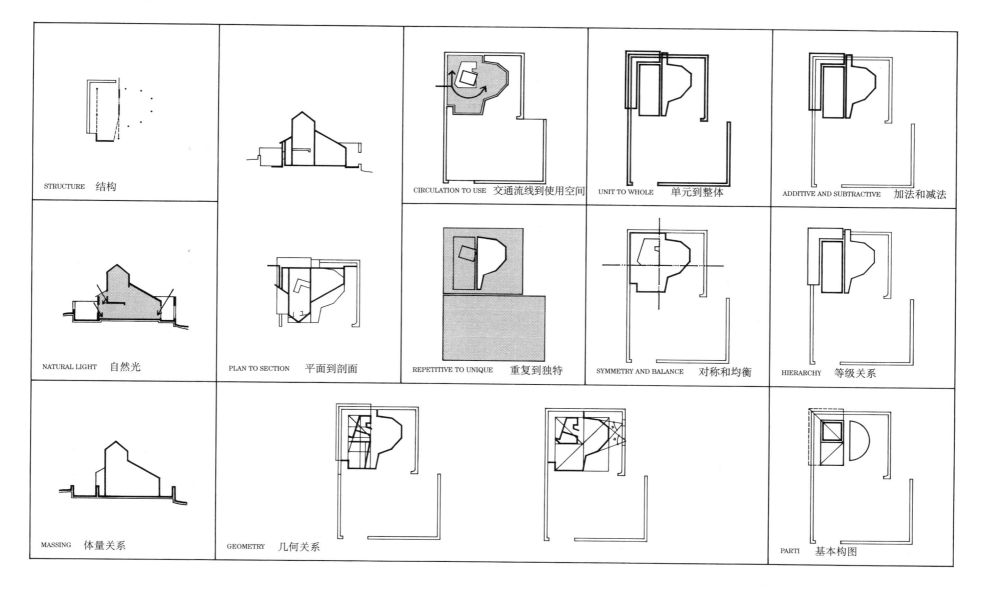

STRUCTURE　结构

NATURAL LIGHT　自然光

MASSING　体量关系

CIRCULATION TO USE　交通流线到使用空间

PLAN TO SECTION　平面到剖面

GEOMETRY　几何关系

UNIT TO WHOLE　单元到整体

REPETITIVE TO UNIQUE　重复到独特

SYMMETRY AND BALANCE　对称和均衡

ADDITIVE AND SUBTRACTIVE　加法和减法

HIERARCHY　等级关系

PARTI　基本构图

BOHLIN CYWINSKI JACKSON (PETER BOHLIN)

博林·西万斯基·雅克松（彼得·博林）

HOUSE IN THE ADIRONDACKS　　阿狄朗达克的住宅
NEW YORK STATE, USA　　纽约州，美国
1987-1992　　1987-1992 年

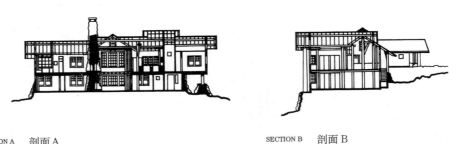

SECTION A　剖面 A

SECTION B　剖面 B

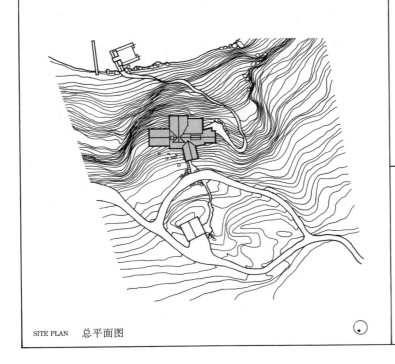

SITE PLAN　总平面图

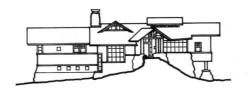

ELEVATION 1　立面 1

ELEVATION 2　立面 2

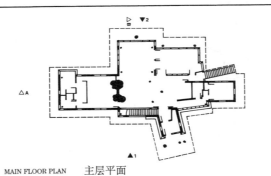

MAIN FLOOR PLAN　主层平面

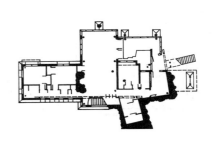

LOWER FLOOR PLAN　底层平面

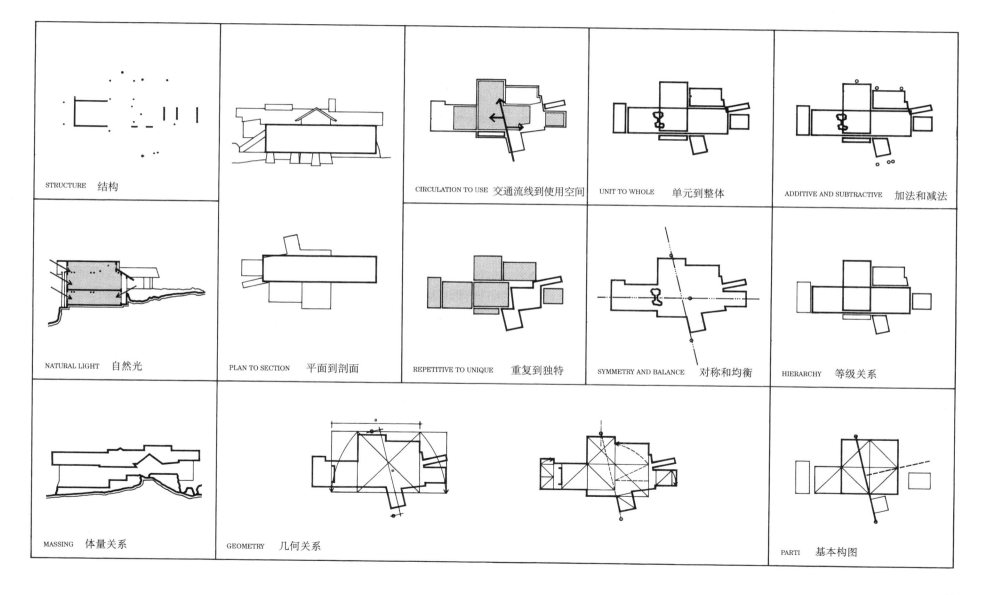

STRUCTURE　结构

CIRCULATION TO USE　交通流线到使用空间

UNIT TO WHOLE　单元到整体

ADDITIVE AND SUBTRACTIVE　加法和减法

NATURAL LIGHT　自然光

PLAN TO SECTION　平面到剖面

REPETITIVE TO UNIQUE　重复到独特

SYMMETRY AND BALANCE　对称和均衡

HIERARCHY　等级关系

MASSING　体量关系

GEOMETRY　几何关系

PARTI　基本构图

43

BOHLIN CYWINSKI JACKSON/JAMES CUTLER ARCHITECTS (PETER BOHLIN)

博林·西万斯基·雅克松／詹姆斯·卡特勒建筑事务所（彼得·博林）

GUEST HOUSE, GATES RESIDENCE 格斯特住宅，盖茨住宅
MEDINA, WASHINGTON, USA 梅迪纳，华盛顿州，美国
1990-1993 1990－1993 年

SECTION A 剖面 A

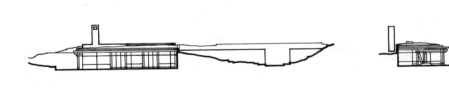

ELEVATION 1 立面 1 ELEVATION 2 立面 2

SITE PLAN 总平面图

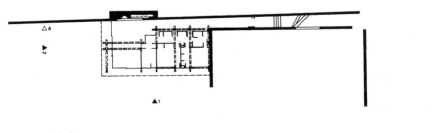

FLOOR PLAN 平面

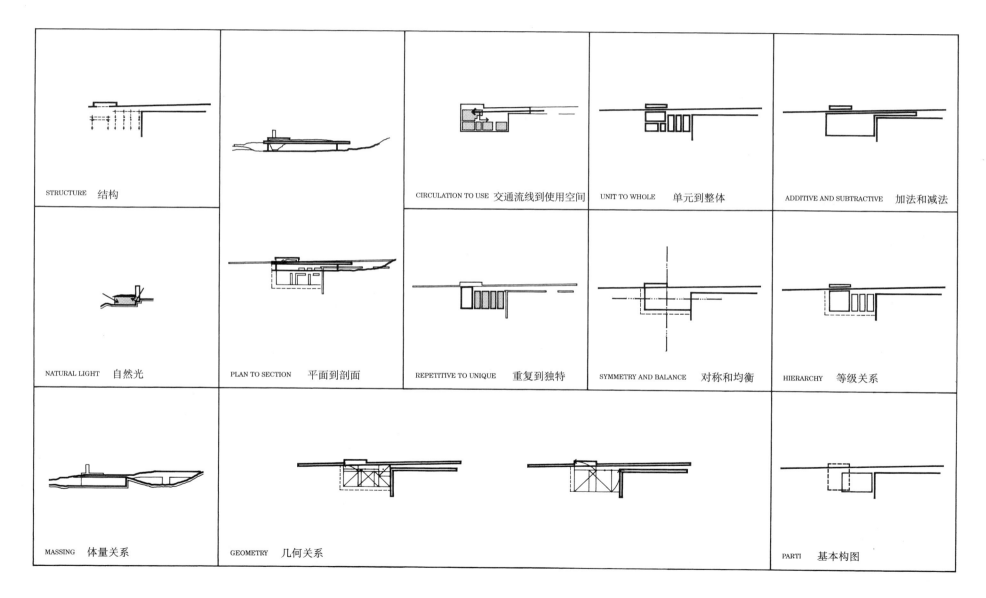

STRUCTURE　结构

CIRCULATION TO USE 交通流线到使用空间

UNIT TO WHOLE　单元到整体

ADDITIVE AND SUBTRACTIVE　加法和减法

NATURAL LIGHT　自然光

PLAN TO SECTION　平面到剖面

REPETITIVE TO UNIQUE　重复到独特

SYMMETRY AND BALANCE　对称和均衡

HIERARCHY　等级关系

MASSING　体量关系

GEOMETRY　几何关系

PARTI　基本构图

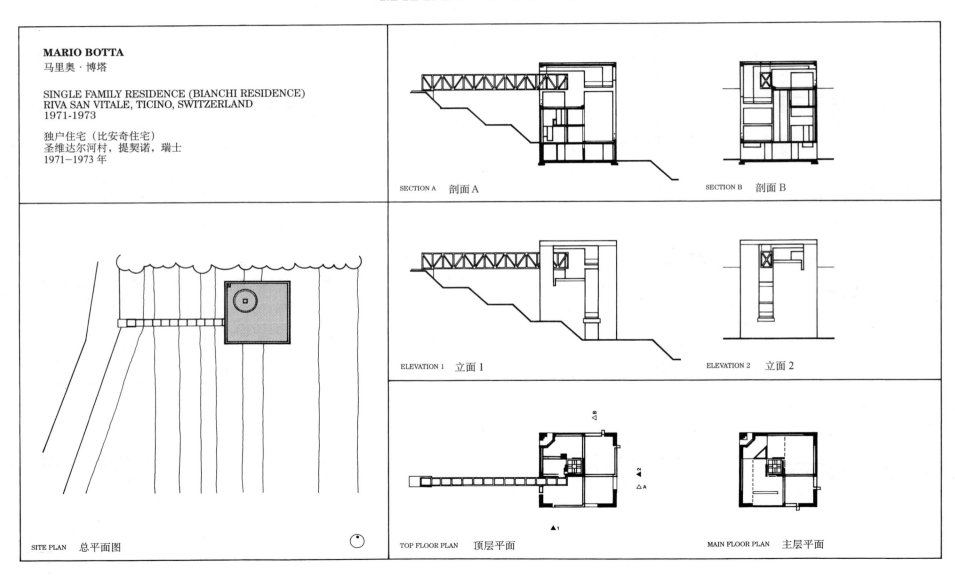

MARIO BOTTA
马里奥·博塔

SINGLE FAMILY RESIDENCE (BIANCHI RESIDENCE)
RIVA SAN VITALE, TICINO, SWITZERLAND
1971-1973

独户住宅（比安奇住宅）
圣维达尔河村，提契诺，瑞士
1971–1973 年

SECTION A　剖面 A

SECTION B　剖面 B

ELEVATION 1　立面 1

ELEVATION 2　立面 2

SITE PLAN　总平面图

TOP FLOOR PLAN　顶层平面

MAIN FLOOR PLAN　主层平面

40

46

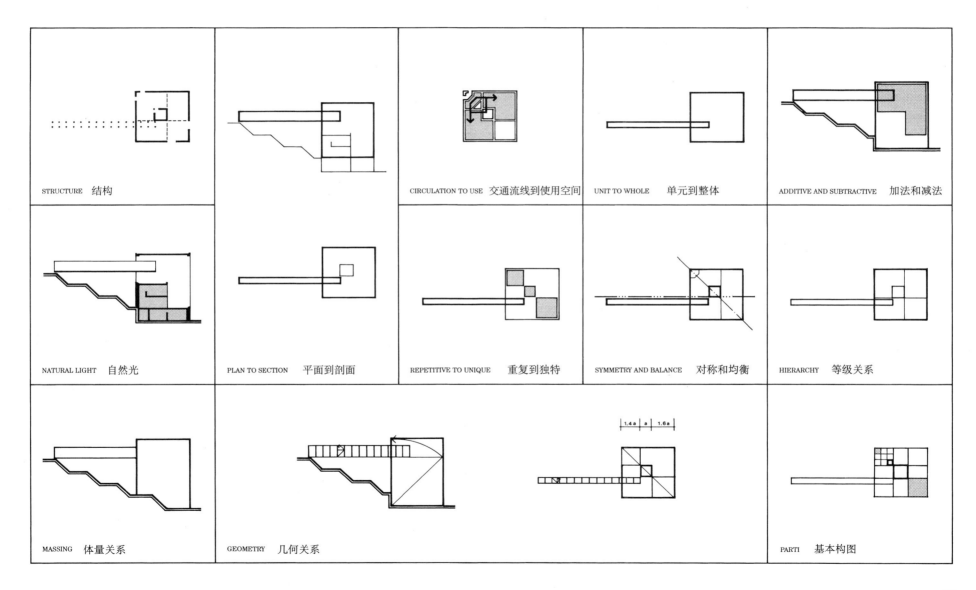

STRUCTURE　结构

CIRCULATION TO USE　交通流线到使用空间

UNIT TO WHOLE　单元到整体

ADDITIVE AND SUBTRACTIVE　加法和减法

NATURAL LIGHT　自然光

PLAN TO SECTION　平面到剖面

REPETITIVE TO UNIQUE　重复到独特

SYMMETRY AND BALANCE　对称和均衡

HIERARCHY　等级关系

MASSING　体量关系

GEOMETRY　几何关系

PARTI　基本构图

MARIO BOTTA
马里奥·博塔

CHURCH OF SAN GIOVANNI BATTISTA (SAINT JOHN THE BAPTIST)
MOGNO, TICINO, SWITZERLAND
1986-1995

圣乔万尼·巴蒂斯塔教堂（圣约翰施洗教堂）
莫尼奥，提契诺，瑞士
1986—1995 年

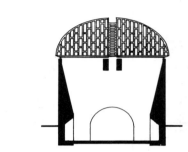

SECTION A　剖面 A

SECTION B　剖面 B

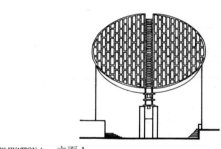

ELEVATION 1　立面 1

ELEVATION 2　立面 2

SITE PLAN　总平面图

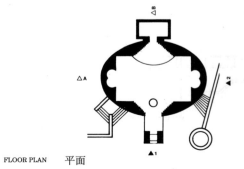

FLOOR PLAN　平面

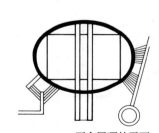

PLAN WITHOUT ROOF　不含屋顶的平面

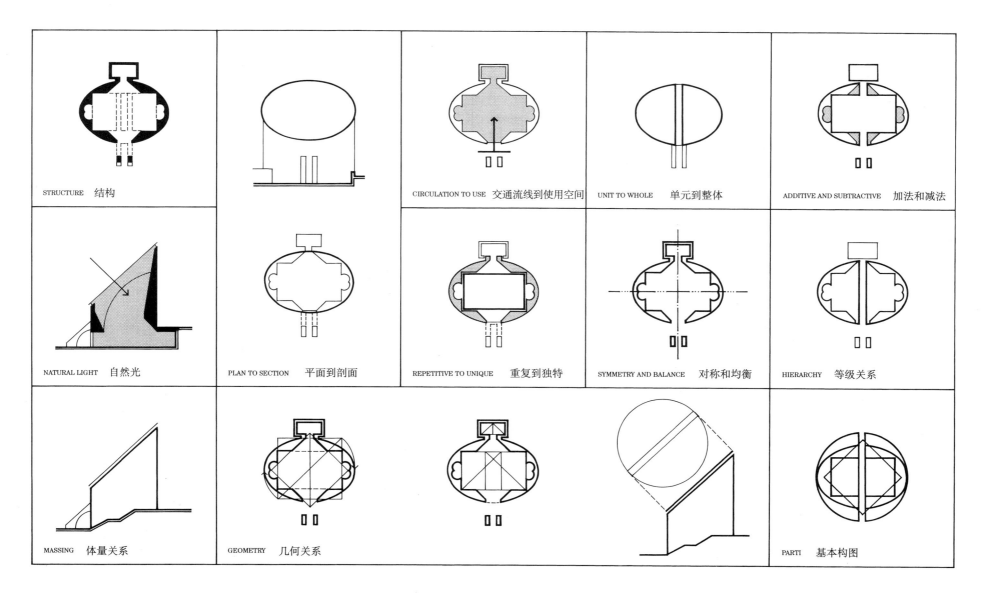

STRUCTURE 结构		CIRCULATION TO USE 交通流线到使用空间	UNIT TO WHOLE 单元到整体	ADDITIVE AND SUBTRACTIVE 加法和减法
NATURAL LIGHT 自然光	PLAN TO SECTION 平面到剖面	REPETITIVE TO UNIQUE 重复到独特	SYMMETRY AND BALANCE 对称和均衡	HIERARCHY 等级关系
MASSING 体量关系	GEOMETRY 几何关系			PARTI 基本构图

MARIO BOTTA　　　　　马里奥·博塔

BIANDA RESIDENCE　　　　比安达住宅
LOSONE, TICINO, SWITZERLAND　　洛索内，提契诺，瑞士
1987-1989　　　　　　　　1987–1989 年

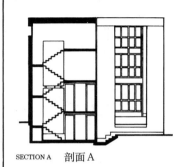

SECTION A　剖面 A

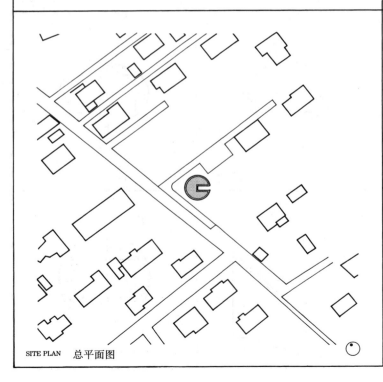

SITE PLAN　总平面图

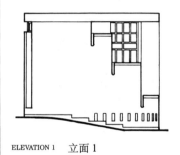

ELEVATION 1　立面 1

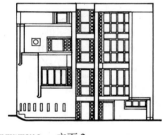

ELEVATION 2　立面 2

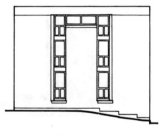

ELEVATION 3　立面 3

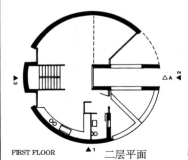

FIRST FLOOR　二层平面

SECOND FLOOR　三层平面

GROUND FLOOR　首层平面

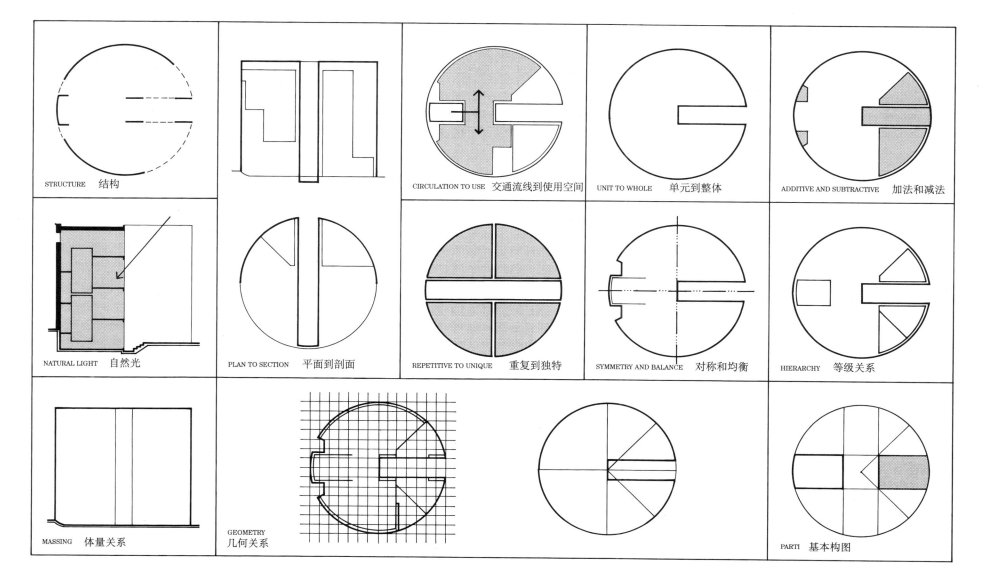

STRUCTURE 结构		CIRCULATION TO USE 交通流线到使用空间	UNIT TO WHOLE 单元到整体	ADDITIVE AND SUBTRACTIVE 加法和减法
NATURAL LIGHT 自然光	PLAN TO SECTION 平面到剖面	REPETITIVE TO UNIQUE 重复到独特	SYMMETRY AND BALANCE 对称和均衡	HIERARCHY 等级关系
MASSING 体量关系	GEOMETRY 几何关系			PARTI 基本构图

MARIO BOTTA　　　　马里奥·博塔

THE CHURCH OF BEATO ODORICO
PORDENONE, ITALY
1987-1992

贝亚托·奥多里科教堂
波代诺内，意大利
1987–1992 年

SECTION A　　剖面 A

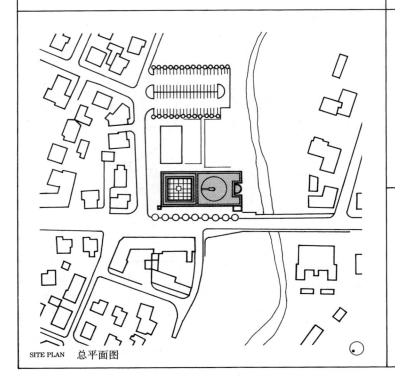

SITE PLAN　　总平面图

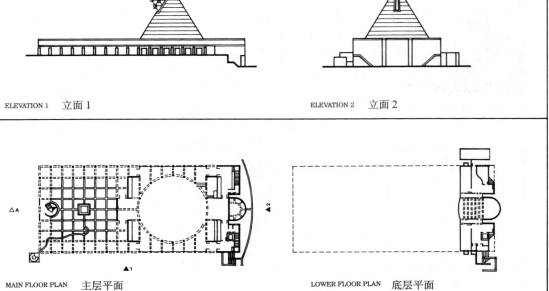

ELEVATION 1　　立面 1

ELEVATION 2　　立面 2

MAIN FLOOR PLAN　　主层平面

LOWER FLOOR PLAN　　底层平面

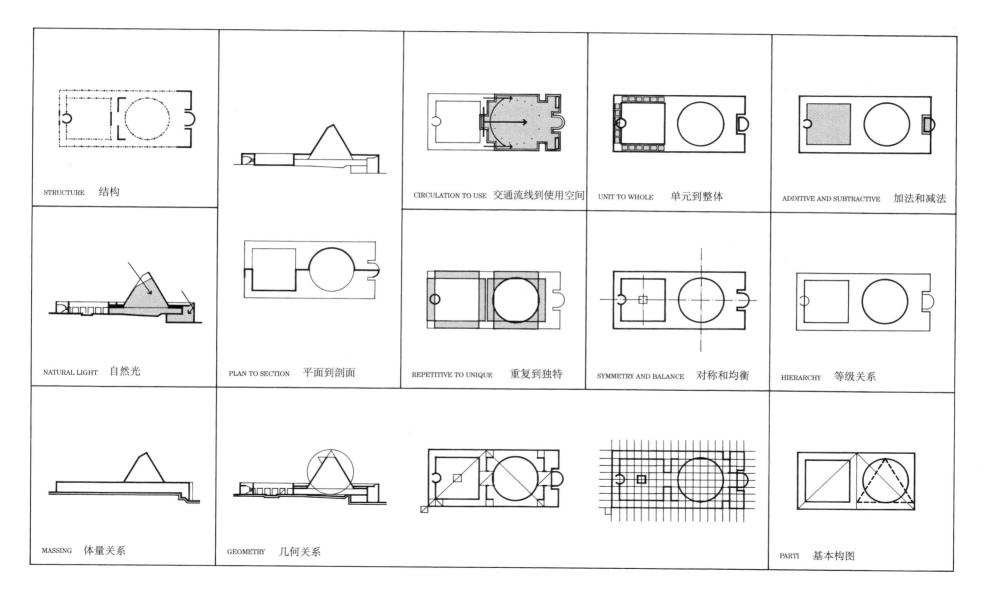

STRUCTURE　结构

CIRCULATION TO USE　交通流线到使用空间

UNIT TO WHOLE　单元到整体

ADDITIVE AND SUBTRACTIVE　加法和减法

NATURAL LIGHT　自然光

PLAN TO SECTION　平面到剖面

REPETITIVE TO UNIQUE　重复到独特

SYMMETRY AND BALANCE　对称和均衡

HIERARCHY　等级关系

MASSING　体量关系

GEOMETRY　几何关系

PARTI　基本构图

菲利波·伯鲁乃列斯基
FILIPPO BRUNELLESCHI

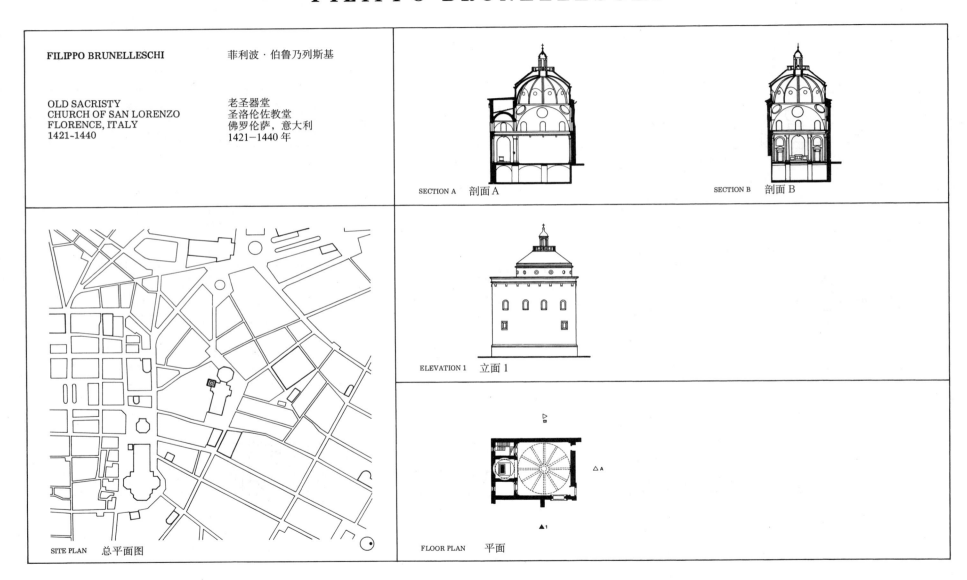

FILIPPO BRUNELLESCHI   菲利波·伯鲁乃列斯基

OLD SACRISTY   老圣器堂
CHURCH OF SAN LORENZO   圣洛伦佐教堂
FLORENCE, ITALY   佛罗伦萨，意大利
1421-1440   1421－1440 年

SECTION A   剖面 A

SECTION B   剖面 B

ELEVATION 1   立面 1

SITE PLAN   总平面图

FLOOR PLAN   平面

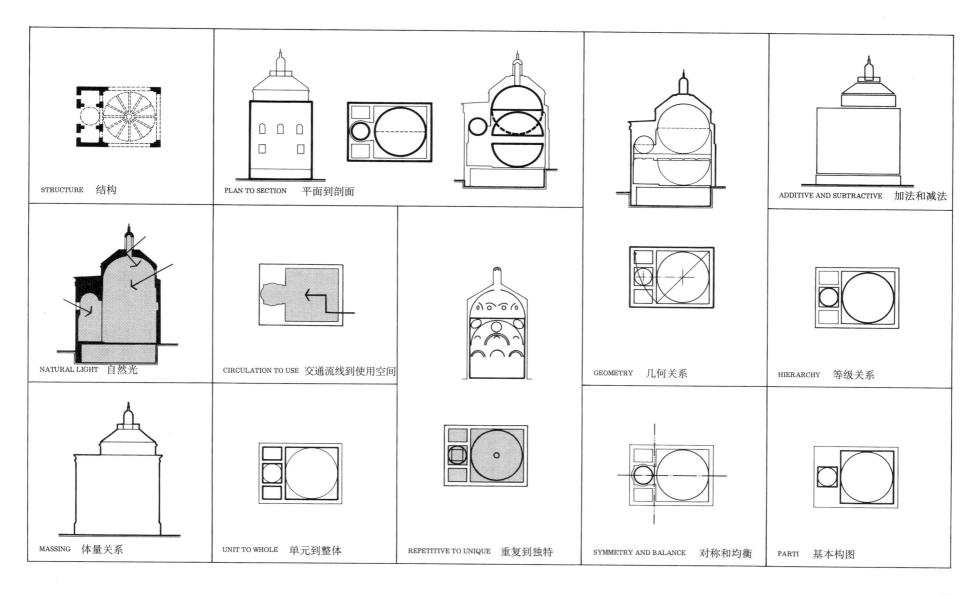

STRUCTURE　结构

PLAN TO SECTION　平面到剖面

ADDITIVE AND SUBTRACTIVE　加法和减法

NATURAL LIGHT　自然光

CIRCULATION TO USE　交通流线到使用空间

GEOMETRY　几何关系

HIERARCHY　等级关系

MASSING　体量关系

UNIT TO WHOLE　单元到整体

REPETITIVE TO UNIQUE　重复到独特

SYMMETRY AND BALANCE　对称和均衡

PARTI　基本构图

FILIPPO BRUNELLESCHI　　　菲利波·伯鲁乃列斯基

OSPEDALE DEGLI INNOCENTI　　孤儿院
FLORENCE, ITALY　　　　　　　佛罗伦萨，意大利
1421-1445　　　　　　　　　　1421-1445 年

SECTION A　剖面 A

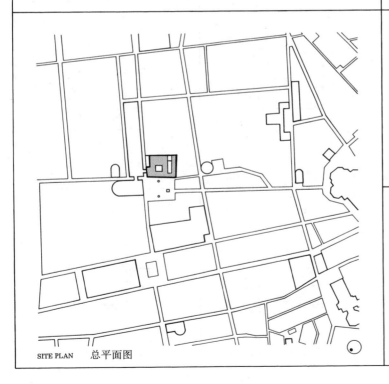

SITE PLAN　总平面图

ELEVATION 1　立面 1

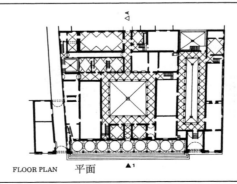

FLOOR PLAN　平面　▲1

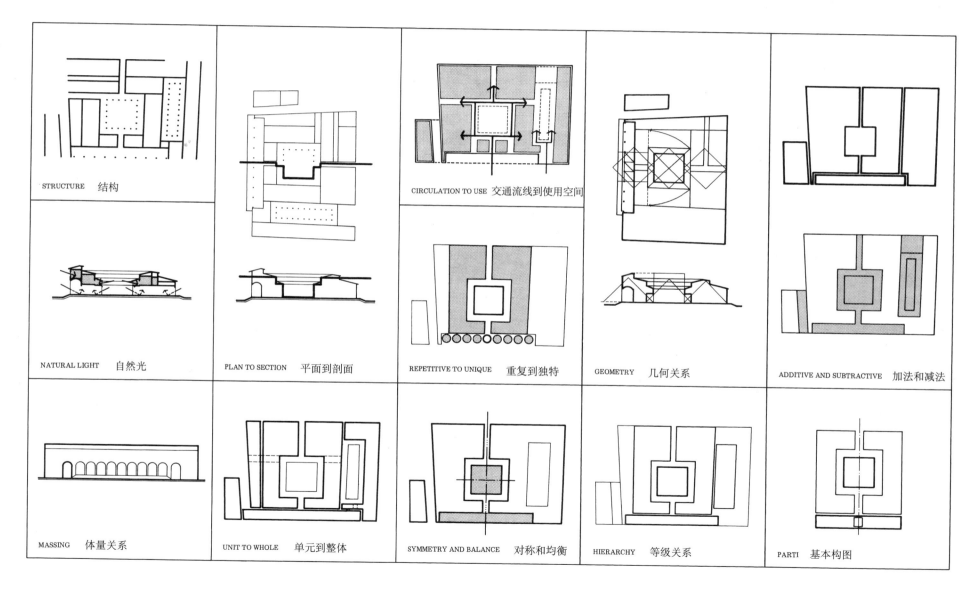

STRUCTURE　结构

CIRCULATION TO USE　交通流线到使用空间

NATURAL LIGHT　自然光

PLAN TO SECTION　平面到剖面

REPETITIVE TO UNIQUE　重复到独特

GEOMETRY　几何关系

ADDITIVE AND SUBTRACTIVE　加法和减法

MASSING　体量关系

UNIT TO WHOLE　单元到整体

SYMMETRY AND BALANCE　对称和均衡

HIERARCHY　等级关系

PARTI　基本构图

FILIPPO BRUNELLESCHI　　　　菲利波·伯鲁乃列斯基

CHURCH OF SAN MARIA DEGLI ANGELI　　圣玛丽亚布道所
FLORENCE, ITALY　　佛罗伦萨，意大利
1434-1436　　1434-1436 年

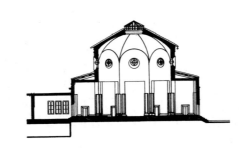

SECTION A　剖面 A

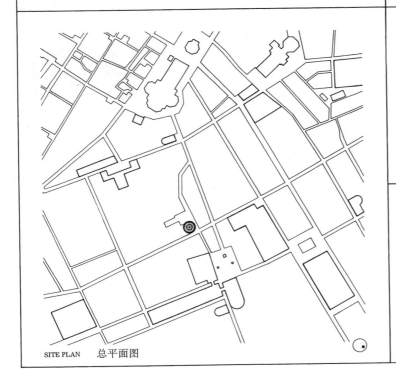

SITE PLAN　总平面图

ELEVATION 1　立面 1

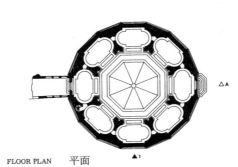

FLOOR PLAN　平面

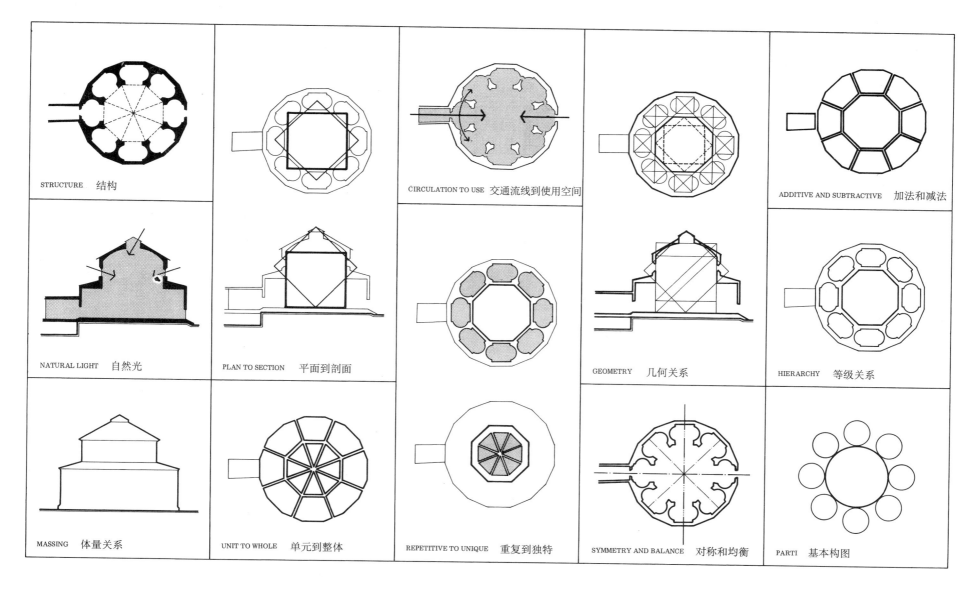

STRUCTURE　结构

CIRCULATION TO USE　交通流线到使用空间

ADDITIVE AND SUBTRACTIVE　加法和减法

NATURAL LIGHT　自然光

PLAN TO SECTION　平面到剖面

GEOMETRY　几何关系

HIERARCHY　等级关系

MASSING　体量关系

UNIT TO WHOLE　单元到整体

REPETITIVE TO UNIQUE　重复到独特

SYMMETRY AND BALANCE　对称和均衡

PARTI　基本构图

FILIPPO BRUNELLESCHI　　　菲利波·伯鲁乃列斯基

CHURCH OF SAN SPIRITO
FLORENCE, ITALY
(Begun) 1434

圣灵教堂
佛罗伦萨，意大利
始于 1434 年

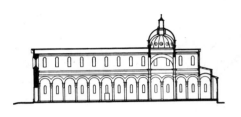

SECTION A　剖面 A

SECTION B　剖面 B

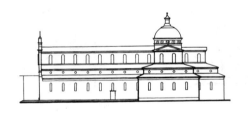

ELEVATION 1　立面 1

ELEVATION 2　立面 2

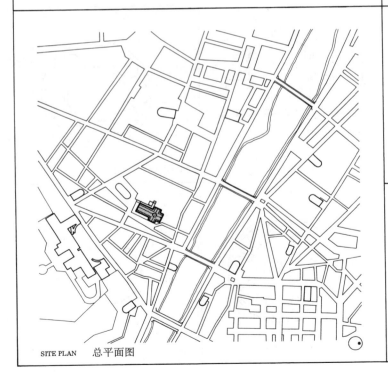

SITE PLAN　总平面图

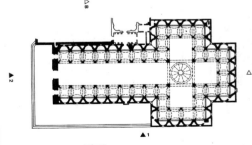

FLOOR PLAN　平面

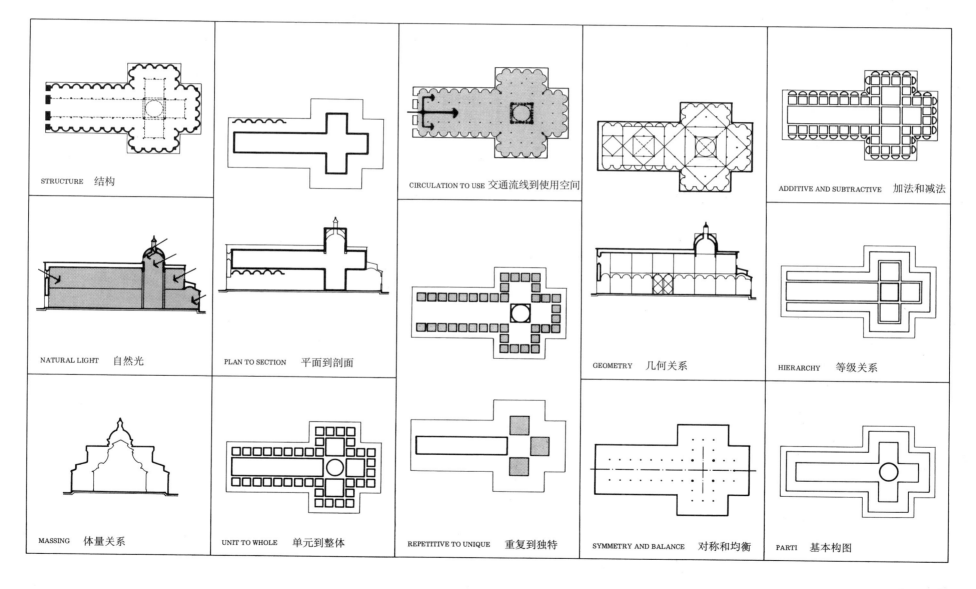

STRUCTURE　结构

CIRCULATION TO USE　交通流线到使用空间

ADDITIVE AND SUBTRACTIVE　加法和减法

NATURAL LIGHT　自然光

PLAN TO SECTION　平面到剖面

GEOMETRY　几何关系

HIERARCHY　等级关系

MASSING　体量关系

UNIT TO WHOLE　单元到整体

REPETITIVE TO UNIQUE　重复到独特

SYMMETRY AND BALANCE　对称和均衡

PARTI　基本构图

戴维·奇普菲尔德
DAVID CHIPPERFIELD

DAVID CHIPPERFIELD　　　　戴维·奇普菲尔德

GALLERY BUILDING AM KUPFERGRABEN 10　　艺廊兼住宅
BERLIN, GERMANY　　柏林，德国
2003 · 2007　　2003－2007 年

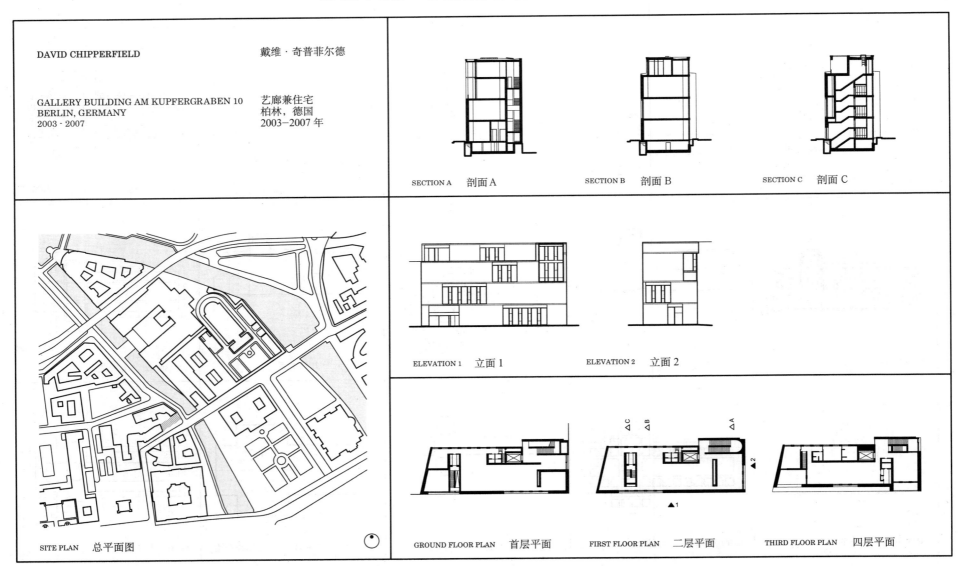

SECTION A　剖面 A　　　SECTION B　剖面 B　　　SECTION C　剖面 C

ELEVATION 1　立面 1　　　ELEVATION 2　立面 2

SITE PLAN　总平面图

GROUND FLOOR PLAN　首层平面　　　FIRST FLOOR PLAN　二层平面　　　THIRD FLOOR PLAN　四层平面

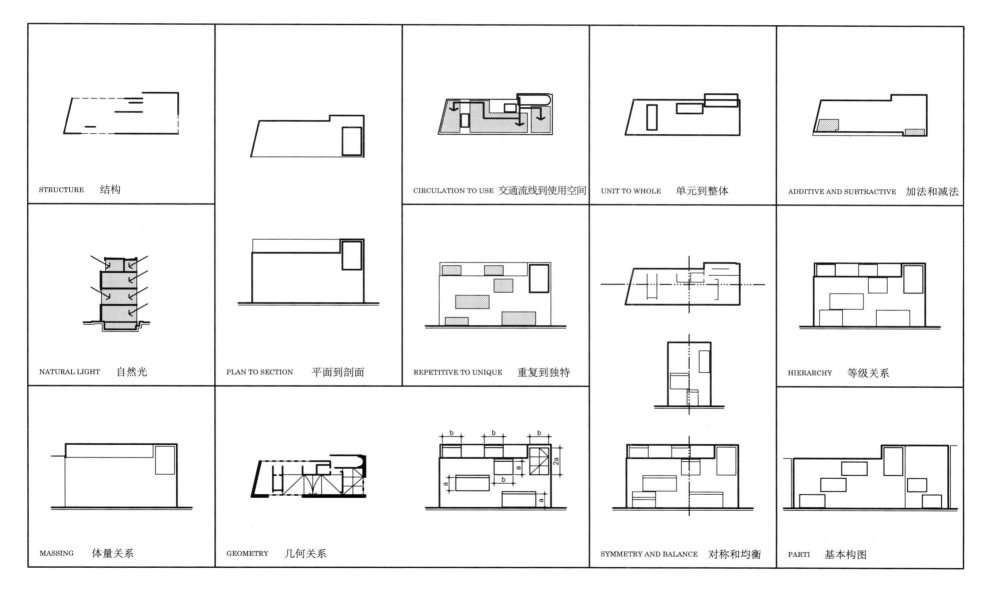

STRUCTURE　　结构

CIRCULATION TO USE　交通流线到使用空间

UNIT TO WHOLE　　单元到整体

ADDITIVE AND SUBTRACTIVE　加法和减法

NATURAL LIGHT　　自然光

PLAN TO SECTION　　平面到剖面

REPETITIVE TO UNIQUE　　重复到独特

HIERARCHY　　等级关系

MASSING　　体量关系

GEOMETRY　　几何关系

SYMMETRY AND BALANCE　对称和均衡

PARTI　　基本构图

DAVID CHIPPERFIELD　　戴维·奇普菲尔德

LIANGZHU CULTURE MUSEUM　　良渚文化博物馆
HANGZHOU, CHINA　　杭州，中国
2000 · 2003　　2000—2003 年

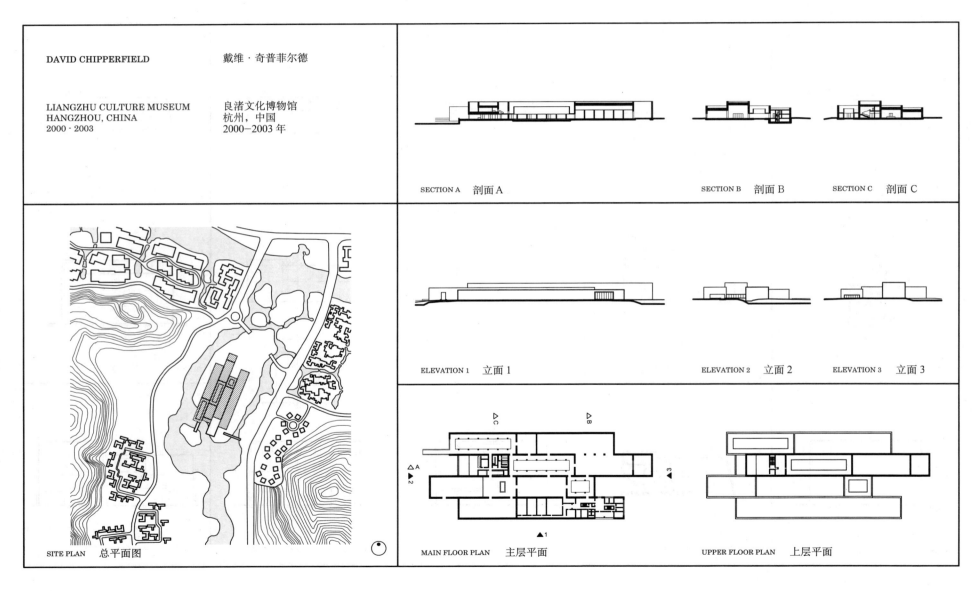

SECTION A 　剖面 A　　　　　SECTION B 　剖面 B　　　SECTION C 　剖面 C

ELEVATION 1 　立面 1　　　　ELEVATION 2 　立面 2　　ELEVATION 3 　立面 3

SITE PLAN 　总平面图

MAIN FLOOR PLAN 　主层平面　　　　UPPER FLOOR PLAN 　上层平面

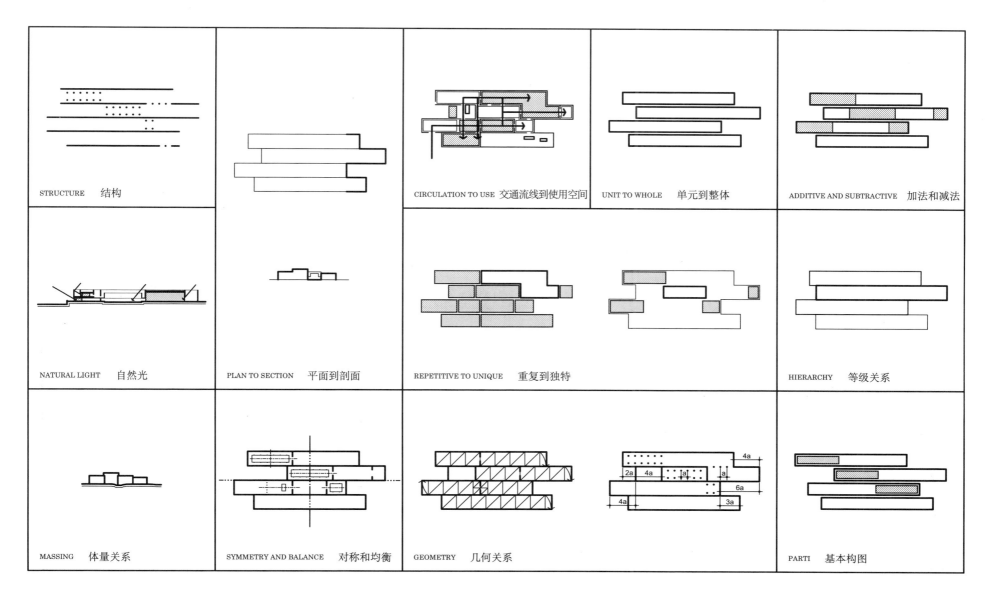

STRUCTURE　　结构

CIRCULATION TO USE　交通流线到使用空间

UNIT TO WHOLE　　单元到整体

ADDITIVE AND SUBTRACTIVE　加法和减法

NATURAL LIGHT　　自然光

PLAN TO SECTION　　平面到剖面

REPETITIVE TO UNIQUE　　重复到独特

HIERARCHY　　等级关系

MASSING　　体量关系

SYMMETRY AND BALANCE　　对称和均衡

GEOMETRY　　几何关系

PARTI　　基本构图

斯韦勒·费恩
S V E R R E   F E H N

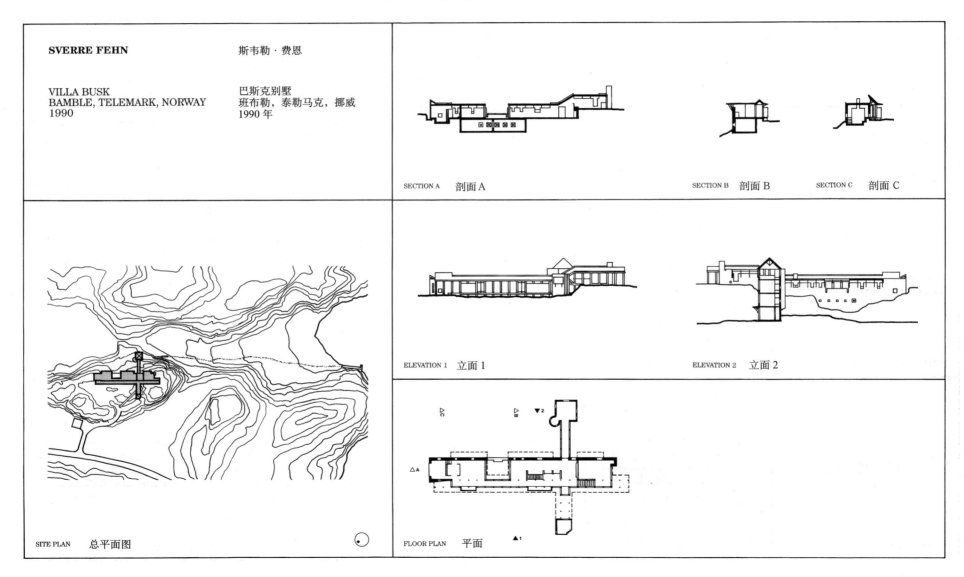

SVERRE FEHN   斯韦勒·费恩

VILLA BUSK   巴斯克别墅
BAMBLE, TELEMARK, NORWAY   班布勒，泰勒马克，挪威
1990   1990 年

SECTION A 剖面 A   SECTION B 剖面 B   SECTION C 剖面 C

ELEVATION 1 立面 1   ELEVATION 2 立面 2

SITE PLAN 总平面图

FLOOR PLAN 平面

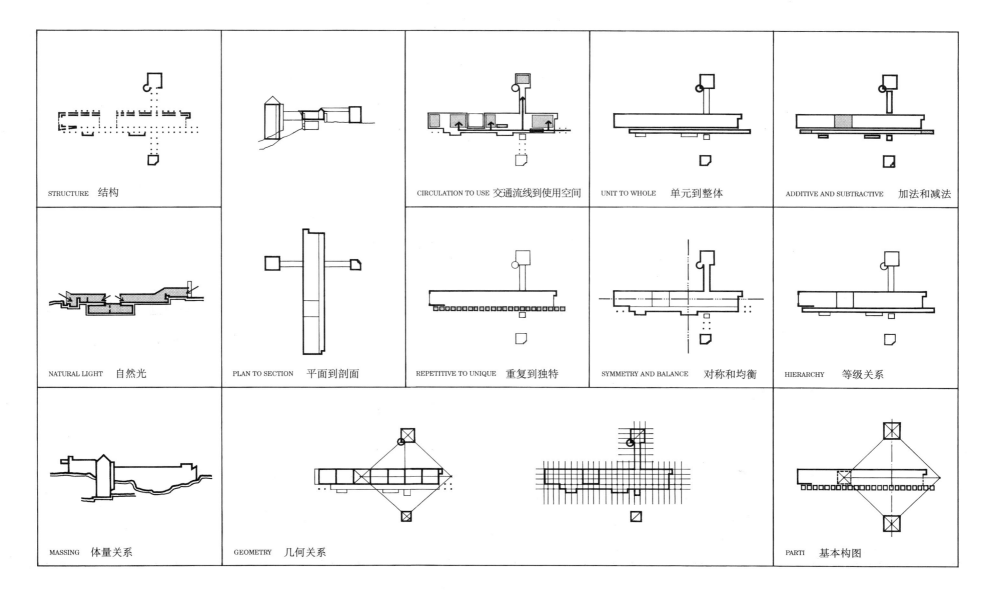

STRUCTURE 结构

CIRCULATION TO USE 交通流线到使用空间

UNIT TO WHOLE 单元到整体

ADDITIVE AND SUBTRACTIVE 加法和减法

NATURAL LIGHT 自然光

PLAN TO SECTION 平面到剖面

REPETITIVE TO UNIQUE 重复到独特

SYMMETRY AND BALANCE 对称和均衡

HIERARCHY 等级关系

MASSING 体量关系

GEOMETRY 几何关系

PARTI 基本构图

SVERRE FEHN　　　　　　斯韦勒·费恩

THE GLACIER MUSEUM　　　格拉西尔博物馆
FJAERLAND, BALESTRAND, NORWAY　菲耶兰，巴莱斯特兰，挪威
1991　　　　　　　　　　　1991 年

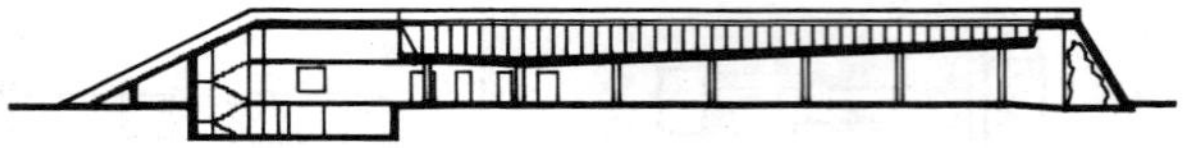

SECTION A　剖面 A

SECTION B　剖面 B

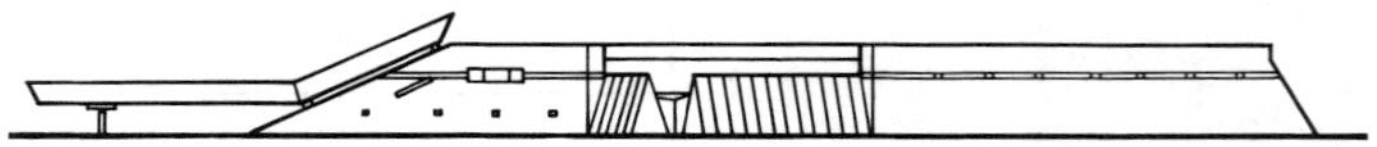

ELEVATION 1　立面 1

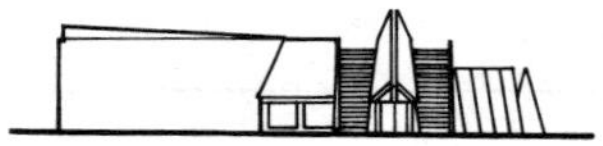

ELEVATION 2　立面 2

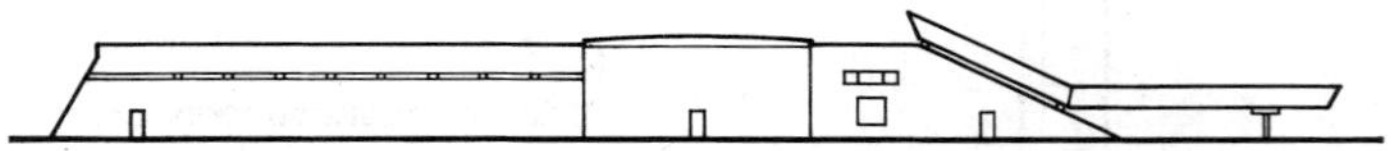

ELEVATION 3　立面 3

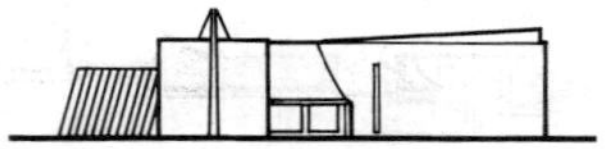

ELEVATION 4　立面 4

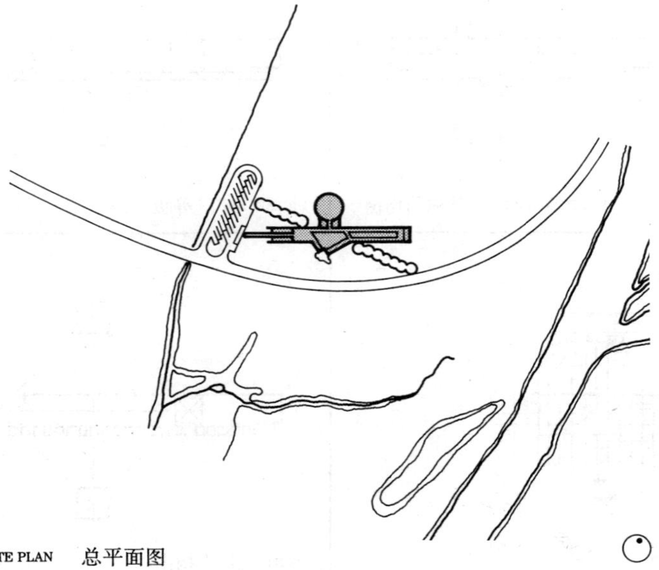

SITE PLAN　总平面图

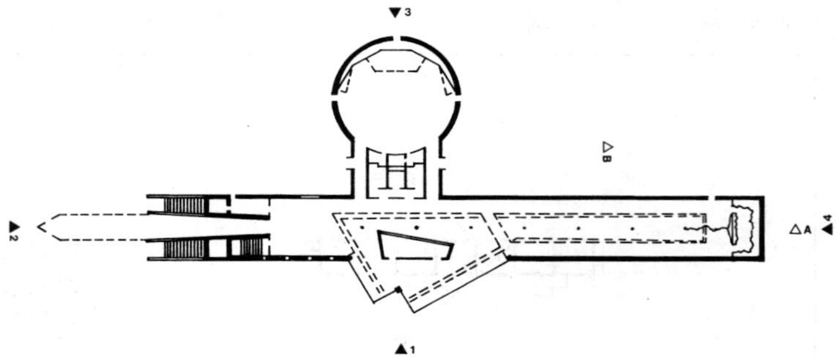

MAIN FLOOR PLAN　主层平面

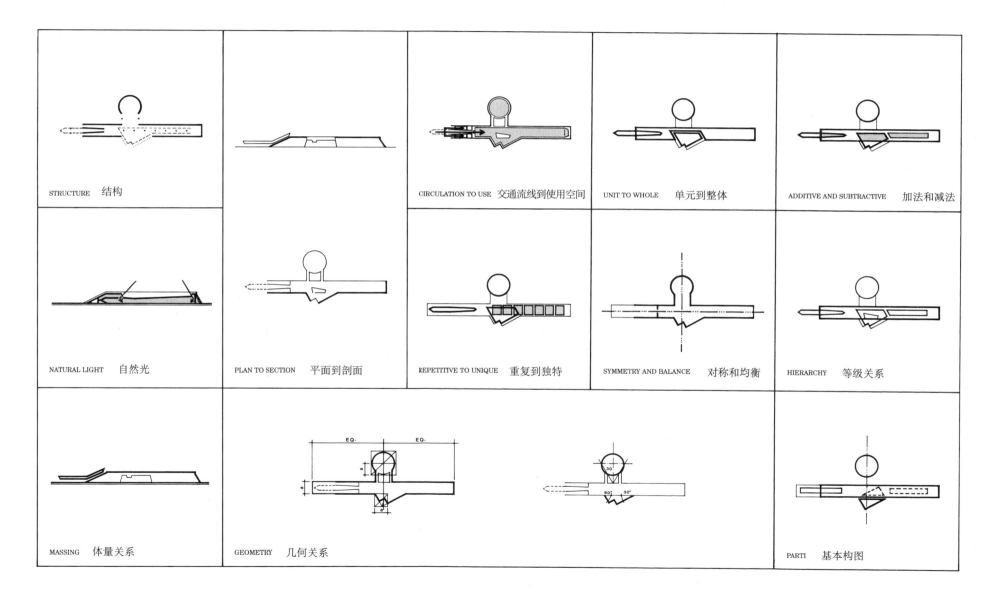

STRUCTURE　结构

CIRCULATION TO USE　交通流线到使用空间

UNIT TO WHOLE　单元到整体

ADDITIVE AND SUBTRACTIVE　加法和减法

NATURAL LIGHT　自然光

PLAN TO SECTION　平面到剖面

REPETITIVE TO UNIQUE　重复到独特

SYMMETRY AND BALANCE　对称和均衡

HIERARCHY　等级关系

MASSING　体量关系

GEOMETRY　几何关系

PARTI　基本构图

罗马尔多·朱尔戈拉
ROMALDO GIURGOLA

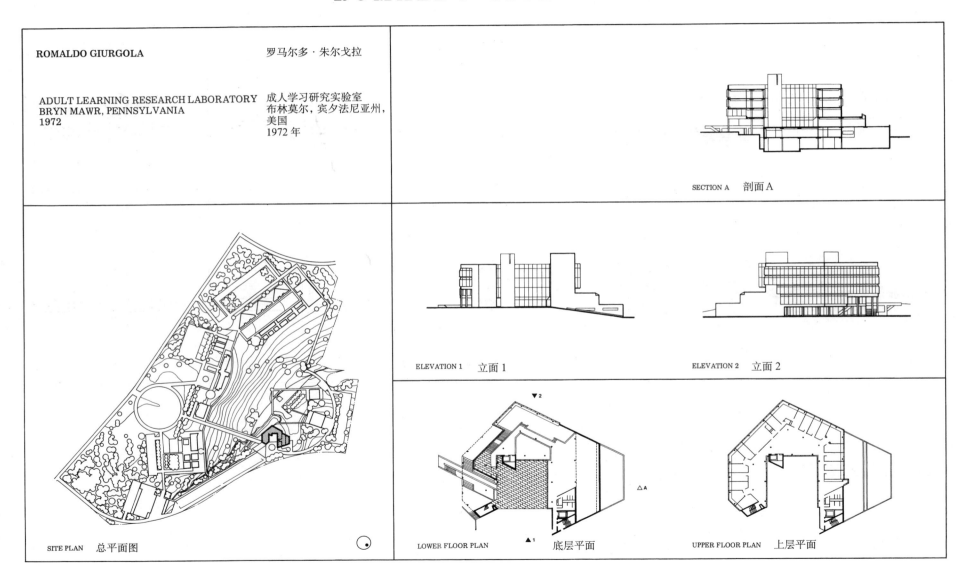

ROMALDO GIURGOLA   罗马尔多·朱尔戈拉

ADULT LEARNING RESEARCH LABORATORY   成人学习研究实验室
BRYN MAWR, PENNSYLVANIA   布林莫尔，宾夕法尼亚州，
1972   美国
                                      1972 年

SECTION A   剖面 A

ELEVATION 1   立面 1   ELEVATION 2   立面 2

SITE PLAN   总平面图

LOWER FLOOR PLAN   底层平面   UPPER FLOOR PLAN   上层平面

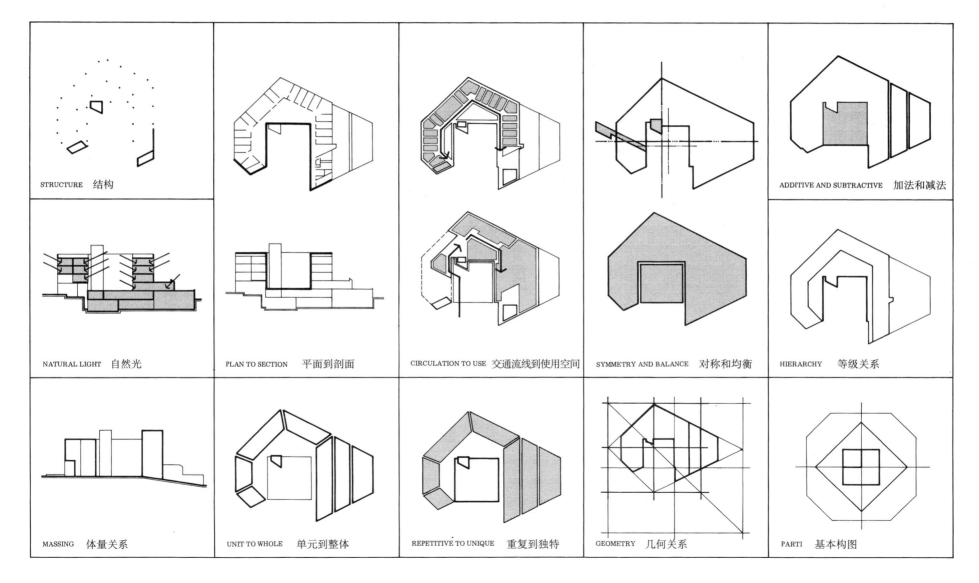

STRUCTURE　结构

NATURAL LIGHT　自然光

PLAN TO SECTION　平面到剖面

CIRCULATION TO USE　交通流线到使用空间

SYMMETRY AND BALANCE　对称和均衡

ADDITIVE AND SUBTRACTIVE　加法和减法

HIERARCHY　等级关系

MASSING　体量关系

UNIT TO WHOLE　单元到整体

REPETITIVE TO UNIQUE　重复到独特

GEOMETRY　几何关系

PARTI　基本构图

ROMALDO GIURGOLA　　　罗马尔多·朱尔戈拉

LANG MUSIC BUILDING　　　兰氏音乐中心
SWARTHMORE COLLEGE　　　斯沃斯莫尔学院
SWARTHMORE, PENNSYLVANIA, USA　斯沃斯莫尔，宾夕法尼亚州，美国
1973　　　　　　　　　　　1973 年

SECTION A　剖面 A

SECTION B　剖面 B

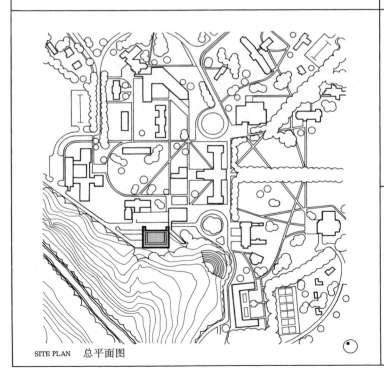

SITE PLAN　总平面图

ELEVATION 1　立面 1

ELEVATION 2　立面 2

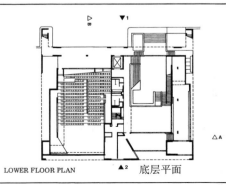

LOWER FLOOR PLAN　底层平面

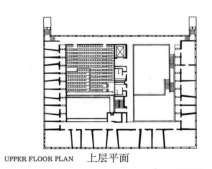

UPPER FLOOR PLAN　上层平面

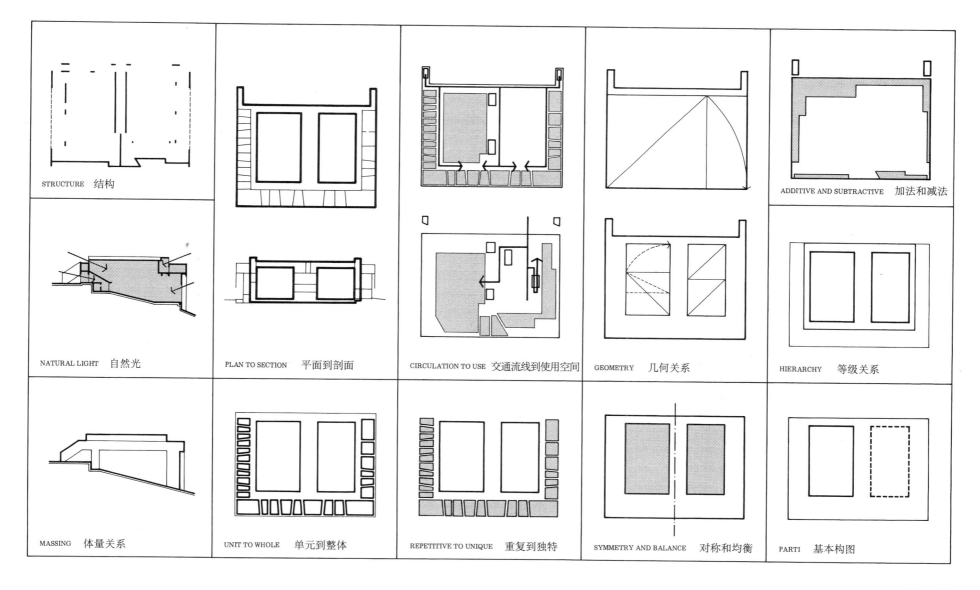

STRUCTURE 结构

NATURAL LIGHT 自然光

MASSING 体量关系

PLAN TO SECTION 平面到剖面

UNIT TO WHOLE 单元到整体

CIRCULATION TO USE 交通流线到使用空间

REPETITIVE TO UNIQUE 重复到独特

GEOMETRY 几何关系

SYMMETRY AND BALANCE 对称和均衡

ADDITIVE AND SUBTRACTIVE 加法和减法

HIERARCHY 等级关系

PARTI 基本构图

ROMALDO GIURGOLA　　　　　　　　　　罗马尔多·朱尔戈拉

STUDENT UNION　　　　　　　　　　　　　学生俱乐部
STATE UNIVERSITY COLLEGE OF NEW YORK　纽约州立大学
PLATTSBURGH, NEW YORK, USA　　　　　　普拉茨堡，纽约州，美国
1974　　　　　　　　　　　　　　　　　　1974 年

SECTION A　剖面 A

SECTION B　剖面 B

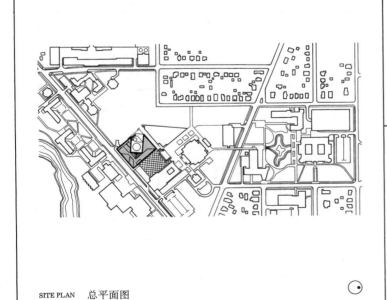

SITE PLAN　总平面图

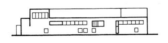

ELEVATION 1　立面 1

ELEVATION 2　立面 2

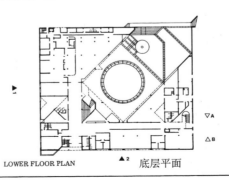

LOWER FLOOR PLAN　▲2　底层平面

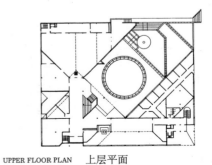

UPPER FLOOR PLAN　上层平面

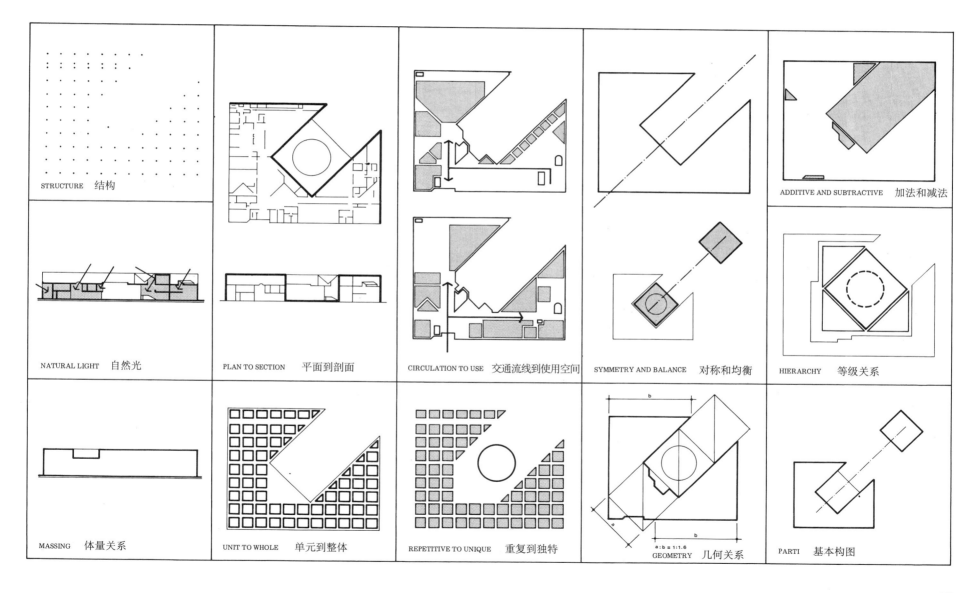

STRUCTURE　结构

ADDITIVE AND SUBTRACTIVE　加法和减法

NATURAL LIGHT　自然光

PLAN TO SECTION　平面到剖面

CIRCULATION TO USE　交通流线到使用空间

SYMMETRY AND BALANCE　对称和均衡

HIERARCHY　等级关系

MASSING　体量关系

UNIT TO WHOLE　单元到整体

REPETITIVE TO UNIQUE　重复到独特

a:b = 1:1.6
GEOMETRY　几何关系

PARTI　基本构图

ROMALDO GIURGOLA　　　　罗马尔多·朱尔戈拉

TREDYFFRIN PUBLIC LIBRARY　　特里迪弗林公共图书馆
STRAFFORD, PENNSYLVANIA, USA　斯特拉福德，宾夕法尼亚州，美国
1976　　　　　　　　　　　　　　1976 年

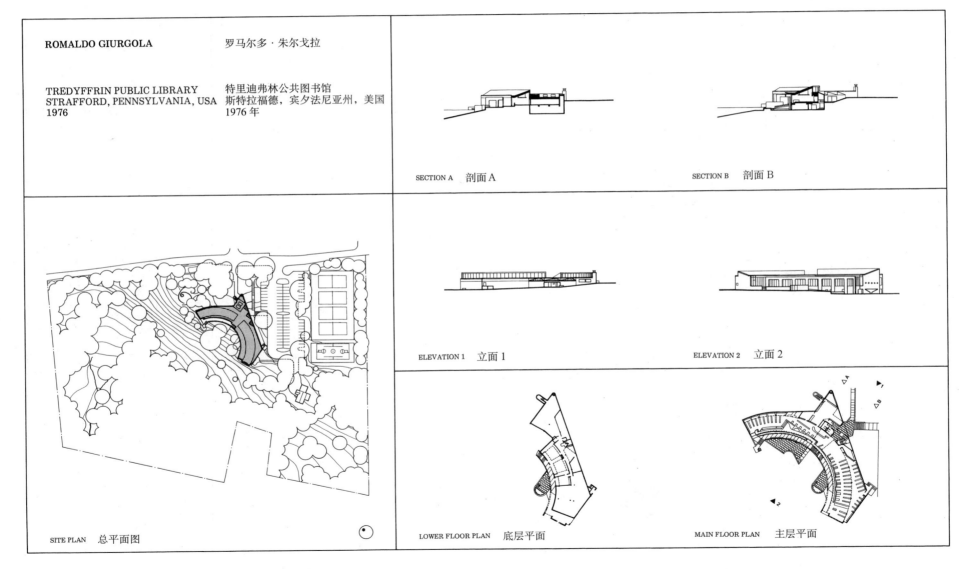

SECTION A　剖面 A

SECTION B　剖面 B

ELEVATION 1　立面 1

ELEVATION 2　立面 2

SITE PLAN　总平面图

LOWER FLOOR PLAN　底层平面

MAIN FLOOR PLAN　主层平面

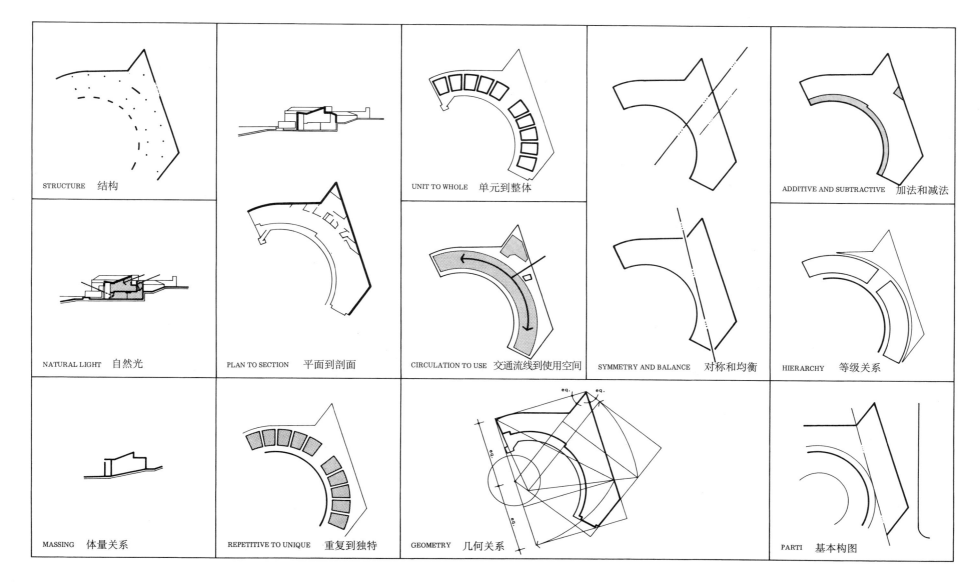

STRUCTURE　结构

NATURAL LIGHT　自然光

MASSING　体量关系

PLAN TO SECTION　平面到剖面

REPETITIVE TO UNIQUE　重复到独特

UNIT TO WHOLE　单元到整体

CIRCULATION TO USE　交通流线到使用空间

GEOMETRY　几何关系

SYMMETRY AND BALANCE　对称和均衡

ADDITIVE AND SUBTRACTIVE　加法和减法

HIERARCHY　等级关系

PARTI　基本构图

尼古拉斯·豪克斯穆尔
NICHOLAS HAWKSMOOR

NICHOLAS HAWKSMOOR   尼古拉斯·豪克斯穆尔

EASTON NESTON   伊斯顿·内斯顿府邸
NORTHAMPTONSHIRE, ENGLAND 北安普敦郡，英国
c. 1695-1710   约 1695-1710 年

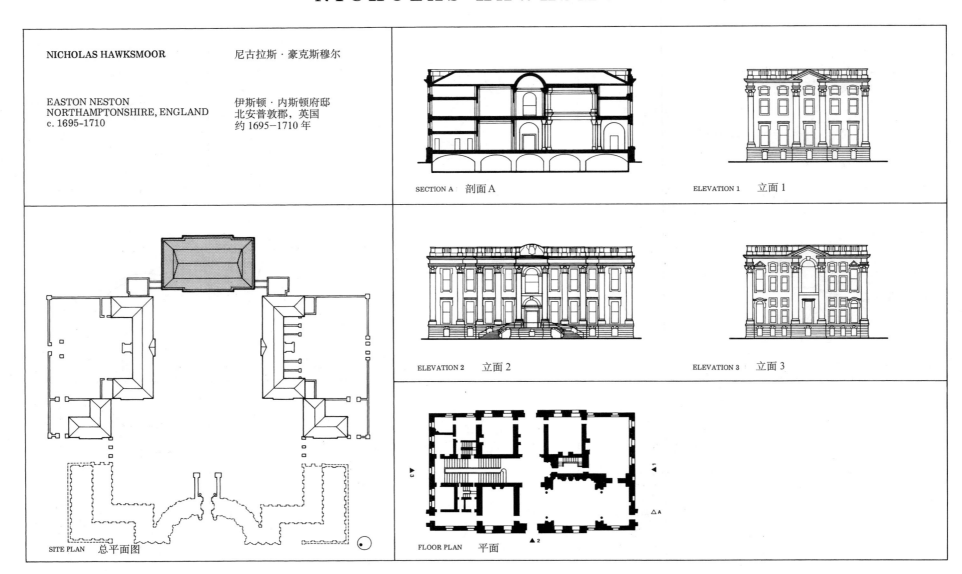

SECTION A   剖面 A

ELEVATION 1   立面 1

ELEVATION 2   立面 2

ELEVATION 3   立面 3

SITE PLAN   总平面图

FLOOR PLAN   平面

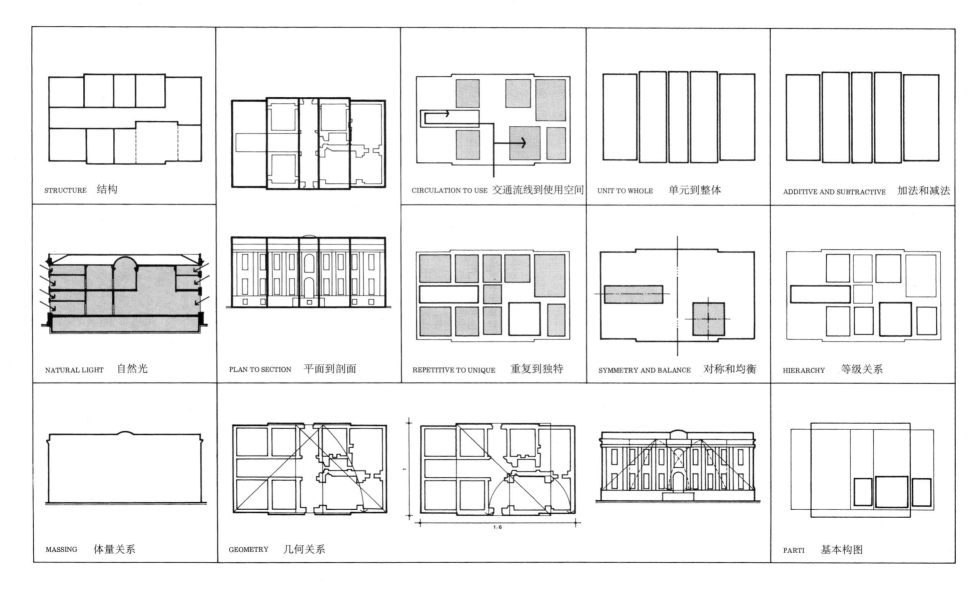

STRUCTURE　结构

CIRCULATION TO USE　交通流线到使用空间

UNIT TO WHOLE　单元到整体

ADDITIVE AND SUBTRACTIVE　加法和减法

NATURAL LIGHT　自然光

PLAN TO SECTION　平面到剖面

REPETITIVE TO UNIQUE　重复到独特

SYMMETRY AND BALANCE　对称和均衡

HIERARCHY　等级关系

MASSING　体量关系

GEOMETRY　几何关系

PARTI　基本构图

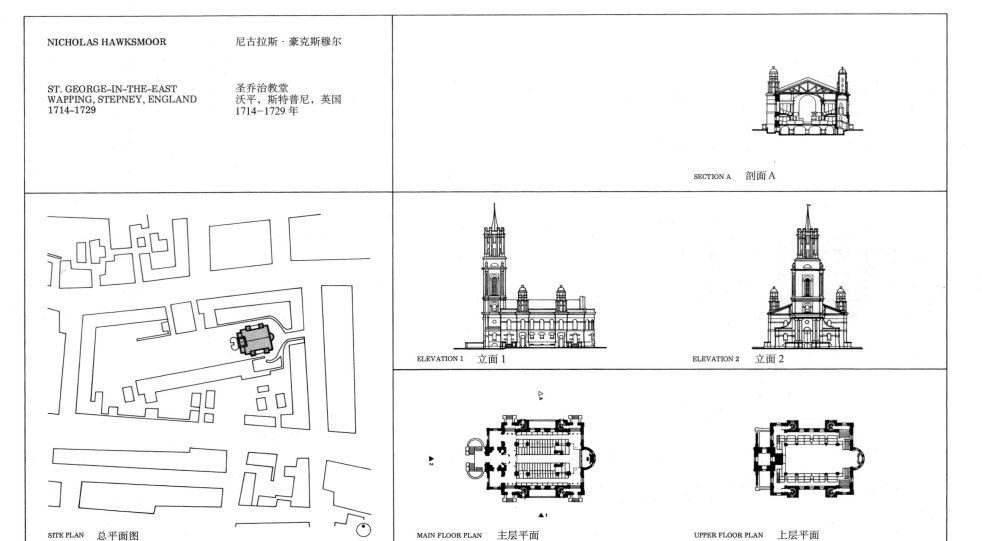

NICHOLAS HAWKSMOOR　　　尼古拉斯·豪克斯穆尔

ST. GEORGE-IN-THE-EAST
WAPPING, STEPNEY, ENGLAND
1714-1729

圣乔治教堂
沃平，斯特普尼，英国
1714-1729 年

SECTION A　剖面 A

SITE PLAN　总平面图

ELEVATION 1　立面 1

ELEVATION 2　立面 2

MAIN FLOOR PLAN　主层平面

UPPER FLOOR PLAN　上层平面

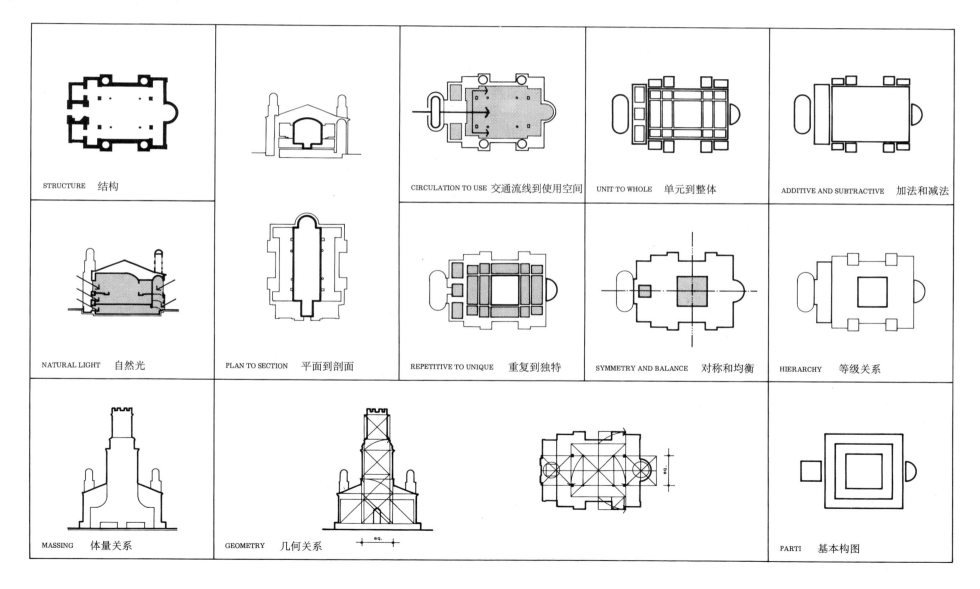

STRUCTURE　结构

CIRCULATION TO USE 交通流线到使用空间

UNIT TO WHOLE　单元到整体

ADDITIVE AND SUBTRACTIVE　加法和减法

NATURAL LIGHT　自然光

PLAN TO SECTION　平面到剖面

REPETITIVE TO UNIQUE　重复到独特

SYMMETRY AND BALANCE　对称和均衡

HIERARCHY　等级关系

MASSING　体量关系

GEOMETRY　几何关系

PARTI　基本构图

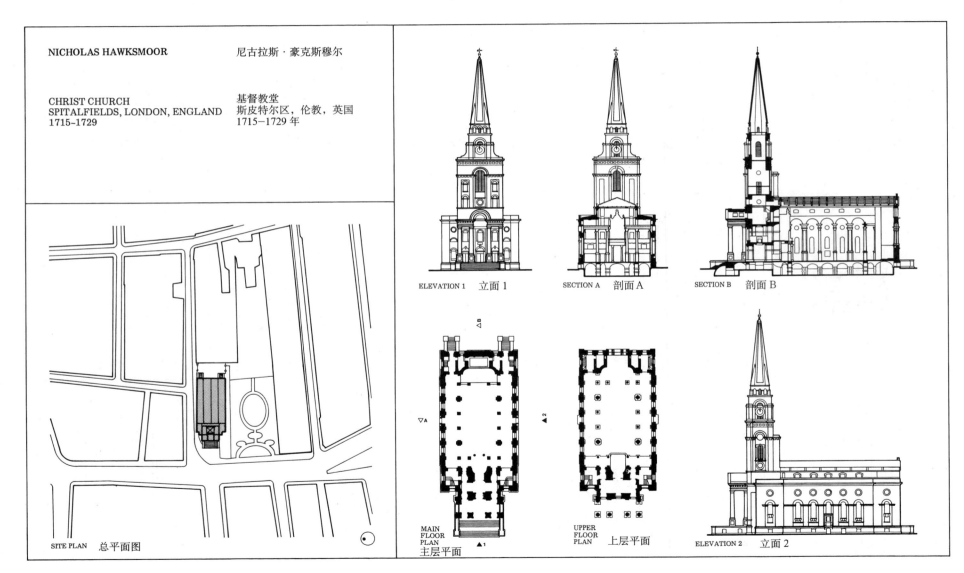

NICHOLAS HAWKSMOOR　　　　尼古拉斯·豪克斯穆尔

CHRIST CHURCH　　　　　　　基督教堂
SPITALFIELDS, LONDON, ENGLAND　斯皮特尔区，伦教，英国
1715~1729　　　　　　　　　　1715~1729 年

SITE PLAN　　总平面图

ELEVATION 1　立面 1

SECTION A　剖面 A

SECTION B　剖面 B

MAIN
FLOOR
PLAN
主层平面

UPPER
FLOOR
PLAN　上层平面

ELEVATION 2　立面 2

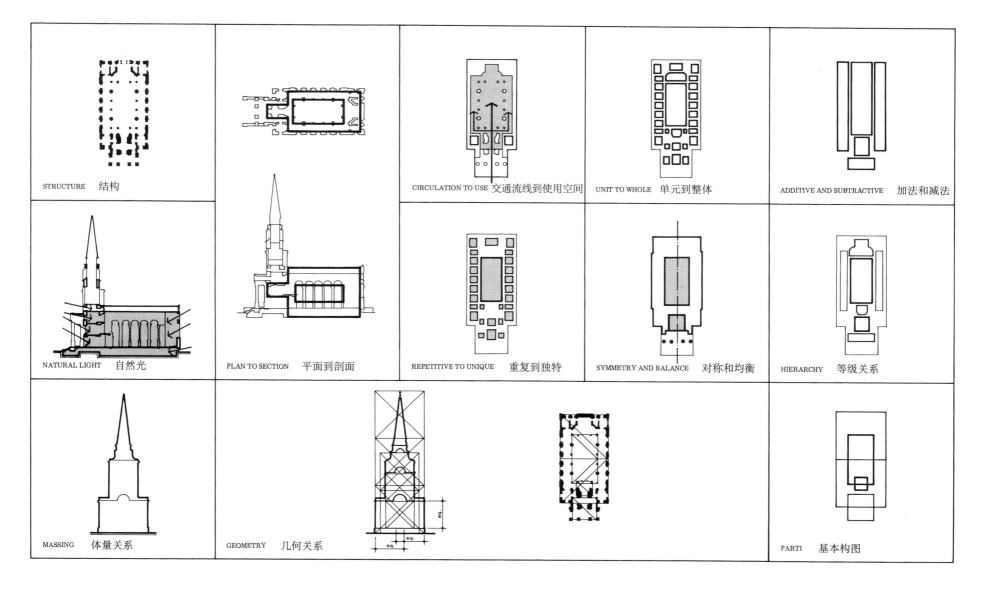

STRUCTURE　结构

CIRCULATION TO USE 交通流线到使用空间

UNIT TO WHOLE　单元到整体

ADDITIVE AND SUBTRACTIVE　加法和减法

NATURAL LIGHT　自然光

PLAN TO SECTION　平面到剖面

REPETITIVE TO UNIQUE　重复到独特

SYMMETRY AND BALANCE　对称和均衡

HIERARCHY　等级关系

MASSING　体量关系

GEOMETRY　几何关系

PARTI　基本构图

83

NICHOLAS HAWKSMOOR　　尼古拉斯·豪克斯穆尔

ST. MARY WOOLNOTH　　圣玛丽·伍尔诺思教堂
LONDON, ENGLAND　　伦教，英国
1716-1724　　1716–1724 年

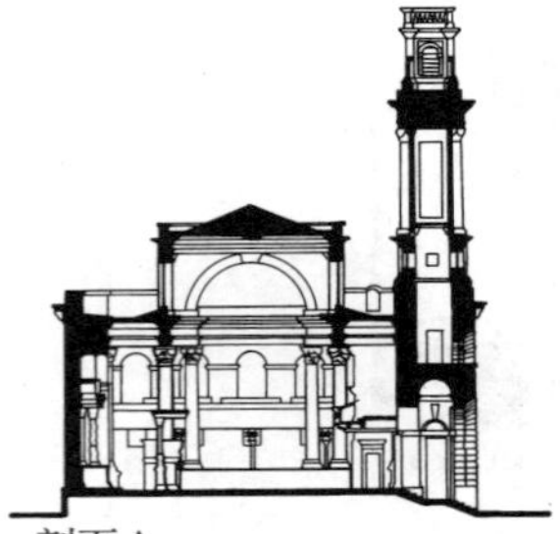

SECTION A　剖面 A

SITE PLAN　总平面图

ELEVATION 1　立面 1

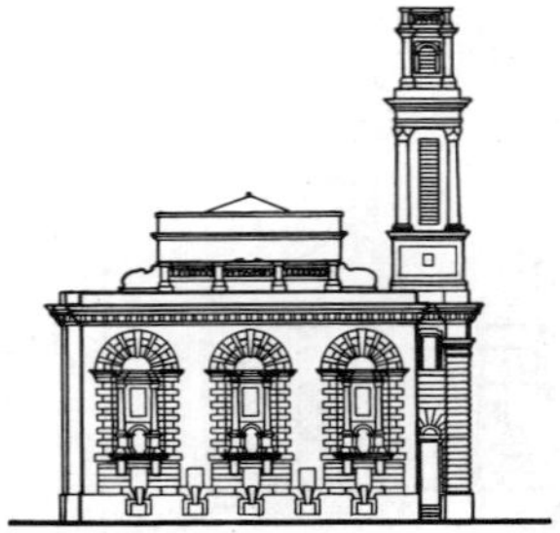

ELEVATION 2　立面 2

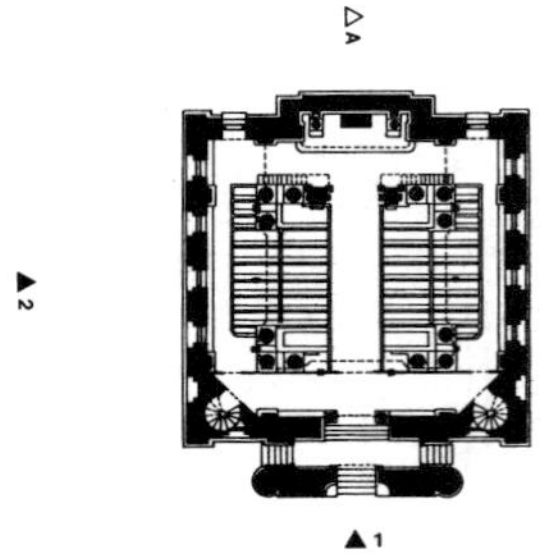

FLOOR PLAN　平面

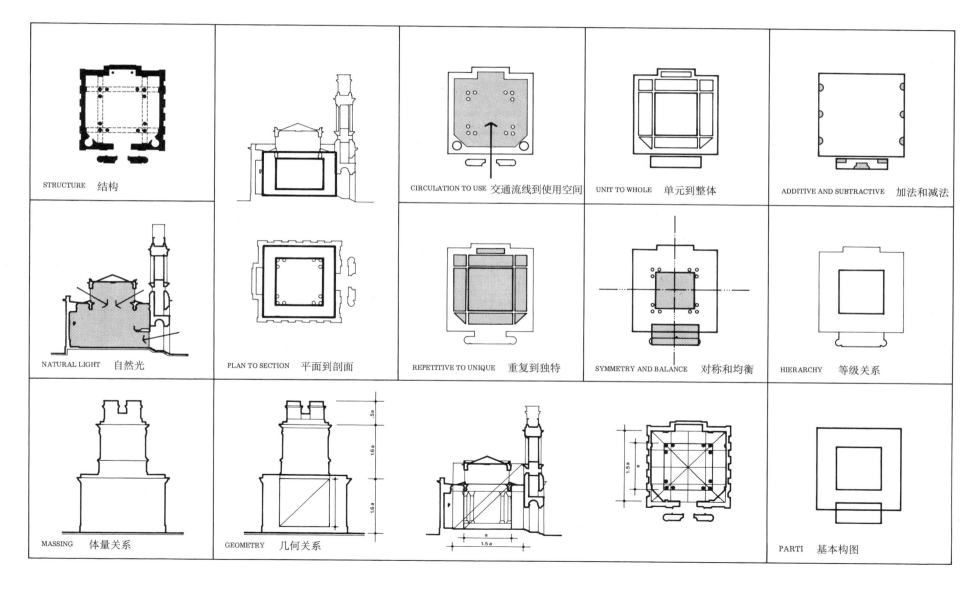

STRUCTURE　结构

CIRCULATION TO USE　交通流线到使用空间

UNIT TO WHOLE　单元到整体

ADDITIVE AND SUBTRACTIVE　加法和减法

NATURAL LIGHT　自然光

PLAN TO SECTION　平面到剖面

REPETITIVE TO UNIQUE　重复到独特

SYMMETRY AND BALANCE　对称和均衡

HIERARCHY　等级关系

MASSING　体量关系

GEOMETRY　几何关系

PARTI　基本构图

赫尔佐格与德梅隆

HERZOG & DE MEURON

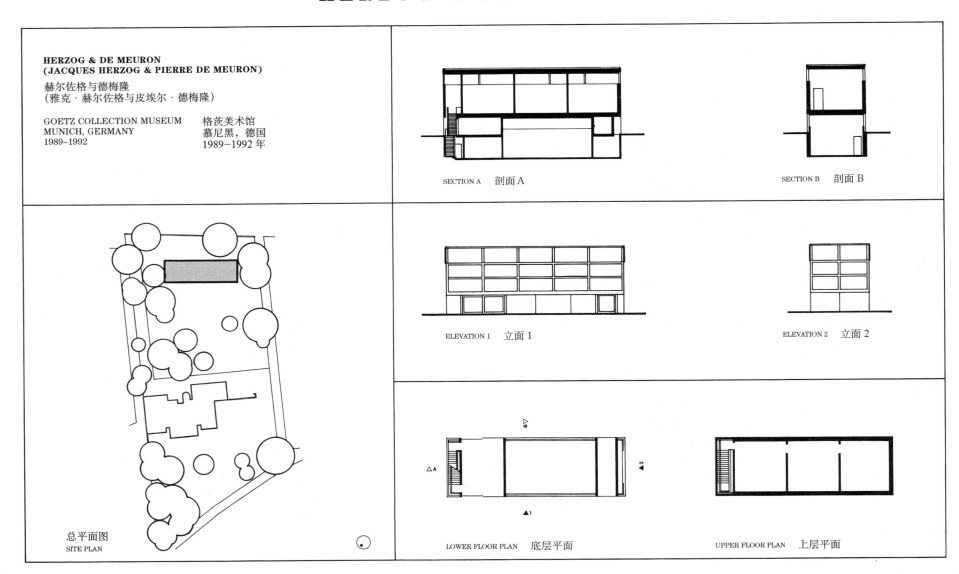

HERZOG & DE MEURON
(JACQUES HERZOG & PIERRE DE MEURON)

赫尔佐格与德梅隆
（雅克·赫尔佐格与皮埃尔·德梅隆）

GOETZ COLLECTION MUSEUM　　格茨美术馆
MUNICH, GERMANY　　　　　　慕尼黑，德国
1989–1992　　　　　　　　　　1989–1992 年

SECTION A　剖面 A

SECTION B　剖面 B

ELEVATION 1　立面 1

ELEVATION 2　立面 2

总平面图
SITE PLAN

LOWER FLOOR PLAN　底层平面

UPPER FLOOR PLAN　上层平面

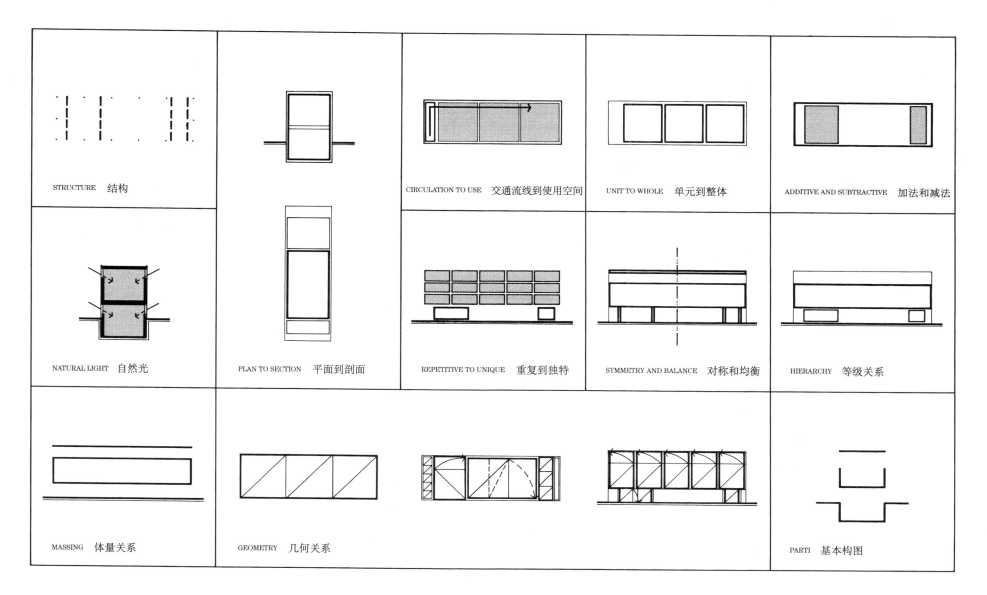

STRUCTURE　结构

CIRCULATION TO USE　交通流线到使用空间

UNIT TO WHOLE　单元到整体

ADDITIVE AND SUBTRACTIVE　加法和减法

NATURAL LIGHT　自然光

PLAN TO SECTION　平面到剖面

REPETITIVE TO UNIQUE　重复到独特

SYMMETRY AND BALANCE　对称和均衡

HIERARCHY　等级关系

MASSING　体量关系

GEOMETRY　几何关系

PARTI　基本构图

**HERZOG & DE MEURON
(JACQUES HERZOG & PIERRE DE MEURON)**

赫尔佐格与德梅隆
（雅克·赫尔佐格与皮埃尔·德梅隆）

DOMINUS WINERY   天主葡萄酒厂
YOUNTVILLE, CALIFORNIA, USA   扬特维尔市，加利福尼亚州，美国
1995–1998   1995–1998 年

SECTION A 剖面 A   SECTION B 剖面 B

ELEVATION 1 立面 1   ELEVATION 2 立面 2

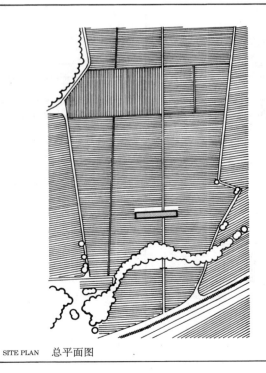

SITE PLAN 总平面图

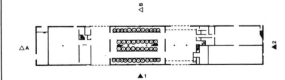

MAIN FLOOR PLAN 主层平面   UPPER FLOOR PLAN 上层平面

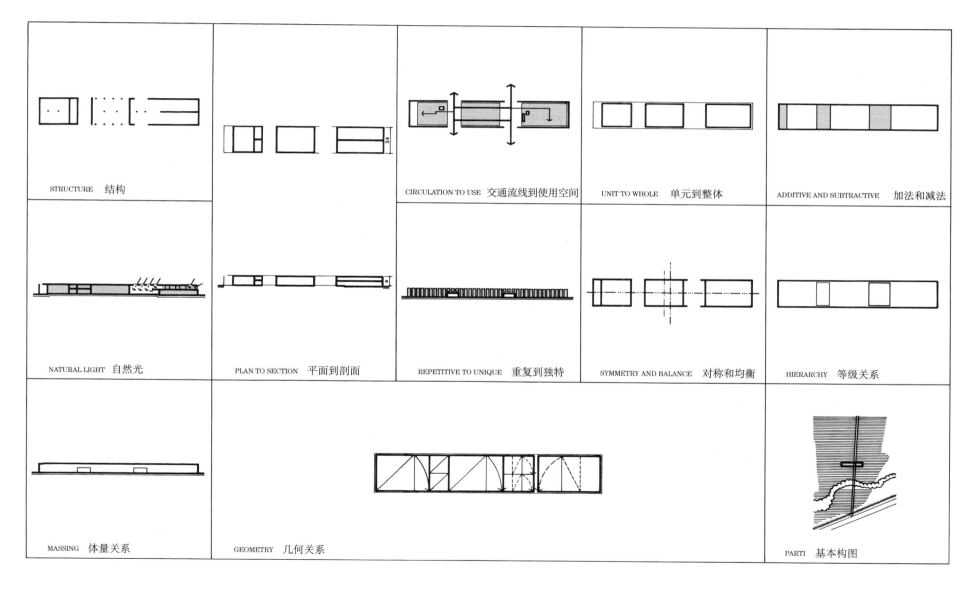

STRUCTURE　结构

CIRCULATION TO USE　交通流线到使用空间

UNIT TO WHOLE　单元到整体

ADDITIVE AND SUBTRACTIVE　加法和减法

NATURAL LIGHT　自然光

PLAN TO SECTION　平面到剖面

REPETITIVE TO UNIQUE　重复到独特

SYMMETRY AND BALANCE　对称和均衡

HIERARCHY　等级关系

MASSING　体量关系

GEOMETRY　几何关系

PARTI　基本构图

斯蒂文·霍尔
STEVEN HOLL

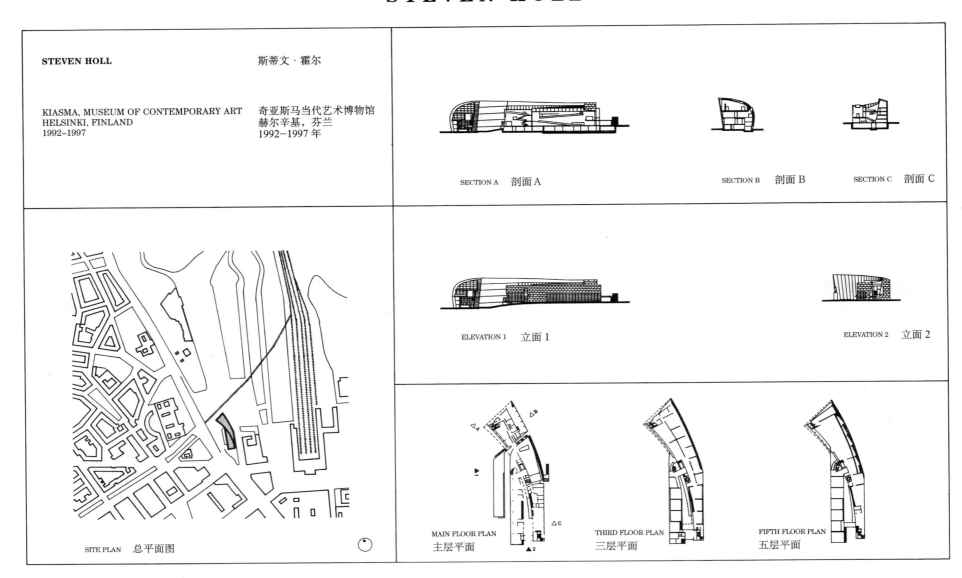

STEVEN HOLL　　　　　　　　　　斯蒂文·霍尔

KIASMA, MUSEUM OF CONTEMPORARY ART　　奇亚斯马当代艺术博物馆
HELSINKI, FINLAND　　　　　　　　赫尔辛基，芬兰
1992–1997　　　　　　　　　　　　1992–1997 年

SECTION A　剖面 A　　　SECTION B　剖面 B　　　SECTION C　剖面 C

ELEVATION 1　立面 1　　　ELEVATION 2　立面 2

SITE PLAN　总平面图

MAIN FLOOR PLAN　　THIRD FLOOR PLAN　　FIFTH FLOOR PLAN
主层平面　　　　　三层平面　　　　　五层平面

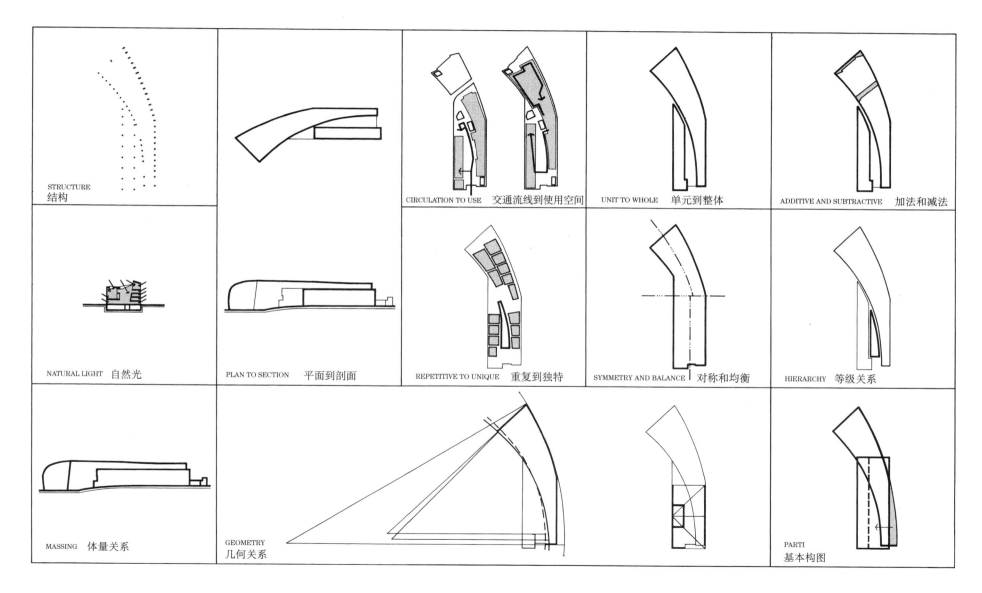

STRUCTURE 结构

CIRCULATION TO USE 交通流线到使用空间

UNIT TO WHOLE 单元到整体

ADDITIVE AND SUBTRACTIVE 加法和减法

NATURAL LIGHT 自然光

PLAN TO SECTION 平面到剖面

REPETITIVE TO UNIQUE 重复到独特

SYMMETRY AND BALANCE 对称和均衡

HIERARCHY 等级关系

MASSING 体量关系

GEOMETRY 几何关系

PARTI 基本构图

STEVEN HOLL　　　　斯蒂文·霍尔

CHAPEL OF ST. IGNATIUS　　圣伊格内修斯小教堂
SEATTLE UNIVERSITY　　　西雅图大学
SEATTLE, WASHINGTON, USA　西雅图，华盛顿州，美国
1994–1997　　　　　　　　1994–1997 年

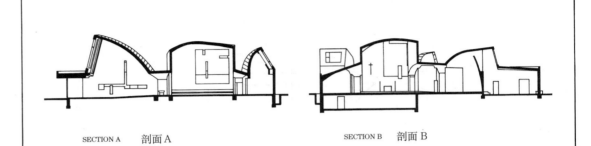

SECTION A　剖面 A　　　　　　SECTION B　剖面 B

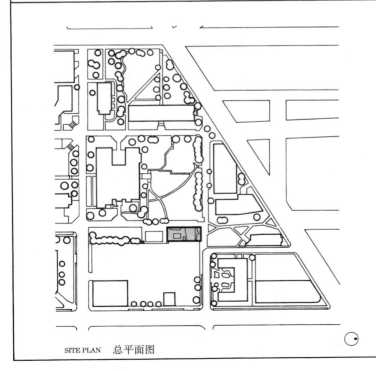

SITE PLAN　总平面图

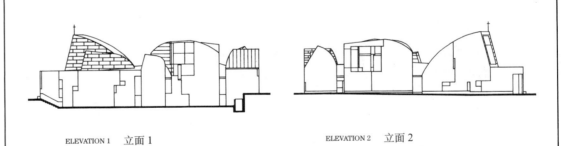

ELEVATION 1　立面 1　　　　　ELEVATION 2　立面 2

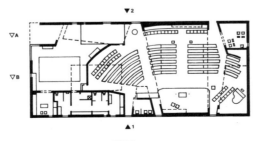

FLOOR PLAN　平面

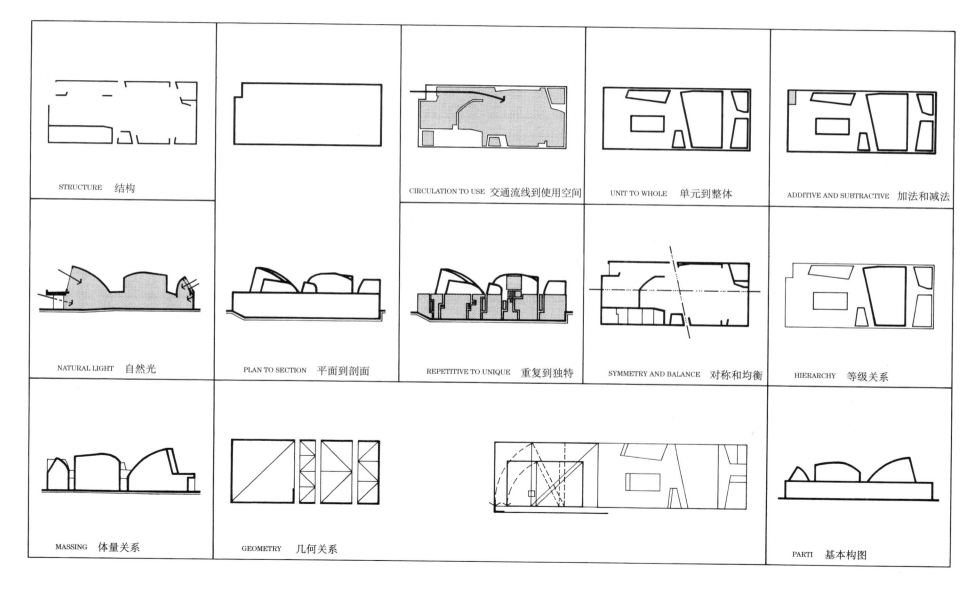

STRUCTURE　结构

CIRCULATION TO USE　交通流线到使用空间

UNIT TO WHOLE　单元到整体

ADDITIVE AND SUBTRACTIVE　加法和减法

NATURAL LIGHT　自然光

PLAN TO SECTION　平面到剖面

REPETITIVE TO UNIQUE　重复到独特

SYMMETRY AND BALANCE　对称和均衡

HIERARCHY　等级关系

MASSING　体量关系

GEOMETRY　几何关系

PARTI　基本构图

伊东丰雄
TOYO ITO

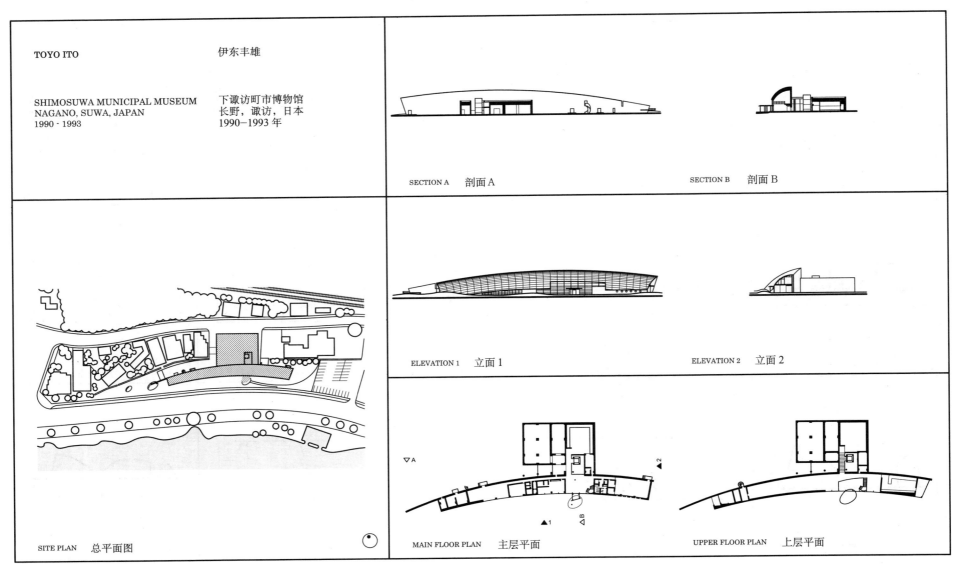

TOYO ITO　　　　　　　　　伊东丰雄

SHIMOSUWA MUNICIPAL MUSEUM　下诹访町市博物馆
NAGANO, SUWA, JAPAN　　　　长野，诹访，日本
1990 - 1993　　　　　　　　　1990—1993 年

SITE PLAN　总平面图

SECTION A　剖面 A

SECTION B　剖面 B

ELEVATION 1　立面 1

ELEVATION 2　立面 2

MAIN FLOOR PLAN　主层平面

UPPER FLOOR PLAN　上层平面

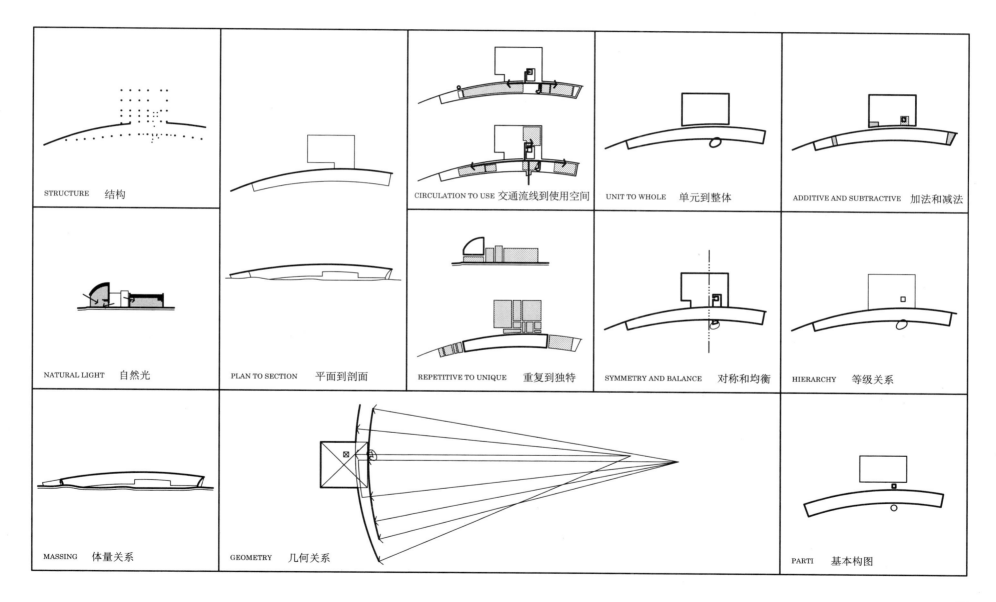

STRUCTURE 结构		CIRCULATION TO USE 交通流线到使用空间	UNIT TO WHOLE 单元到整体	ADDITIVE AND SUBTRACTIVE 加法和减法
NATURAL LIGHT 自然光	PLAN TO SECTION 平面到剖面	REPETITIVE TO UNIQUE 重复到独特	SYMMETRY AND BALANCE 对称和均衡	HIERARCHY 等级关系
MASSING 体量关系	GEOMETRY 几何关系			PARTI 基本构图

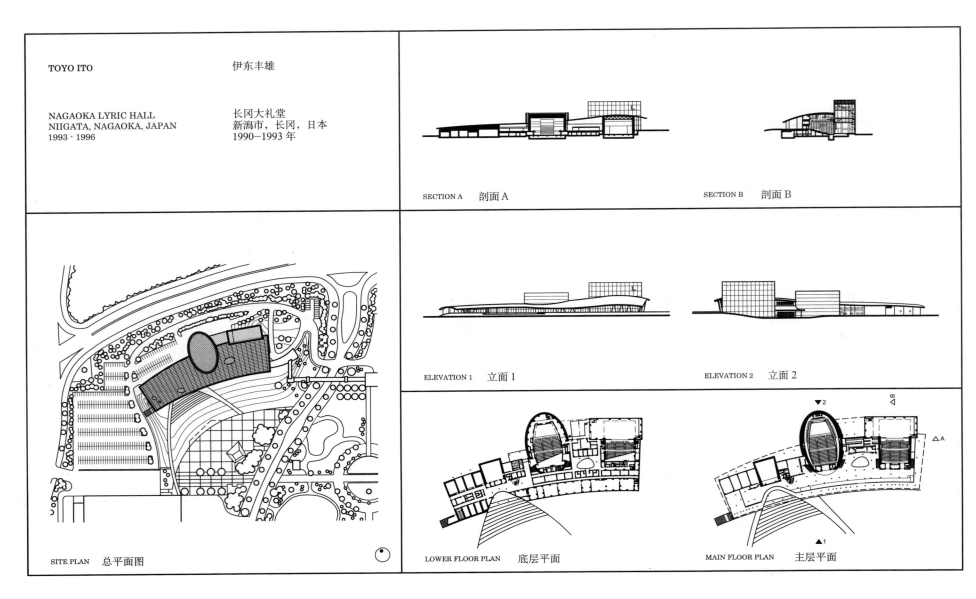

TOYO ITO   伊东丰雄

NAGAOKA LYRIC HALL   长冈大礼堂
NIIGATA, NAGAOKA, JAPAN   新潟市，长冈，日本
1993 - 1996   1990—1993 年

SECTION A   剖面 A   SECTION B   剖面 B

ELEVATION 1   立面 1   ELEVATION 2   立面 2

SITE PLAN   总平面图

LOWER FLOOR PLAN   底层平面   MAIN FLOOR PLAN   主层平面

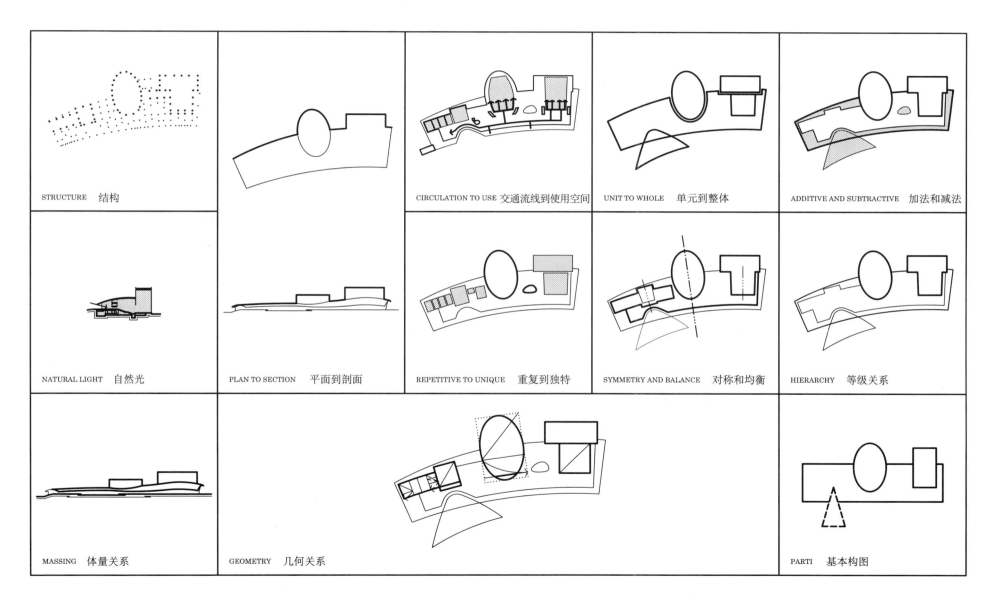

STRUCTURE　结构

CIRCULATION TO USE　交通流线到使用空间

UNIT TO WHOLE　单元到整体

ADDITIVE AND SUBTRACTIVE　加法和减法

NATURAL LIGHT　自然光

PLAN TO SECTION　平面到剖面

REPETITIVE TO UNIQUE　重复到独特

SYMMETRY AND BALANCE　对称和均衡

HIERARCHY　等级关系

MASSING　体量关系

GEOMETRY　几何关系

PARTI　基本构图

路易斯·I·康
LOUIS I. KAHN

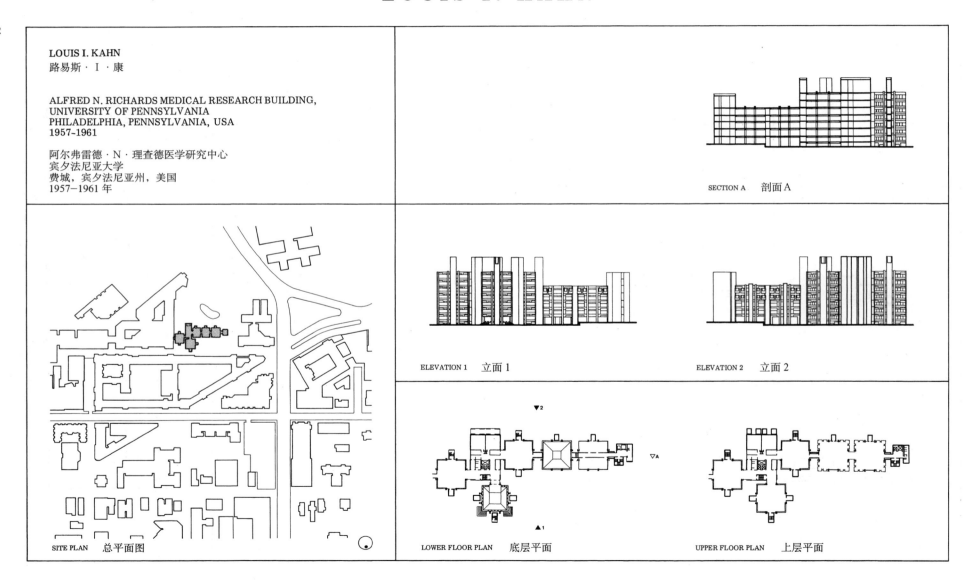

LOUIS I. KAHN
路易斯·I·康

ALFRED N. RICHARDS MEDICAL RESEARCH BUILDING,
UNIVERSITY OF PENNSYLVANIA
PHILADELPHIA, PENNSYLVANIA, USA
1957-1961

阿尔弗雷德·N·理查德医学研究中心
宾夕法尼亚大学
费城，宾夕法尼亚州，美国
1957-1961 年

SECTION A   剖面A

ELEVATION 1   立面 1

ELEVATION 2   立面 2

SITE PLAN   总平面图

LOWER FLOOR PLAN   底层平面

UPPER FLOOR PLAN   上层平面

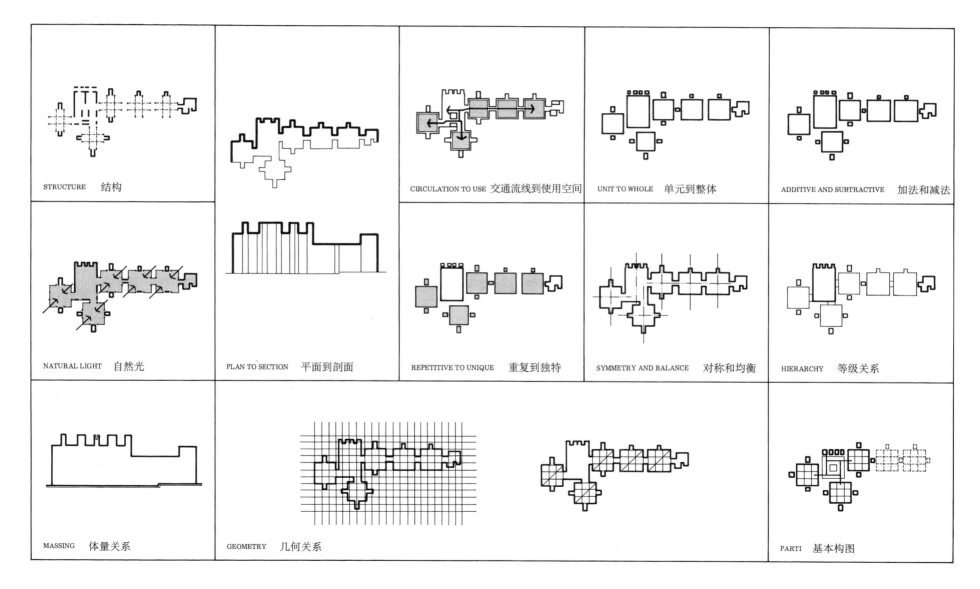

STRUCTURE　结构

CIRCULATION TO USE　交通流线到使用空间

UNIT TO WHOLE　单元到整体

ADDITIVE AND SUBTRACTIVE　加法和减法

NATURAL LIGHT　自然光

PLAN TO SECTION　平面到剖面

REPETITIVE TO UNIQUE　重复到独特

SYMMETRY AND BALANCE　对称和均衡

HIERARCHY　等级关系

MASSING　体量关系

GEOMETRY　几何关系

PARTI　基本构图

LOUIS I. KAHN
路易斯·Ⅰ·康

SALK INSTITUTE OF BIOLOGICAL STUDIES
LA JOLLA, CALIFORNIA, USA
1959-1965

萨尔克生物研究所
拉霍亚，加利福尼亚州，美国
1959-1965 年

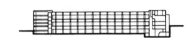

SECTION A   剖面 A

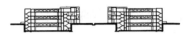

SECTION B   剖面 B

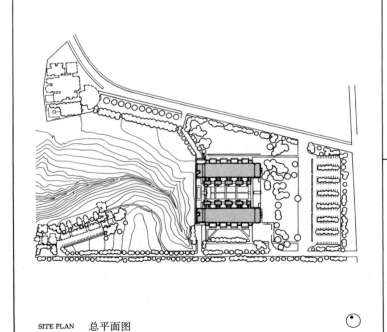

SITE PLAN   总平面图

ELEVATION 1   立面 1

ELEVATION 2   立面 2

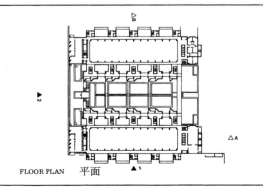

FLOOR PLAN   平面

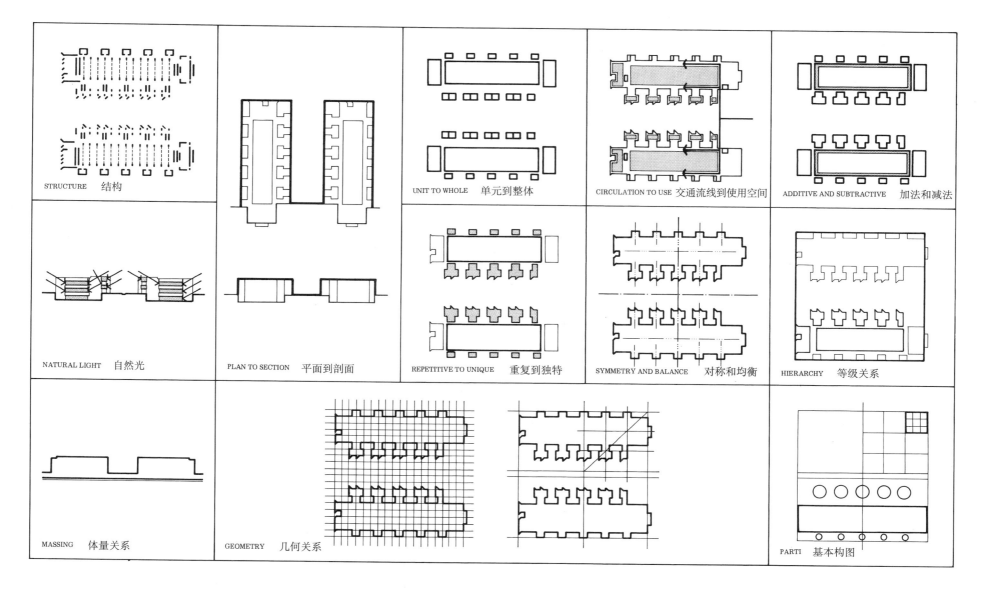

STRUCTURE　结构

UNIT TO WHOLE　单元到整体

CIRCULATION TO USE　交通流线到使用空间

ADDITIVE AND SUBTRACTIVE　加法和减法

NATURAL LIGHT　自然光

PLAN TO SECTION　平面到剖面

REPETITIVE TO UNIQUE　重复到独特

SYMMETRY AND BALANCE　对称和均衡

HIERARCHY　等级关系

MASSING　体量关系

GEOMETRY　几何关系

PARTI　基本构图

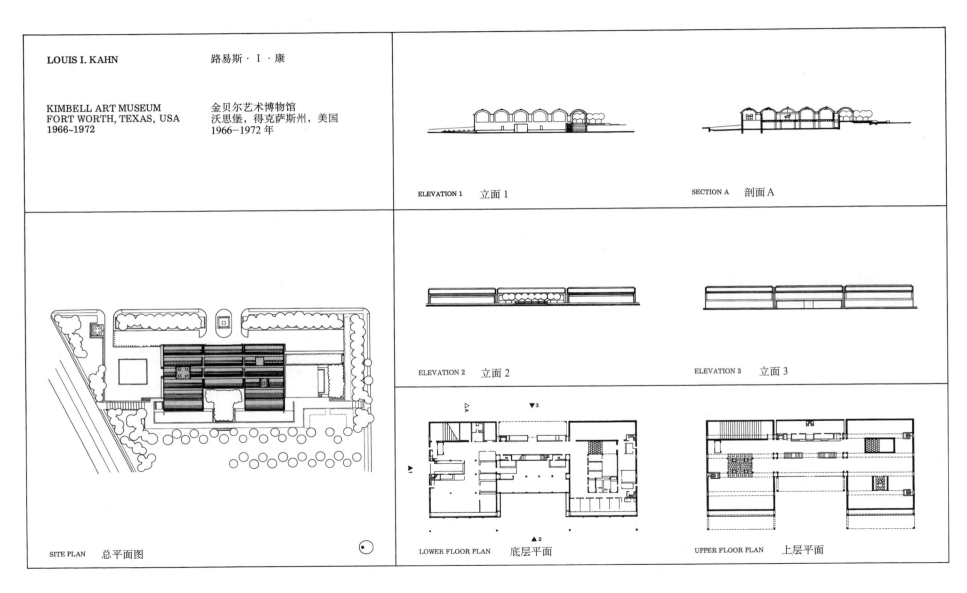

LOUIS I. KAHN　　　路易斯·Ⅰ·康

KIMBELL ART MUSEUM　　金贝尔艺术博物馆
FORT WORTH, TEXAS, USA　沃思堡，得克萨斯州，美国
1966-1972　　　　　　　1966-1972 年

ELEVATION 1　立面 1

SECTION A　剖面 A

ELEVATION 2　立面 2

ELEVATION 3　立面 3

SITE PLAN　总平面图

LOWER FLOOR PLAN　底层平面

UPPER FLOOR PLAN　上层平面

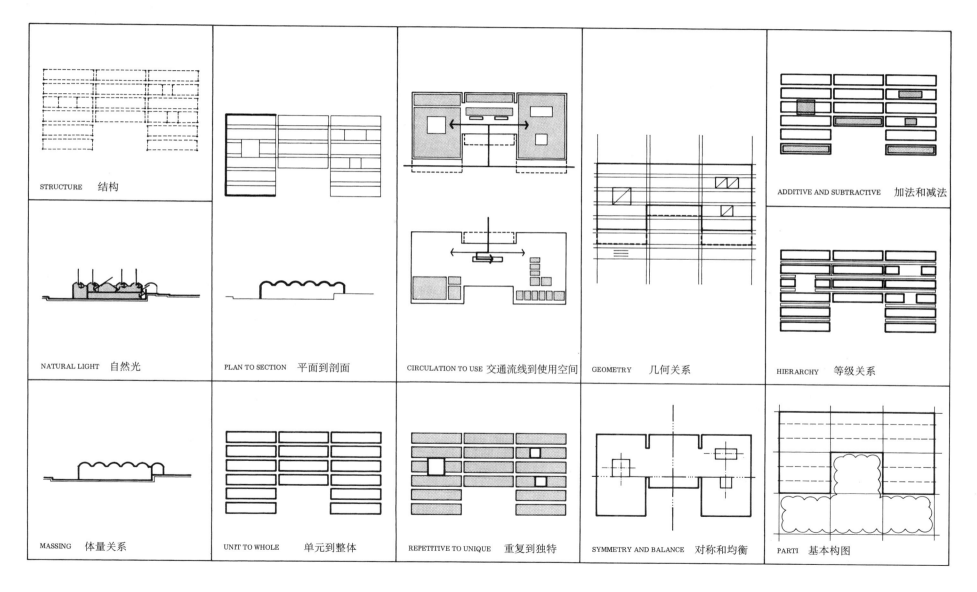

STRUCTURE　结构

NATURAL LIGHT　自然光

MASSING　体量关系

PLAN TO SECTION　平面到剖面

UNIT TO WHOLE　单元到整体

CIRCULATION TO USE 交通流线到使用空间

REPETITIVE TO UNIQUE　重复到独特

GEOMETRY　几何关系

SYMMETRY AND BALANCE　对称和均衡

ADDITIVE AND SUBTRACTIVE　加法和减法

HIERARCHY　等级关系

PARTI　基本构图

LOUIS I. KAHN　　　　　路易斯·Ⅰ·康

LIBRARY　　　　　　　图书馆
PHILIP EXETER ACADEMY　菲利浦·埃克塞特学院
EXETER, NEW HAMPSHIRE, USA　埃克塞特，新罕布什尔州，美国
1967–1972　　　　　　1967–1972 年

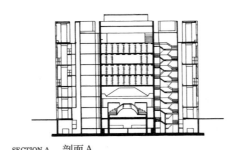

SECTION A　剖面 A

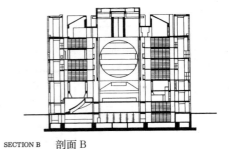

SECTION B　剖面 B

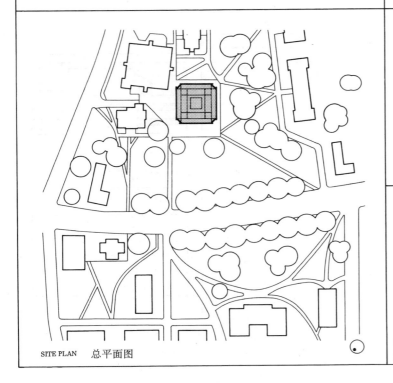

SITE PLAN　总平面图

ELEVATION 1　立面 1

ELEVATION 2　立面 2

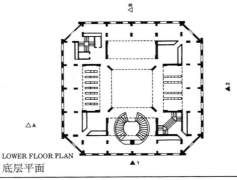

LOWER FLOOR PLAN
底层平面

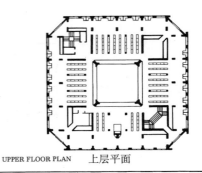

UPPER FLOOR PLAN　上层平面

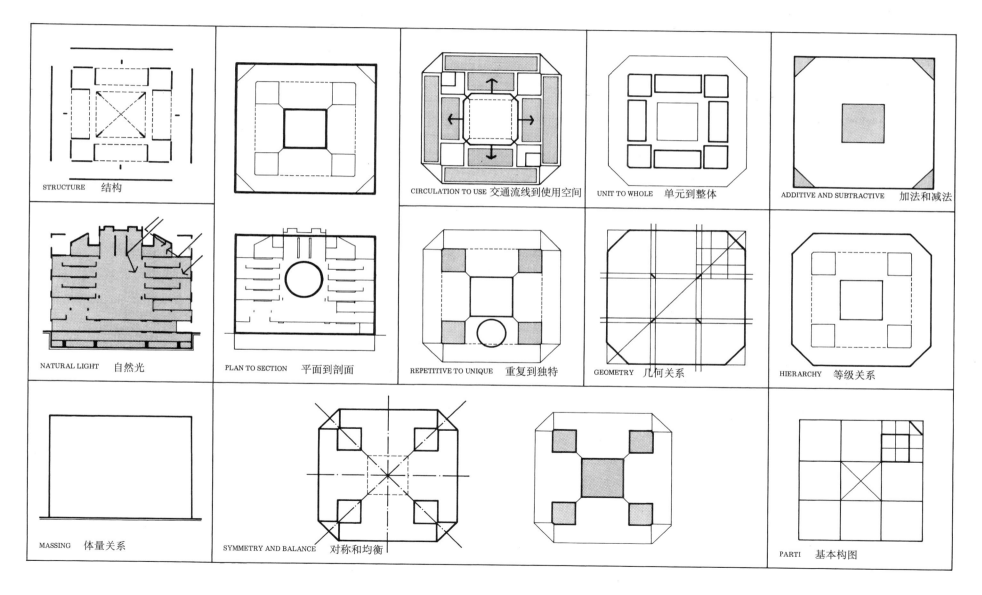

STRUCTURE　结构

CIRCULATION TO USE 交通流线到使用空间

UNIT TO WHOLE　单元到整体

ADDITIVE AND SUBTRACTIVE　加法和减法

NATURAL LIGHT　自然光

PLAN TO SECTION　平面到剖面

REPETITIVE TO UNIQUE　重复到独特

GEOMETRY　几何关系

HIERARCHY　等级关系

MASSING　体量关系

SYMMETRY AND BALANCE　对称和均衡

PARTI　基本构图

105

汤姆·库迪格
TOM KUNDIG

TOM KUNDIG   汤姆·库迪格

DELTA SHELTER   德尔塔别墅
MAZAMA, WASHINGTON, USA   马扎马，华盛顿州，美国
1998-2002   1998−2002 年

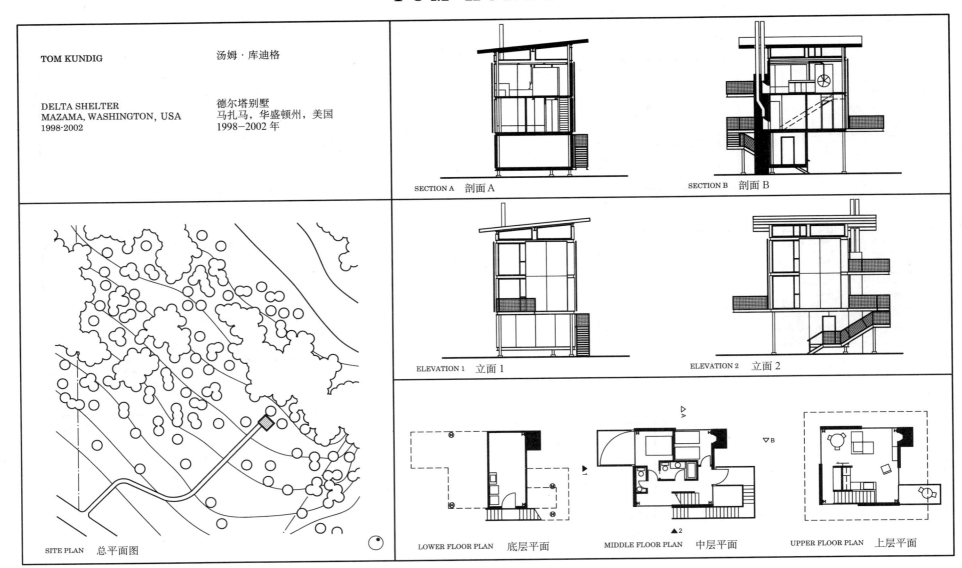

SECTION A 剖面 A   SECTION B 剖面 B

ELEVATION 1 立面 1   ELEVATION 2 立面 2

SITE PLAN 总平面图

LOWER FLOOR PLAN 底层平面   MIDDLE FLOOR PLAN 中层平面   UPPER FLOOR PLAN 上层平面

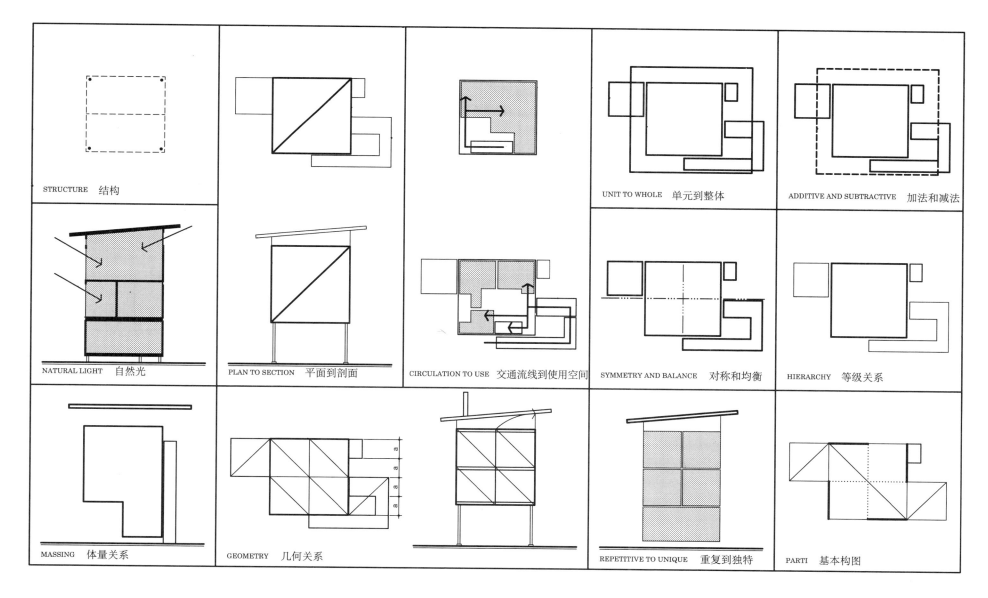

STRUCTURE　结构

UNIT TO WHOLE　单元到整体

ADDITIVE AND SUBTRACTIVE　加法和减法

NATURAL LIGHT　自然光

PLAN TO SECTION　平面到剖面

CIRCULATION TO USE　交通流线到使用空间

SYMMETRY AND BALANCE　对称和均衡

HIERARCHY　等级关系

MASSING　体量关系

GEOMETRY　几何关系

REPETITIVE TO UNIQUE　重复到独特

PARTI　基本构图

107

TOM KUNDIG　　　　　　汤姆·库迪格

CHICKEN POINT CABIN　　　住宅
HAYDEN LAKE, IDAHO, USA　海登湖，爱达荷州，美国
2000-2003　　　　　　　　2000—2003 年

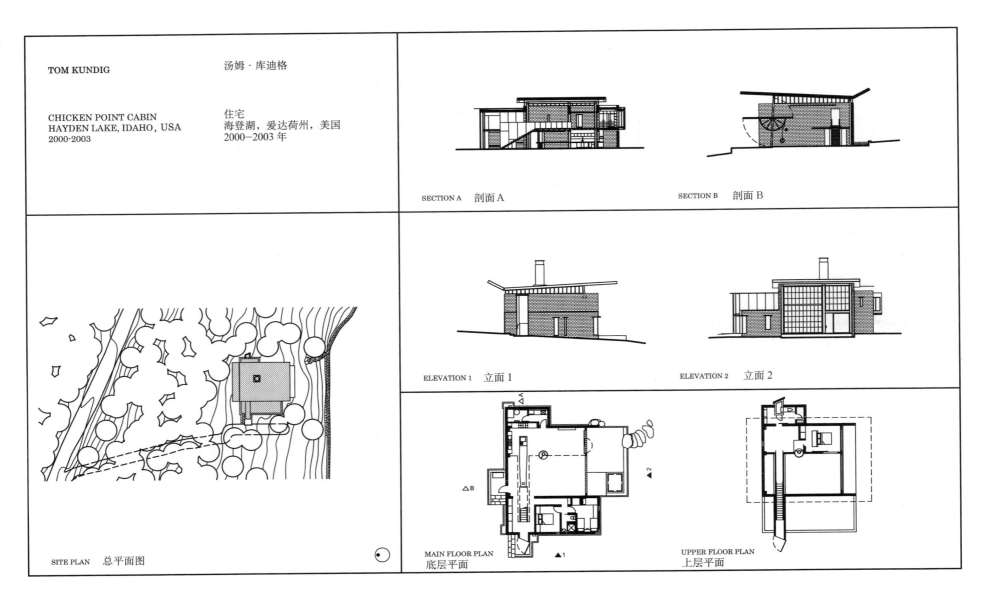

SECTION A　剖面 A

SECTION B　剖面 B

ELEVATION 1　立面 1

ELEVATION 2　立面 2

SITE PLAN　总平面图

MAIN FLOOR PLAN
底层平面

UPPER FLOOR PLAN
上层平面

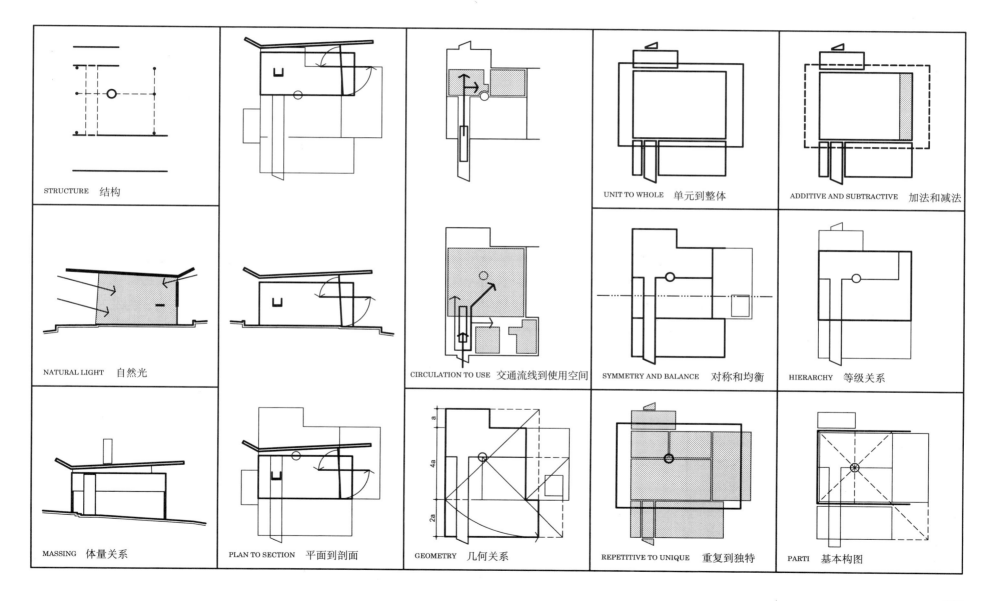

STRUCTURE　结构

NATURAL LIGHT　自然光

MASSING　体量关系

PLAN TO SECTION　平面到剖面

CIRCULATION TO USE　交通流线到使用空间

GEOMETRY　几何关系

UNIT TO WHOLE　单元到整体

SYMMETRY AND BALANCE　对称和均衡

REPETITIVE TO UNIQUE　重复到独特

ADDITIVE AND SUBTRACTIVE　加法和减法

HIERARCHY　等级关系

PARTI　基本构图

109

勒·柯布西耶
LE CORBUSIER

LE CORBUSIER　　　　勒·柯布西耶

VILLA SAVOYE　　　　萨伏伊别墅
POISSY, FRANCE　　　普瓦西，法国
1928-1931　　　　　　1928-1931 年

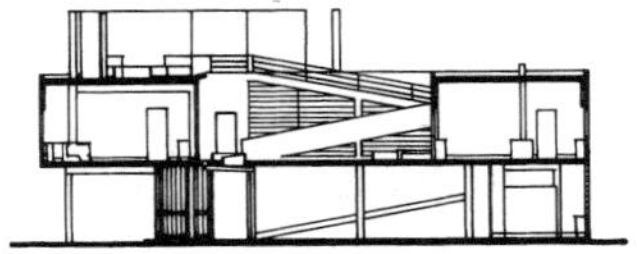

SECTION A　剖面 A

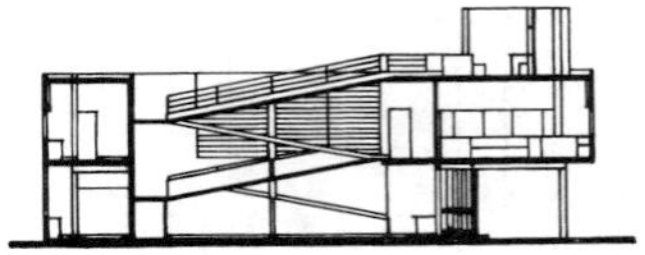

SECTION B　剖面 B

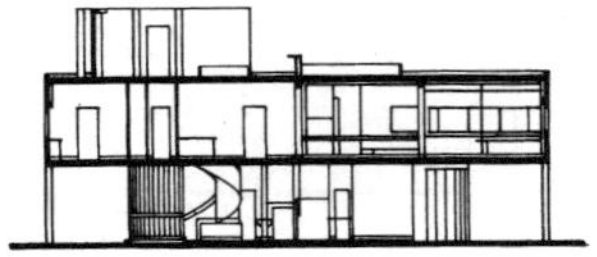

SECTION C　剖面 C

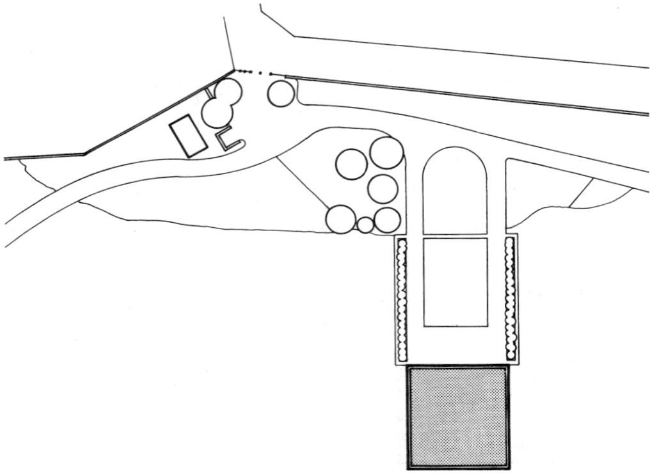

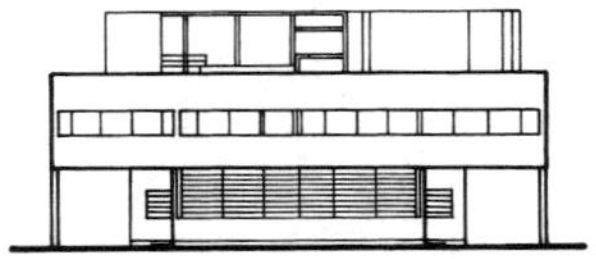

ELEVATION 1　立面 1

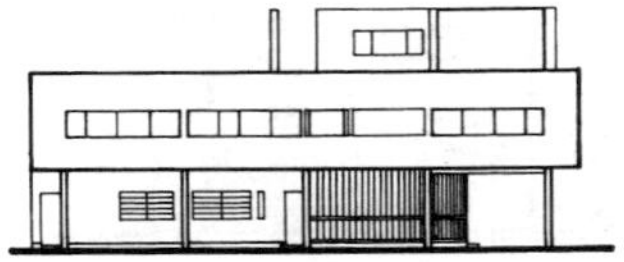

ELEVATION 2　立面 2

ELEVATION 3　立面 3

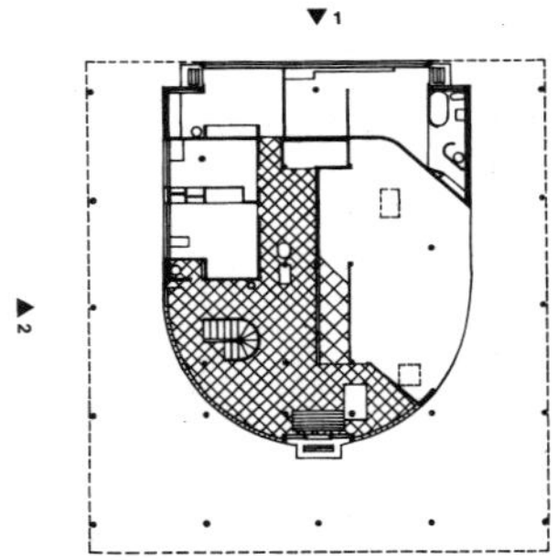

SITE PLAN　总平面图

LOWER FLOOR PLAN　▲3　底层平面

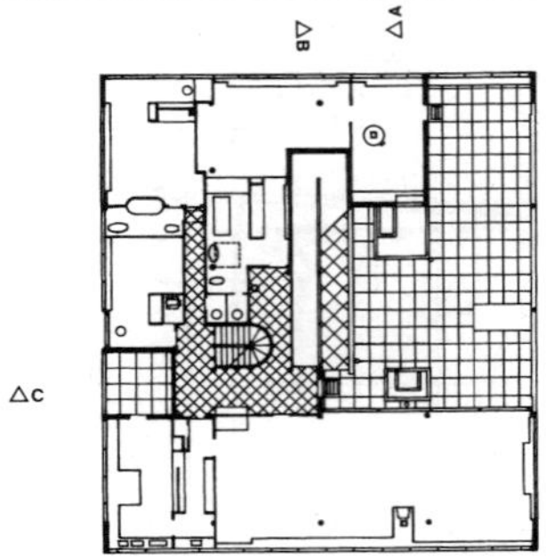

MIDDLE FLOOR PLAN　中层平面

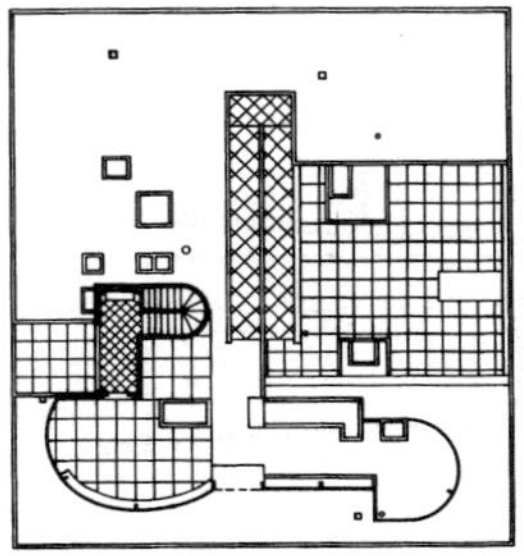

UPPER FLOOR PLAN　上层平面

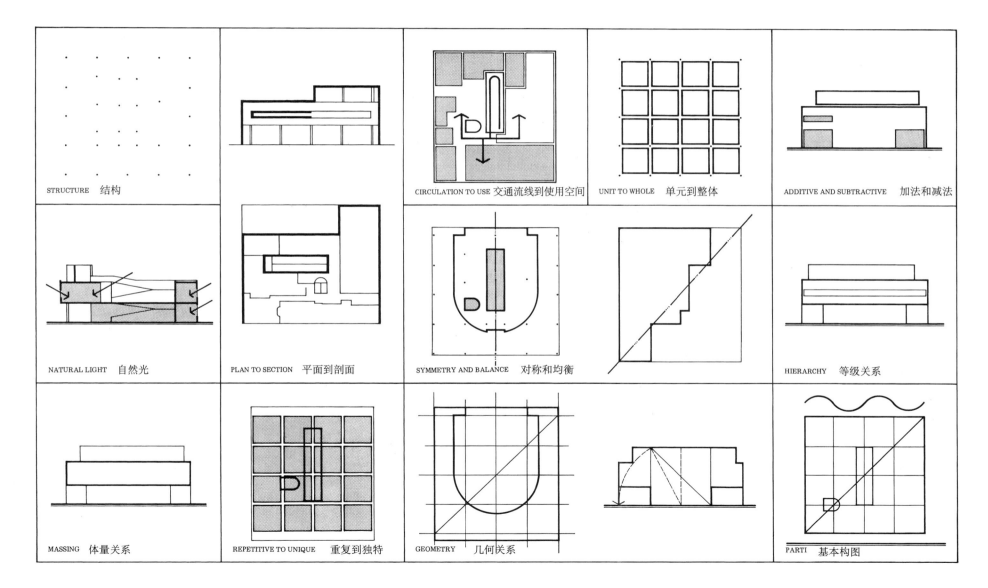

STRUCTURE　结构

CIRCULATION TO USE 交通流线到使用空间

UNIT TO WHOLE　单元到整体

ADDITIVE AND SUBTRACTIVE　加法和减法

NATURAL LIGHT　自然光

PLAN TO SECTION　平面到剖面

SYMMETRY AND BALANCE　对称和均衡

HIERARCHY　等级关系

MASSING　体量关系

REPETITIVE TO UNIQUE　重复到独特

GEOMETRY　几何关系

PARTI　基本构图

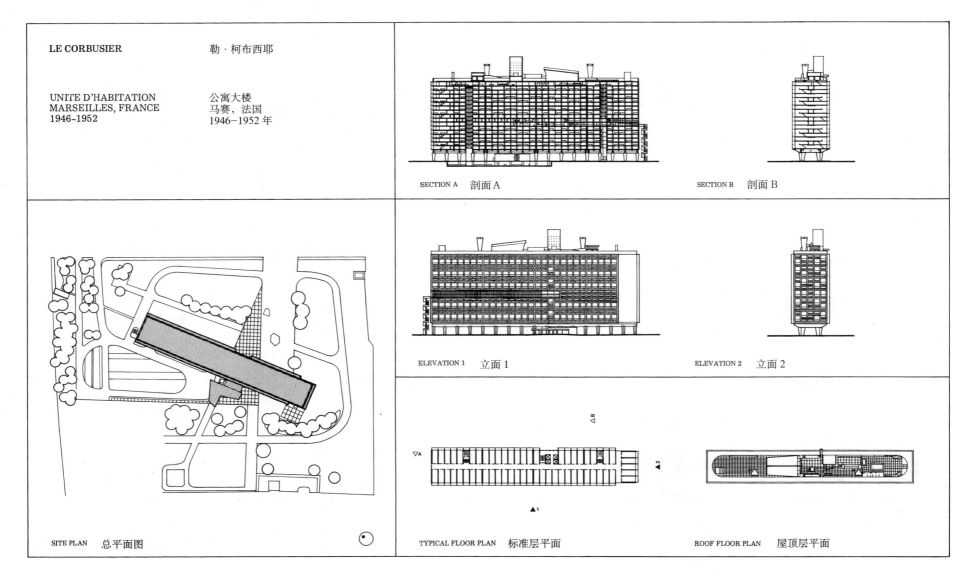

LE CORBUSIER　　　　勒·柯布西耶

UNITE D'HABITATION　　公寓大楼
MARSEILLES, FRANCE　　马赛，法国
1946-1952　　　　　　1946-1952 年

SECTION A　剖面 A

SECTION B　剖面 B

ELEVATION 1　立面 1

ELEVATION 2　立面 2

SITE PLAN　　总平面图

TYPICAL FLOOR PLAN　标准层平面

ROOF FLOOR PLAN　屋顶层平面

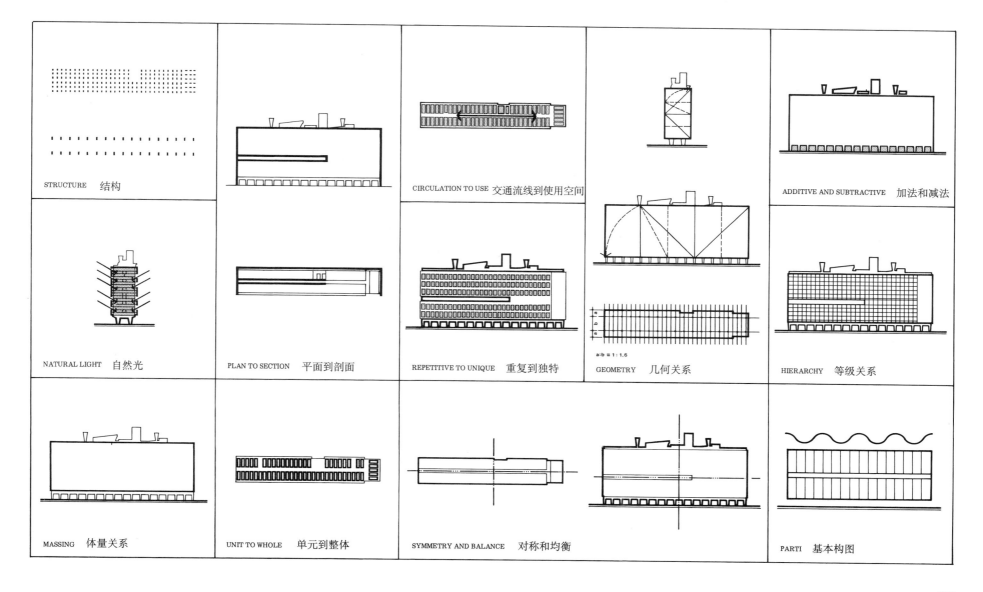

STRUCTURE　结构

CIRCULATION TO USE　交通流线到使用空间

ADDITIVE AND SUBTRACTIVE　加法和减法

NATURAL LIGHT　自然光

PLAN TO SECTION　平面到剖面

REPETITIVE TO UNIQUE　重复到独特

a:b = 1 : 1.6

GEOMETRY　几何关系

HIERARCHY　等级关系

MASSING　体量关系

UNIT TO WHOLE　单元到整体

SYMMETRY AND BALANCE　对称和均衡

PARTI　基本构图

LE CORBUSIER 勒·柯布西耶

NOTRE DAME DU HAUT CHAPEL
RONCHAMP, FRANCE
1950-1955

高地圣母教堂（朗香教堂）
龙尚，法国
1950-1955 年

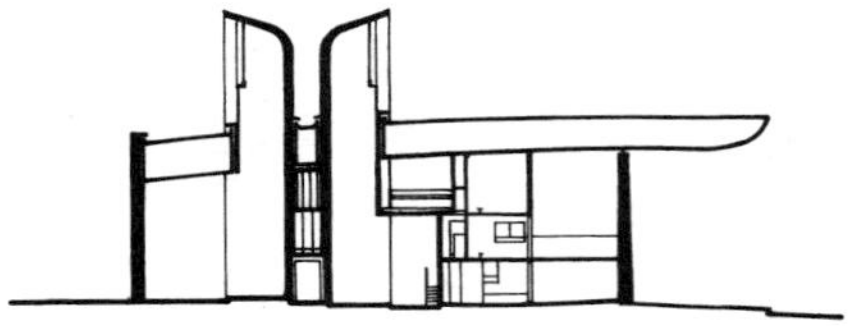

SECTION A 剖面 A

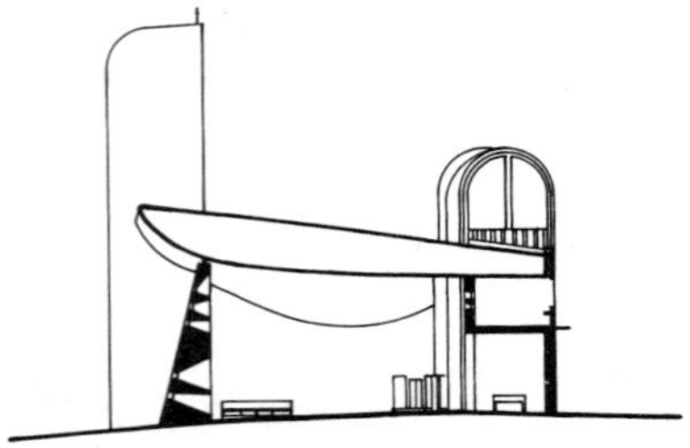

SECTION B 剖面 B

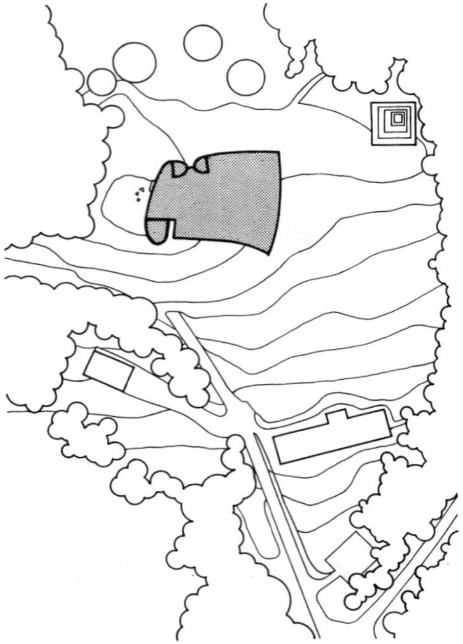

SITE PLAN 总平面图

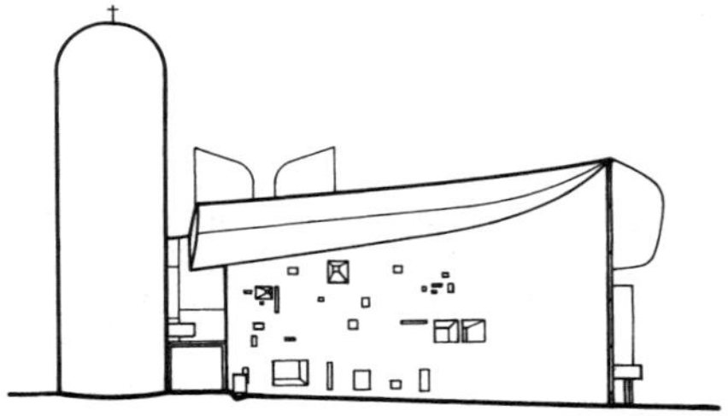

ELEVATION 1 立面 1

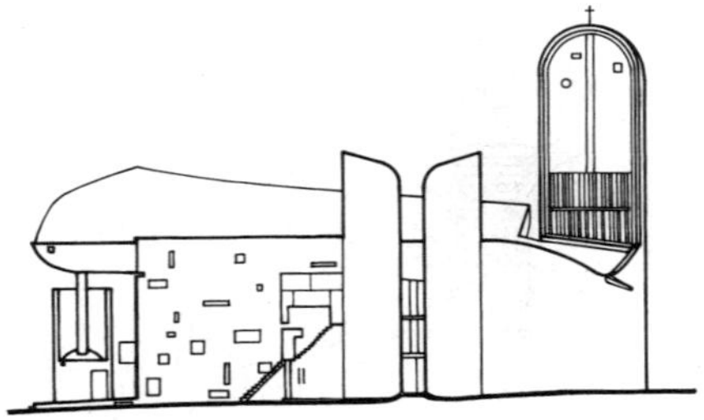

ELEVATION 2 立面 2

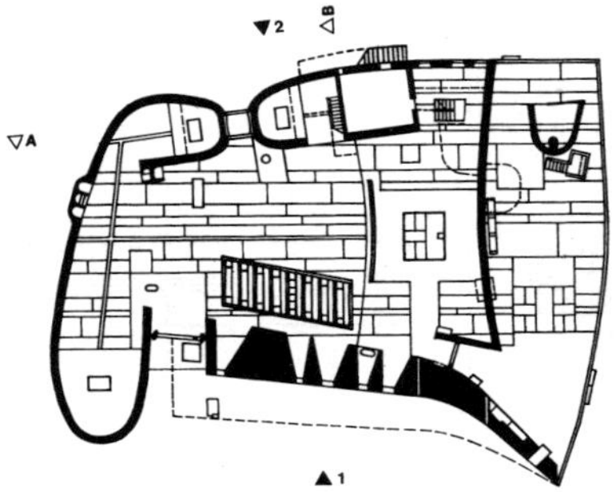

FLOOR PLAN 平面

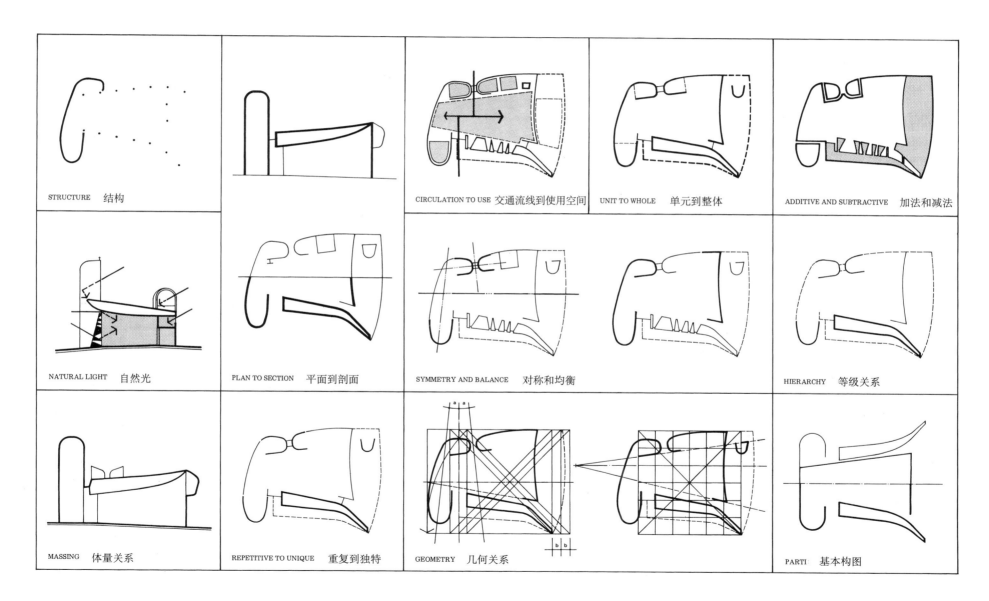

STRUCTURE　结构

CIRCULATION TO USE 交通流线到使用空间

UNIT TO WHOLE　单元到整体

ADDITIVE AND SUBTRACTIVE　加法和减法

NATURAL LIGHT　自然光

PLAN TO SECTION　平面到剖面

SYMMETRY AND BALANCE　对称和均衡

HIERARCHY　等级关系

MASSING　体量关系

REPETITIVE TO UNIQUE　重复到独特

GEOMETRY　几何关系

PARTI　基本构图

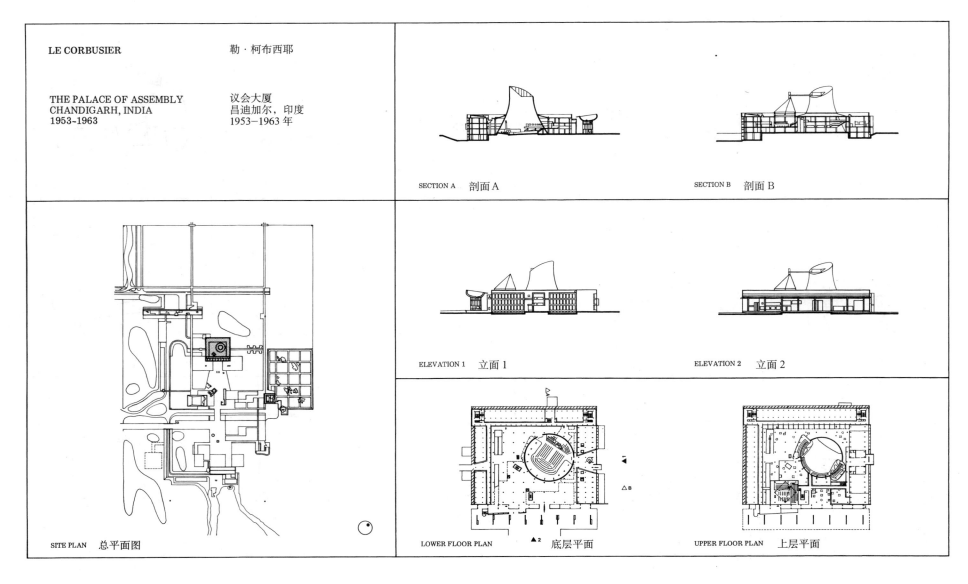

LE CORBUSIER　　　　　　勒·柯布西耶

THE PALACE OF ASSEMBLY　　议会大厦
CHANDIGARH, INDIA　　　　昌迪加尔，印度
1953~1963　　　　　　　　1953~1963 年

SECTION A　剖面 A

SECTION B　剖面 B

ELEVATION 1　立面 1

ELEVATION 2　立面 2

SITE PLAN　总平面图

LOWER FLOOR PLAN　底层平面

UPPER FLOOR PLAN　上层平面

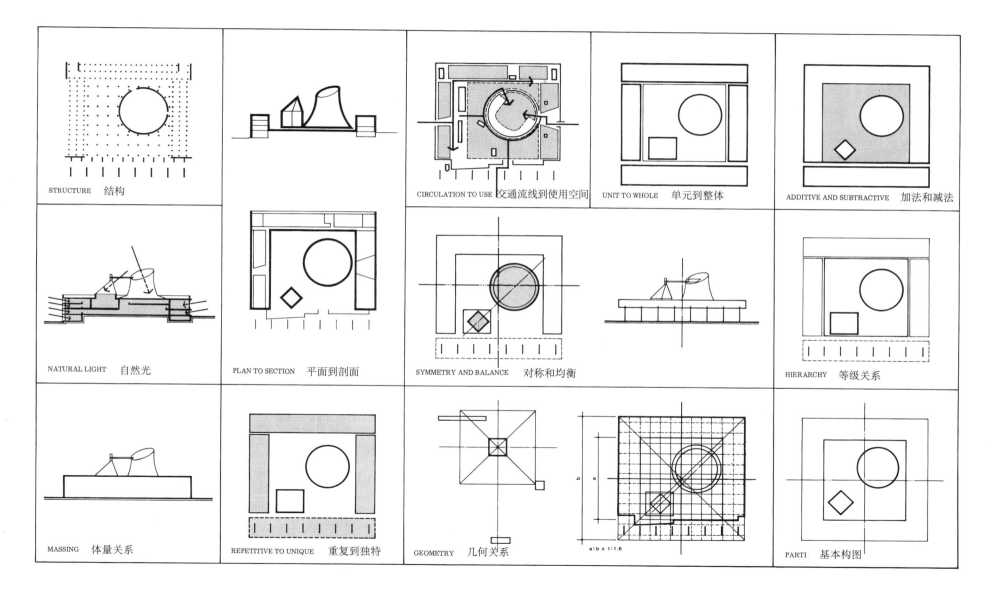

STRUCTURE 结构

CIRCULATION TO USE 交通流线到使用空间

UNIT TO WHOLE 单元到整体

ADDITIVE AND SUBTRACTIVE 加法和减法

NATURAL LIGHT 自然光

PLAN TO SECTION 平面到剖面

SYMMETRY AND BALANCE 对称和均衡

HIERARCHY 等级关系

MASSING 体量关系

REPETITIVE TO UNIQUE 重复到独特

GEOMETRY 几何关系

a:b = 1:1.6

PARTI 基本构图

克洛德·尼古拉·勒杜
CLAUDE NICHOLAS LEDOUX

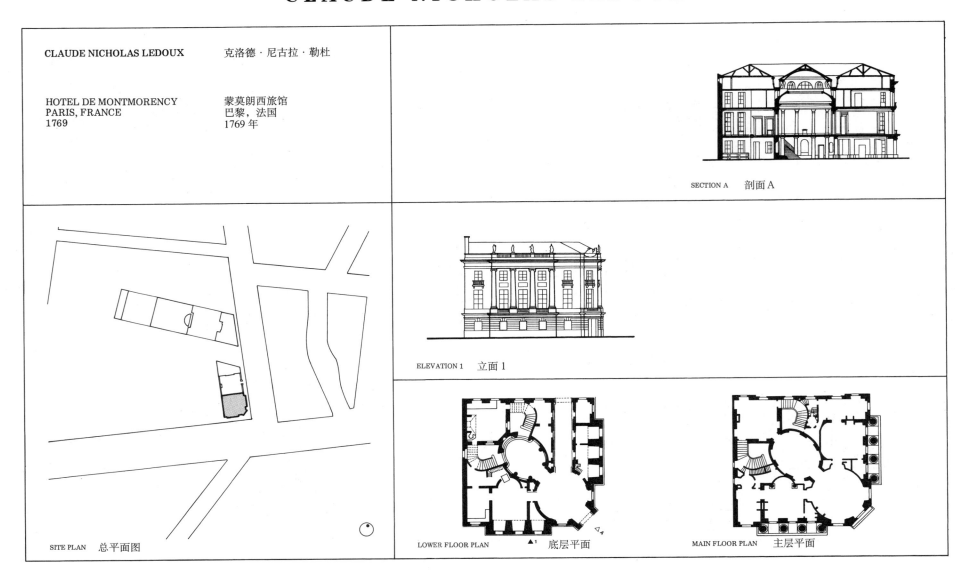

CLAUDE NICHOLAS LEDOUX　　克洛德·尼古拉·勒杜

HOTEL DE MONTMORENCY　　蒙莫朗西旅馆
PARIS, FRANCE　　巴黎，法国
1769　　1769 年

SECTION A　　剖面 A

ELEVATION 1　　立面 1

SITE PLAN　　总平面图

LOWER FLOOR PLAN　　底层平面

MAIN FLOOR PLAN　　主层平面

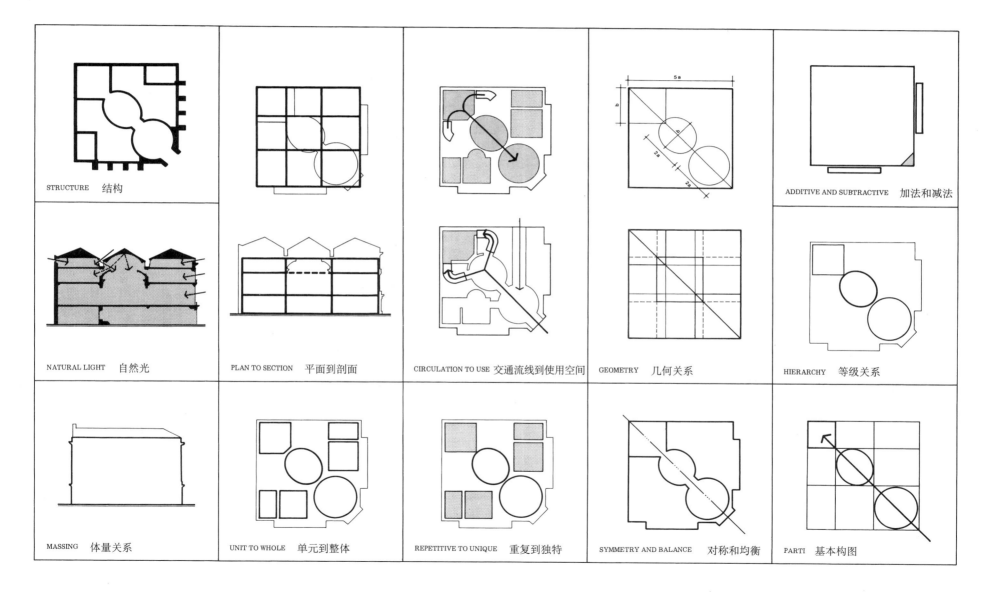

STRUCTURE　结构

NATURAL LIGHT　自然光

MASSING　体量关系

PLAN TO SECTION　平面到剖面

UNIT TO WHOLE　单元到整体

CIRCULATION TO USE 交通流线到使用空间

REPETITIVE TO UNIQUE　重复到独特

GEOMETRY　几何关系

SYMMETRY AND BALANCE　对称和均衡

ADDITIVE AND SUBTRACTIVE　加法和减法

HIERARCHY　等级关系

PARTI　基本构图

CLAUDE NICHOLAS LEDOUX　　　克洛德·尼古拉·勒杜

HOTEL GUIMARD　　　　　　　圭马尔旅馆
PARIS, FRANCE　　　　　　　　巴黎，法国
1770　　　　　　　　　　　　　1770 年

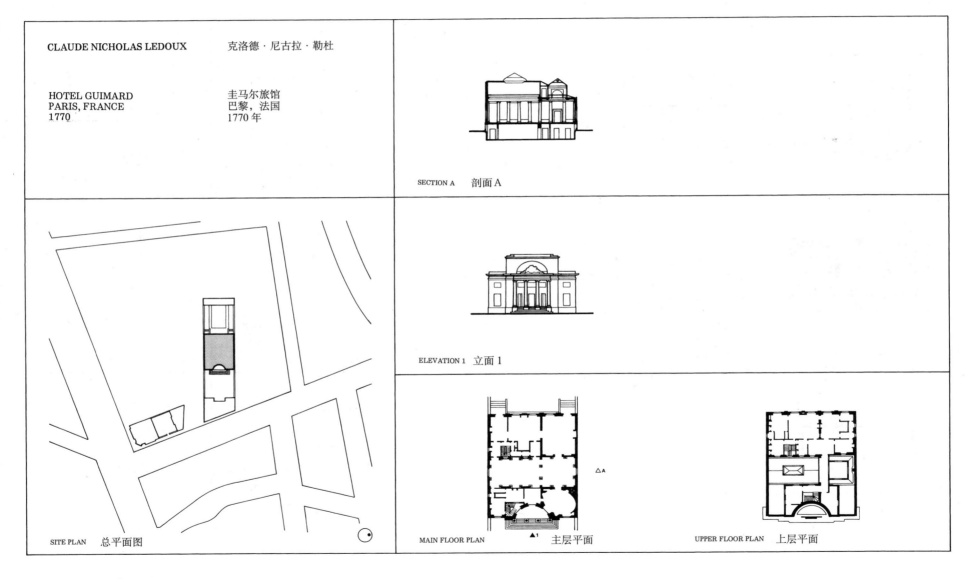

SECTION A　剖面 A

ELEVATION 1　立面 1

SITE PLAN　总平面图

MAIN FLOOR PLAN　　▲1　主层平面

UPPER FLOOR PLAN　上层平面

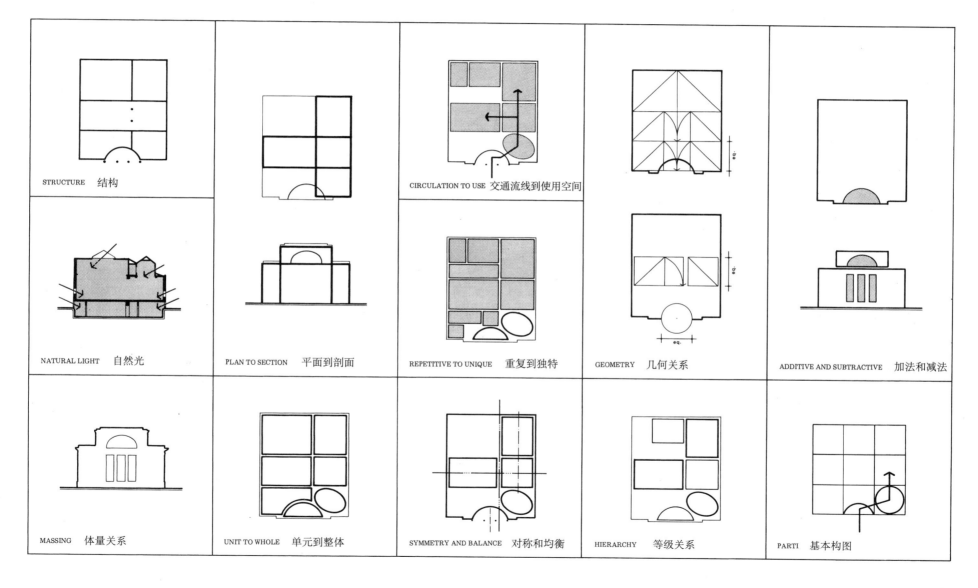

STRUCTURE　结构

NATURAL LIGHT　自然光

MASSING　体量关系

PLAN TO SECTION　平面到剖面

UNIT TO WHOLE　单元到整体

CIRCULATION TO USE　交通流线到使用空间

REPETITIVE TO UNIQUE　重复到独特

SYMMETRY AND BALANCE　对称和均衡

GEOMETRY　几何关系

HIERARCHY　等级关系

ADDITIVE AND SUBTRACTIVE　加法和减法

PARTI　基本构图

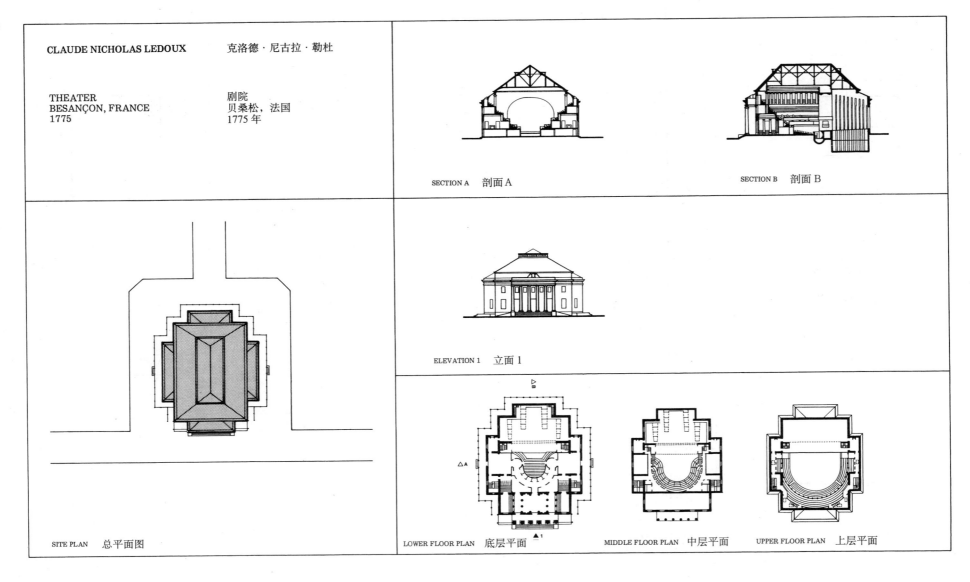

CLAUDE NICHOLAS LEDOUX   克洛德·尼古拉·勒杜

THEATER   剧院
BESANÇON, FRANCE   贝桑松，法国
1775   1775 年

SECTION A   剖面 A   SECTION B   剖面 B

ELEVATION 1   立面 1

SITE PLAN   总平面图

LOWER FLOOR PLAN   底层平面   MIDDLE FLOOR PLAN   中层平面   UPPER FLOOR PLAN   上层平面

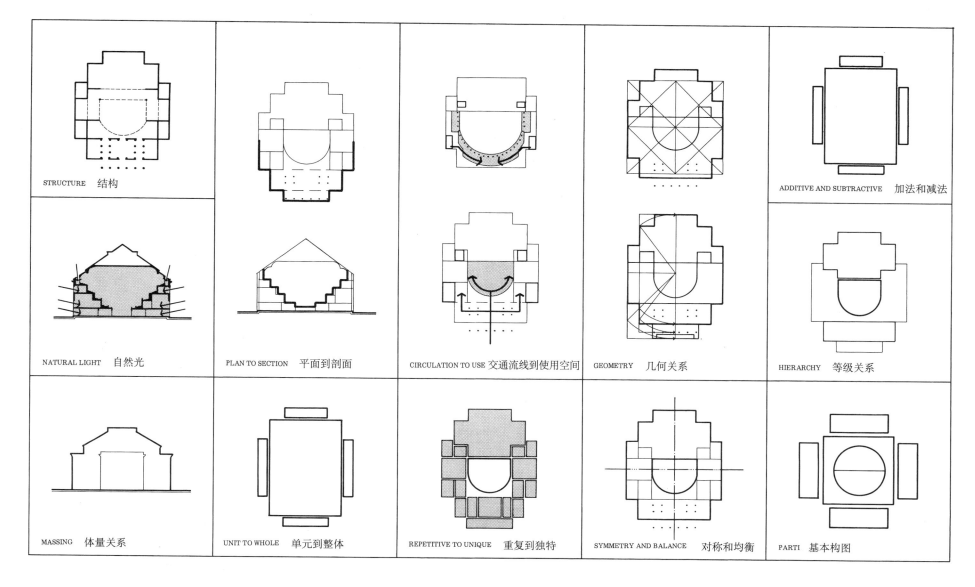

STRUCTURE　结构

NATURAL LIGHT　自然光

MASSING　体量关系

PLAN TO SECTION　平面到剖面

UNIT TO WHOLE　单元到整体

CIRCULATION TO USE 交通流线到使用空间

REPETITIVE TO UNIQUE　重复到独特

GEOMETRY　几何关系

SYMMETRY AND BALANCE　对称和均衡

ADDITIVE AND SUBTRACTIVE　加法和减法

HIERARCHY　等级关系

PARTI　基本构图

CLAUDE NICHOLAS LEDOUX　　　克洛德·尼古拉·勒杜

DIRECTOR'S HOUSE　　　　　　场长住宅
SALTWORKS OF ARC AND SENANS　阿克和塞南斯盐场
NEAR BESANÇON, FRANCE　　　贝桑松附近，法国
1775-1779　　　　　　　　　　1775-1779 年

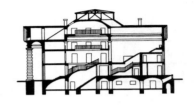

SECTION A　剖面 A

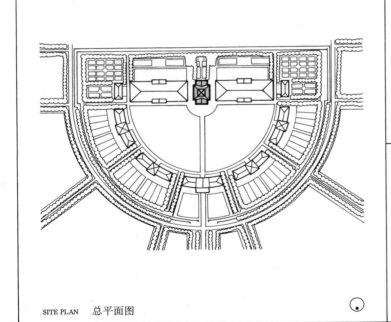

SITE PLAN　总平面图

ELEVATION 1　立面 1

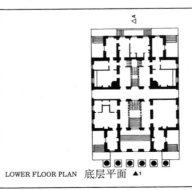

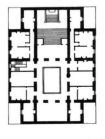

LOWER FLOOR PLAN　底层平面　▲1　　MIDDLE FLOOR PLAN　中层平面　　UPPER FLOOR PLAN　上层平面

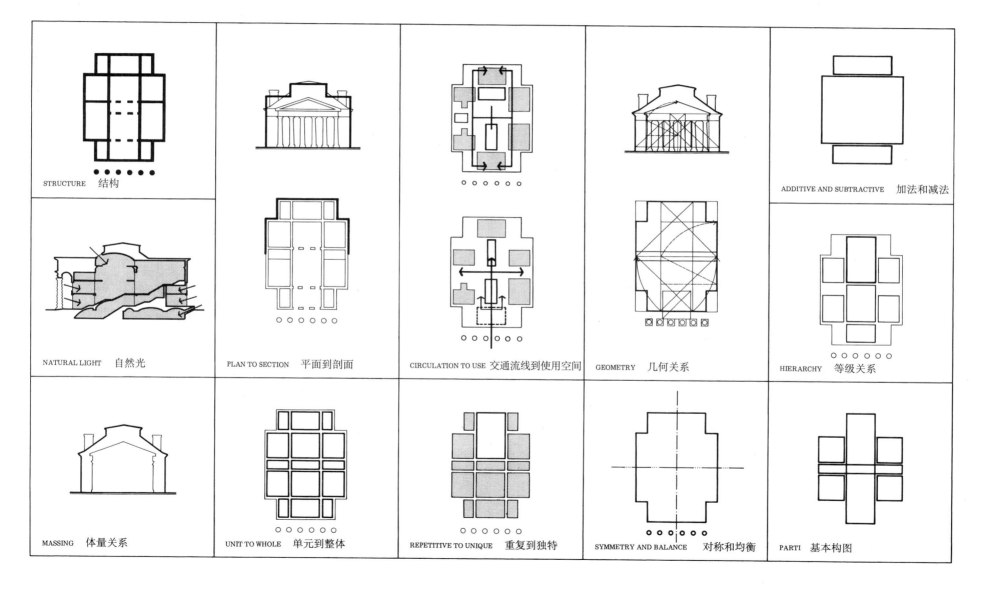

STRUCTURE　结构

NATURAL LIGHT　自然光

MASSING　体量关系

PLAN TO SECTION　平面到剖面

UNIT TO WHOLE　单元到整体

CIRCULATION TO USE　交通流线到使用空间

REPETITIVE TO UNIQUE　重复到独特

GEOMETRY　几何关系

SYMMETRY AND BALANCE　对称和均衡

ADDITIVE AND SUBTRACTIVE　加法和减法

HIERARCHY　等级关系

PARTI　基本构图

西古德·莱韦伦茨
SIGURD LEWERENTZ

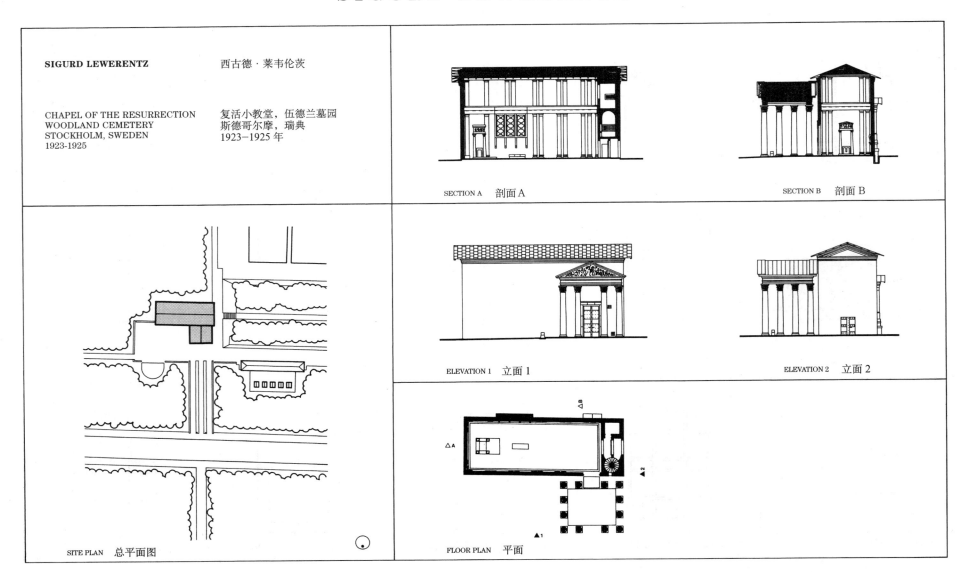

SIGURD LEWERENTZ　　　西古德·莱韦伦茨

CHAPEL OF THE RESURRECTION　　复活小教堂，伍德兰墓园
WOODLAND CEMETERY　　　　斯德哥尔摩，瑞典
STOCKHOLM, SWEDEN　　　　1923-1925 年
1923-1925

SECTION A　剖面 A

SECTION B　剖面 B

ELEVATION 1　立面 1

ELEVATION 2　立面 2

SITE PLAN　总平面图

FLOOR PLAN　平面

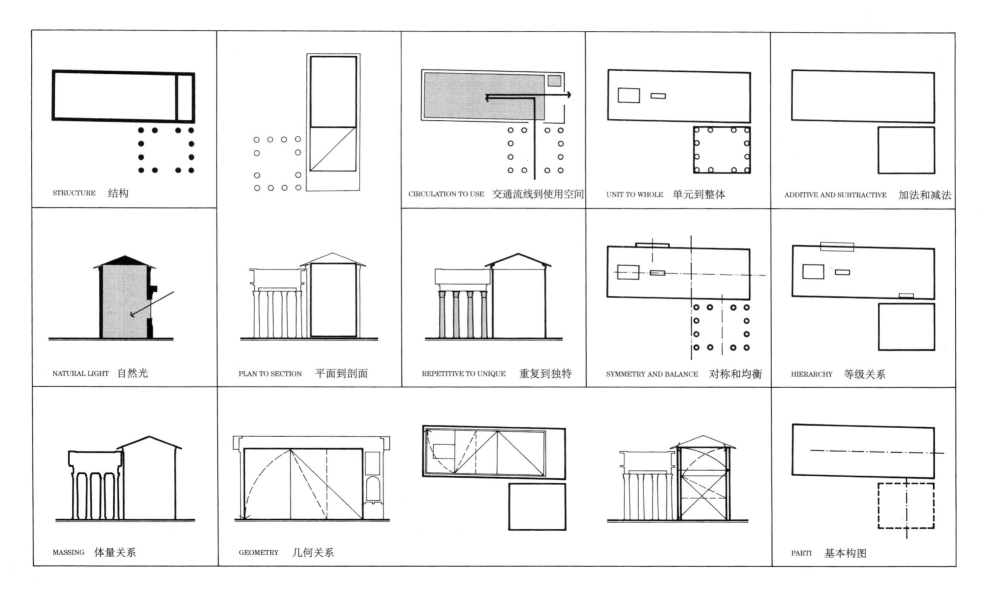

STRUCTURE　结构

CIRCULATION TO USE　交通流线到使用空间

UNIT TO WHOLE　单元到整体

ADDITIVE AND SUBTRACTIVE　加法和减法

NATURAL LIGHT　自然光

PLAN TO SECTION　平面到剖面

REPETITIVE TO UNIQUE　重复到独特

SYMMETRY AND BALANCE　对称和均衡

HIERARCHY　等级关系

MASSING　体量关系

GEOMETRY　几何关系

PARTI　基本构图

SIGURD LEWERENTZ   西古德·莱韦伦茨

CHURCH OF ST. PETER   圣彼得教堂
KLIPPAN, SWEDEN   克利潘，瑞典
1963-1966   1963-1966 年

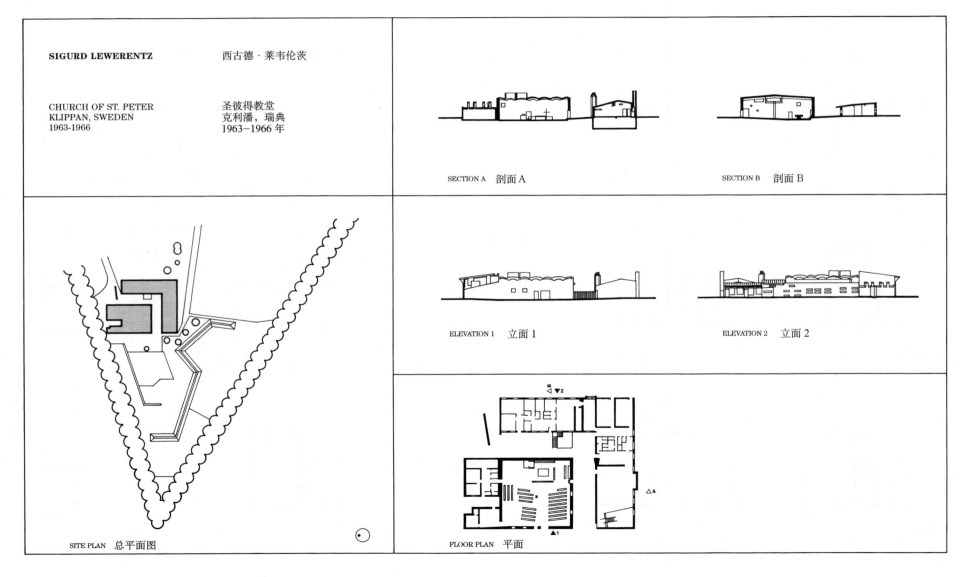

SECTION A   剖面 A

SECTION B   剖面 B

ELEVATION 1   立面 1

ELEVATION 2   立面 2

SITE PLAN   总平面图

FLOOR PLAN   平面

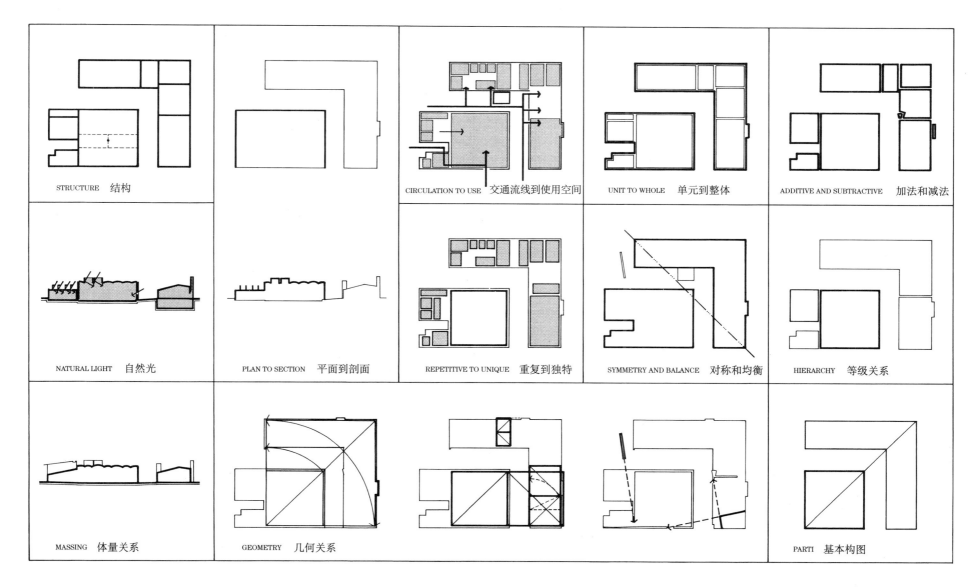

STRUCTURE　结构

CIRCULATION TO USE　交通流线到使用空间

UNIT TO WHOLE　单元到整体

ADDITIVE AND SUBTRACTIVE　加法和减法

NATURAL LIGHT　自然光

PLAN TO SECTION　平面到剖面

REPETITIVE TO UNIQUE　重复到独特

SYMMETRY AND BALANCE　对称和均衡

HIERARCHY　等级关系

MASSING　体量关系

GEOMETRY　几何关系

PARTI　基本构图

埃德温·勒琴斯
EDWIN LUTYENS

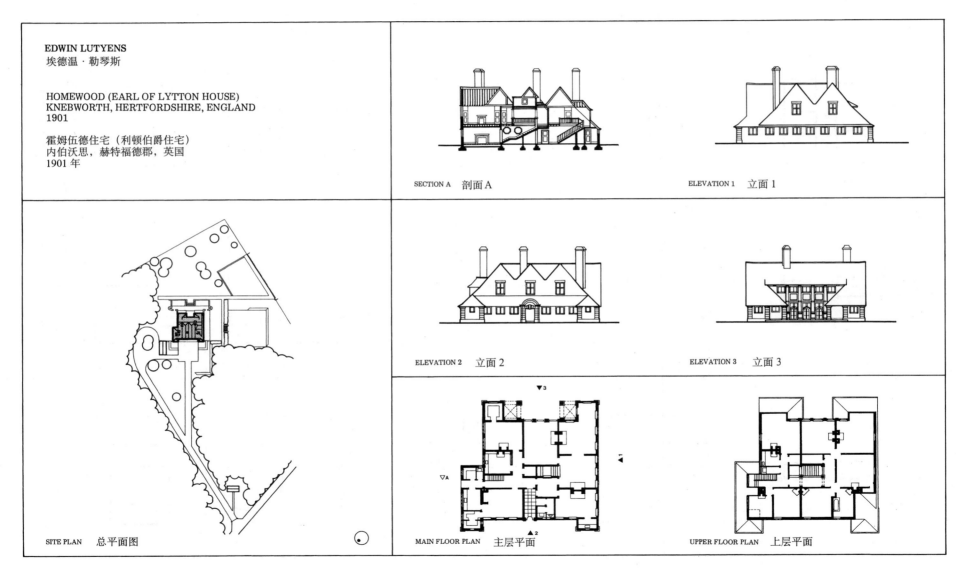

EDWIN LUTYENS
埃德温·勒琴斯

HOMEWOOD (EARL OF LYTTON HOUSE)
KNEBWORTH, HERTFORDSHIRE, ENGLAND
1901

霍姆伍德住宅（利顿伯爵住宅）
内伯沃思，赫特福德郡，英国
1901 年

SITE PLAN　总平面图

SECTION A　剖面 A

ELEVATION 1　立面 1

ELEVATION 2　立面 2

ELEVATION 3　立面 3

MAIN FLOOR PLAN　主层平面

UPPER FLOOR PLAN　上层平面

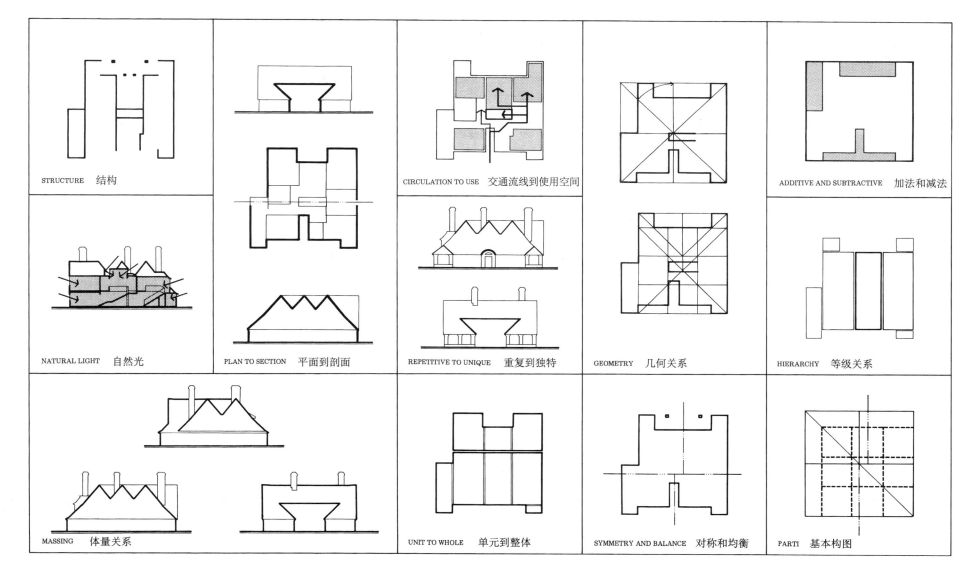

STRUCTURE　结构

CIRCULATION TO USE　交通流线到使用空间

ADDITIVE AND SUBTRACTIVE　加法和减法

NATURAL LIGHT　自然光

PLAN TO SECTION　平面到剖面

REPETITIVE TO UNIQUE　重复到独特

GEOMETRY　几何关系

HIERARCHY　等级关系

MASSING　体量关系

UNIT TO WHOLE　单元到整体

SYMMETRY AND BALANCE　对称和均衡

PARTI　基本构图

EDWIN LUTYENS
埃德温·勒琴斯

NASHDOM (PRINCESS ALEXIS DOLGORONKI HOUSE)
TAPLOW, BUCKINGHAMSHIRE, ENGLAND
1905~1909

纳希顿府邸（亚历克西斯·多尔戈朗墓公主府邸）
泰浦罗，白金汉郡，英国
1905~1909 年

SECTION A　剖面 A

SECTION B　剖面 B

ELEVATION 1　立面 1

ELEVATION 2　立面 2

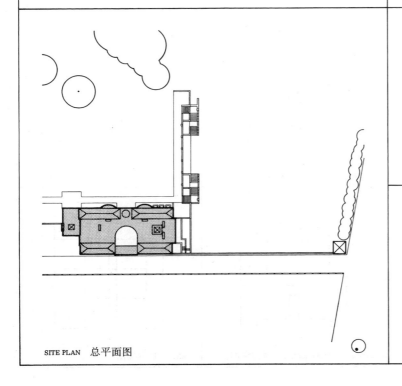

SITE PLAN　总平面图

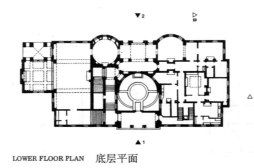

LOWER FLOOR PLAN　底层平面

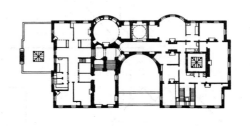

UPPER FLOOR PLAN　上层平面

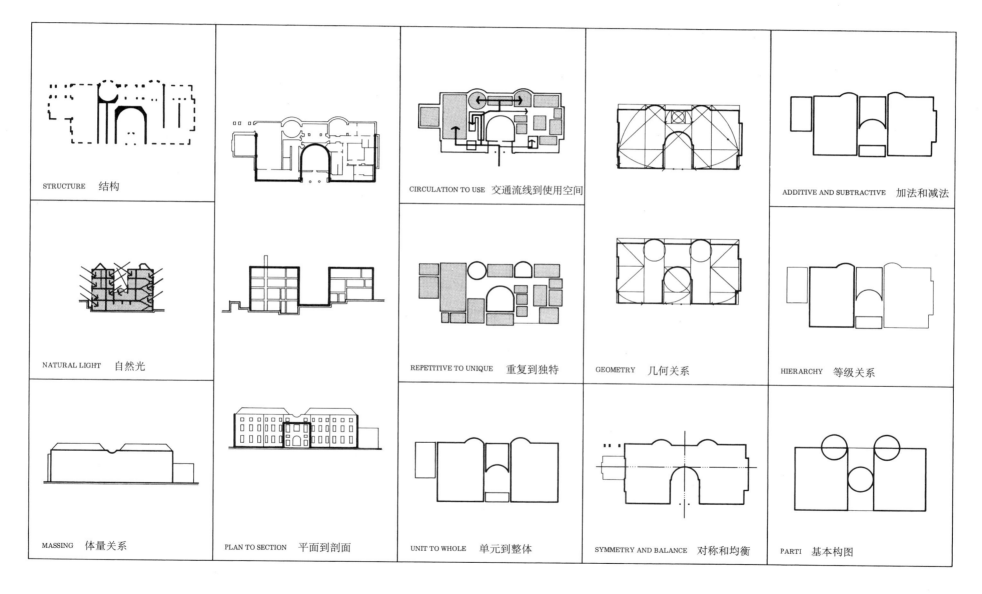

STRUCTURE　结构

CIRCULATION TO USE　交通流线到使用空间

ADDITIVE AND SUBTRACTIVE　加法和减法

NATURAL LIGHT　自然光

REPETITIVE TO UNIQUE　重复到独特

GEOMETRY　几何关系

HIERARCHY　等级关系

MASSING　体量关系

PLAN TO SECTION　平面到剖面

UNIT TO WHOLE　单元到整体

SYMMETRY AND BALANCE　对称和均衡

PARTI　基本构图

EDWIN LUTYENS　　　　埃德温·勒琴斯

HEATHCOTE (HEMINGWAY HOUSE)　　希思科特住宅（汉明威住宅）
ILKLEY, YORKSHIRE, ENGLAND　　伊尔克利，约克郡，英国
1906　　　　1906 年

SECTION A　剖面 A

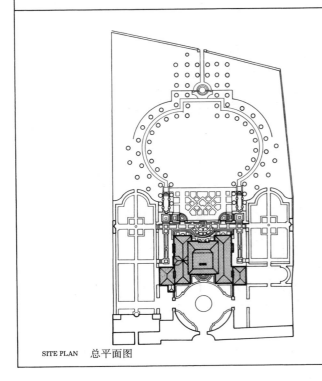

SITE PLAN　总平面图

ELEVATION 1　立面 1

ELEVATION 2　立面 2

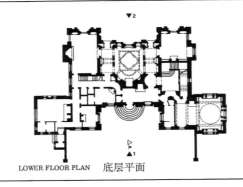

LOWER FLOOR PLAN　底层平面

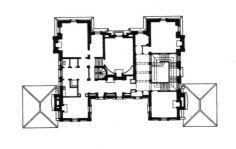

UPPER FLOOR PLAN　上层平面

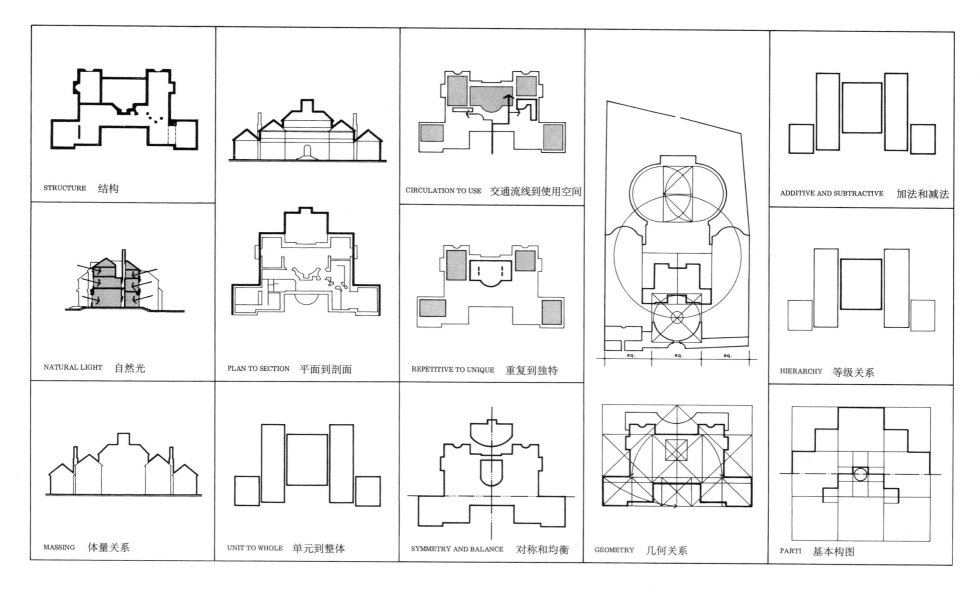

STRUCTURE　结构

CIRCULATION TO USE　交通流线到使用空间

ADDITIVE AND SUBTRACTIVE　加法和减法

NATURAL LIGHT　自然光

PLAN TO SECTION　平面到剖面

REPETITIVE TO UNIQUE　重复到独特

HIERARCHY　等级关系

MASSING　体量关系

UNIT TO WHOLE　单元到整体

SYMMETRY AND BALANCE　对称和均衡

GEOMETRY　几何关系

PARTI　基本构图

EDWIN LUTYENS
埃德温·勒琴斯

THE SALUTATION (HENRY FARRER HOUSE)
SANDWICH, KENT, ENGLAND
1911

萨吕泰兴府邸（亨利·法勒住宅）
桑威奇，肯特郡，英国
1911 年

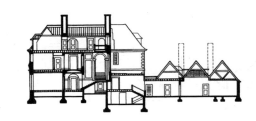

SECTION A　剖面 A

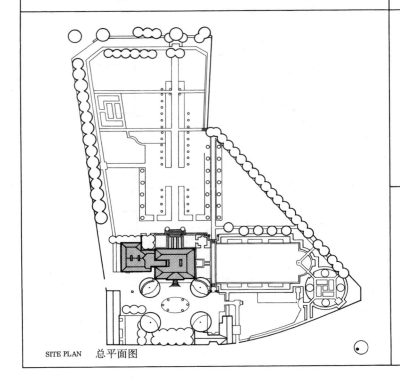

SITE PLAN　总平面图

ELEVATION 1　立面 1　　　　ELEVATION 2　立面 2

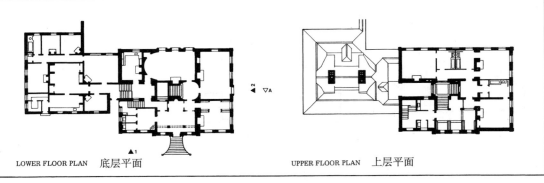

LOWER FLOOR PLAN　底层平面　　　　UPPER FLOOR PLAN　上层平面

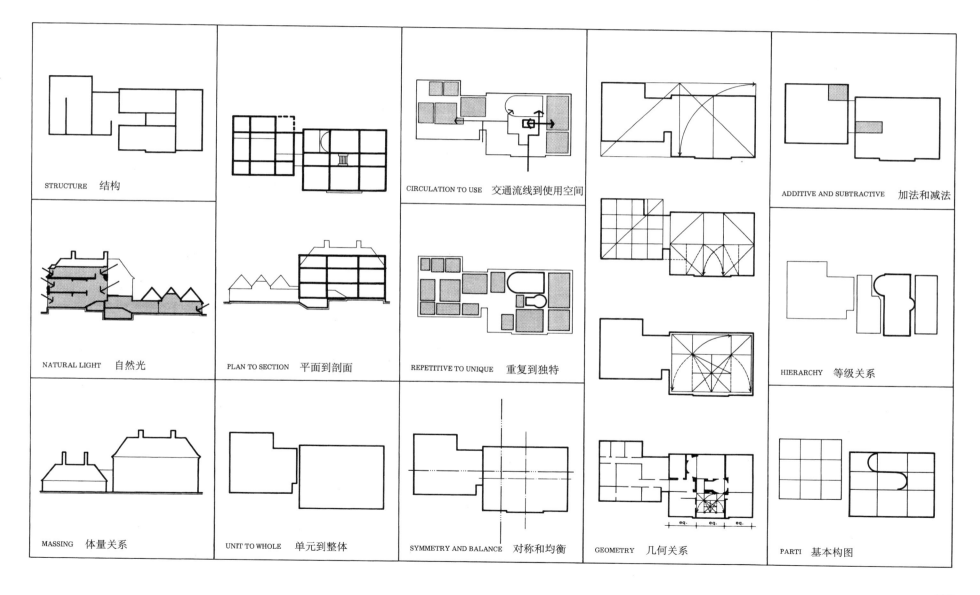

STRUCTURE　结构

CIRCULATION TO USE　交通流线到使用空间

ADDITIVE AND SUBTRACTIVE　加法和减法

NATURAL LIGHT　自然光

PLAN TO SECTION　平面到剖面

REPETITIVE TO UNIQUE　重复到独特

HIERARCHY　等级关系

MASSING　体量关系

UNIT TO WHOLE　单元到整体

SYMMETRY AND BALANCE　对称和均衡

GEOMETRY　几何关系

PARTI　基本构图

布赖恩·迈克－里昂
BRIAN MACKAY–LYONS

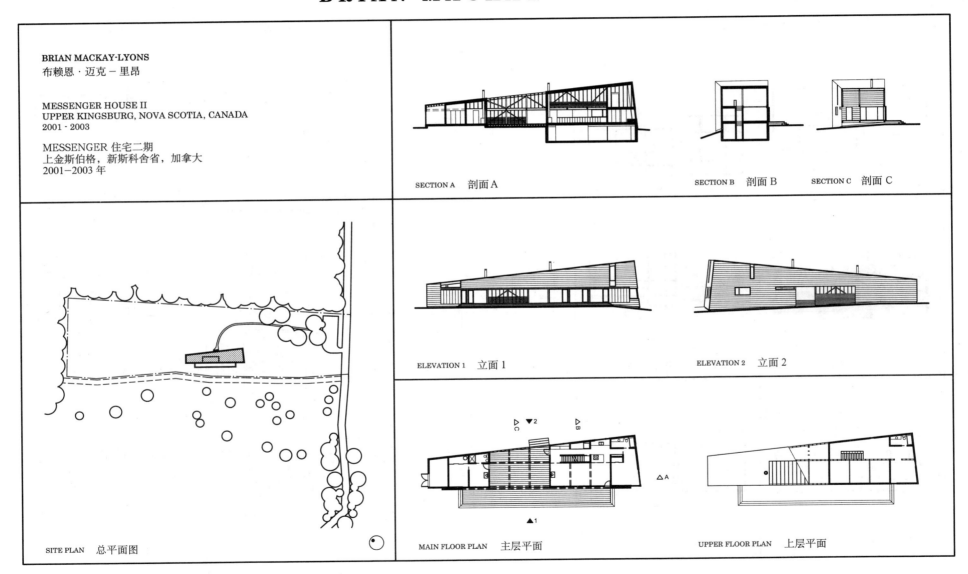

BRIAN MACKAY-LYONS
布赖恩·迈克－里昂

MESSENGER HOUSE II
UPPER KINGSBURG, NOVA SCOTIA, CANADA
2001 · 2003

MESSENGER 住宅二期
上金斯伯格，新斯科舍省，加拿大
2001－2003 年

SECTION A　剖面 A

SECTION B　剖面 B　　SECTION C　剖面 C

ELEVATION 1　立面 1

ELEVATION 2　立面 2

SITE PLAN　总平面图

MAIN FLOOR PLAN　主层平面

UPPER FLOOR PLAN　上层平面

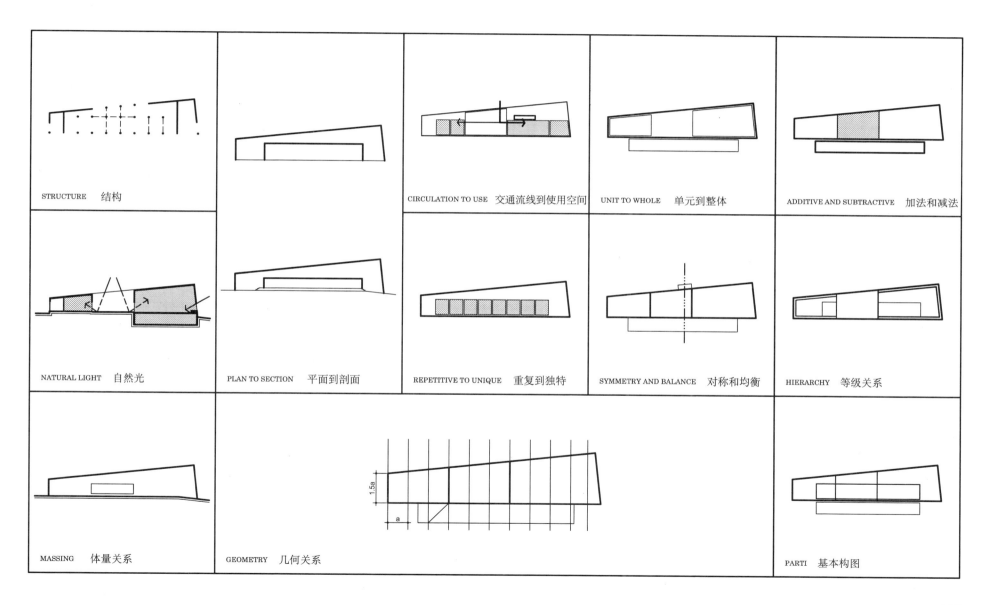

STRUCTURE　结构

CIRCULATION TO USE　交通流线到使用空间

UNIT TO WHOLE　单元到整体

ADDITIVE AND SUBTRACTIVE　加法和减法

NATURAL LIGHT　自然光

PLAN TO SECTION　平面到剖面

REPETITIVE TO UNIQUE　重复到独特

SYMMETRY AND BALANCE　对称和均衡

HIERARCHY　等级关系

MASSING　体量关系

GEOMETRY　几何关系

PARTI　基本构图

BRIAN MACKAY-LYONS　　　　　　布赖恩·迈克－里昂

HILL HOUSE　　　　　　　　　　　希尔住宅
KINGSBURG, NOVA SCOTIA, CANADA　金斯伯格，新斯科舍省，加拿大
2002 - 2004　　　　　　　　　　　2002－2004 年

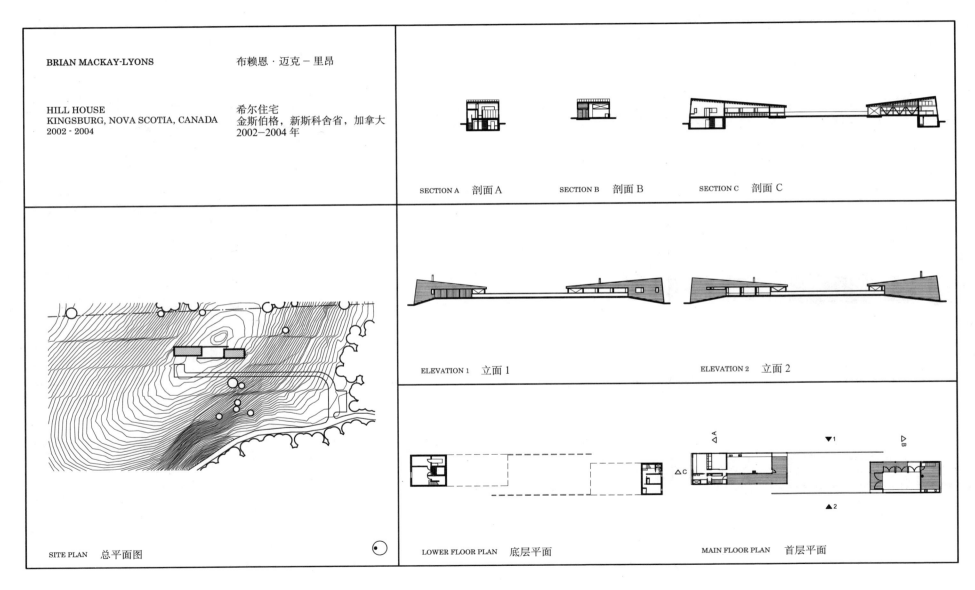

SECTION A　剖面 A　　　　　SECTION B　剖面 B　　　　　SECTION C　剖面 C

ELEVATION 1　立面 1　　　　　　　ELEVATION 2　立面 2

SITE PLAN　总平面图

LOWER FLOOR PLAN　底层平面　　　　MAIN FLOOR PLAN　首层平面

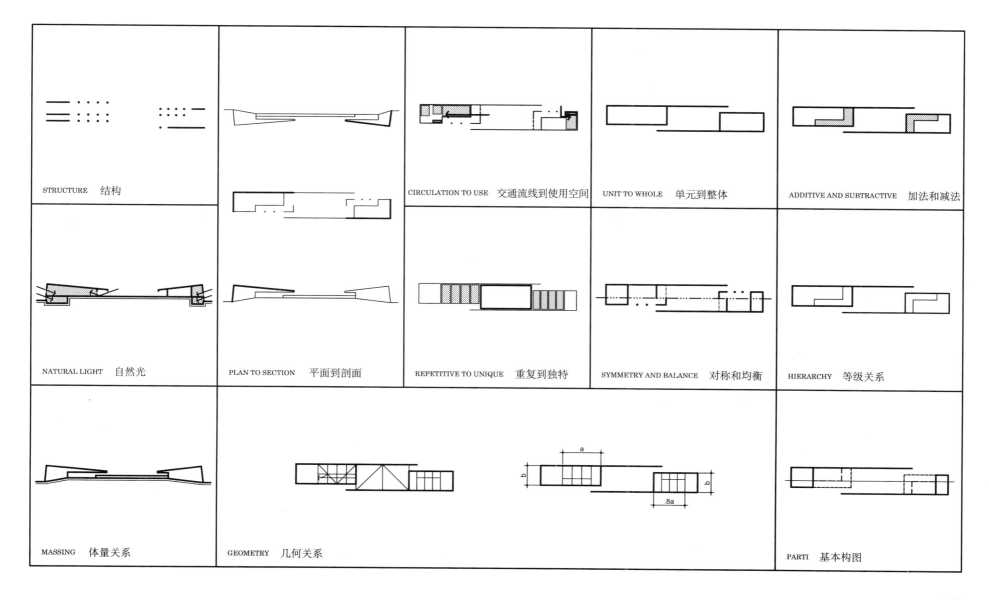

STRUCTURE　结构

CIRCULATION TO USE　交通流线到使用空间

UNIT TO WHOLE　单元到整体

ADDITIVE AND SUBTRACTIVE　加法和减法

NATURAL LIGHT　自然光

PLAN TO SECTION　平面到剖面

REPETITIVE TO UNIQUE　重复到独特

SYMMETRY AND BALANCE　对称和均衡

HIERARCHY　等级关系

MASSING　体量关系

GEOMETRY　几何关系

PARTI　基本构图

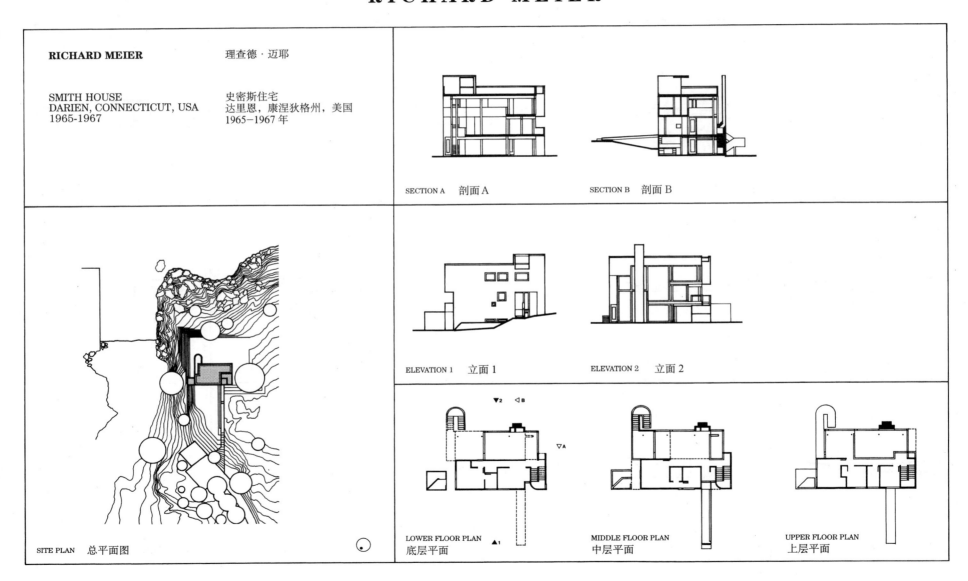

RICHARD MEIER　　　　　理查德·迈耶

SMITH HOUSE　　　　　史密斯住宅
DARIEN, CONNECTICUT, USA　　达里恩，康涅狄格州，美国
1965-1967　　　　　　1965－1967 年

SECTION A　剖面 A　　　　SECTION B　剖面 B

ELEVATION 1　立面 1　　　　ELEVATION 2　立面 2

SITE PLAN　总平面图

LOWER FLOOR PLAN　　MIDDLE FLOOR PLAN　　UPPER FLOOR PLAN
底层平面　　　　中层平面　　　　上层平面

136

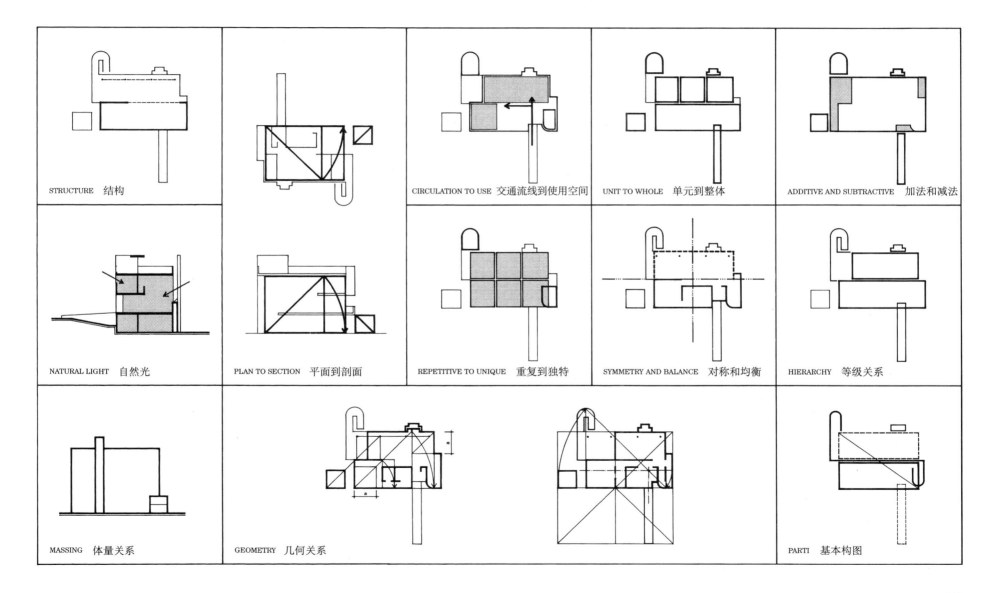

STRUCTURE　结构

CIRCULATION TO USE　交通流线到使用空间

UNIT TO WHOLE　单元到整体

ADDITIVE AND SUBTRACTIVE　加法和减法

NATURAL LIGHT　自然光

PLAN TO SECTION　平面到剖面

REPETITIVE TO UNIQUE　重复到独特

SYMMETRY AND BALANCE　对称和均衡

HIERARCHY　等级关系

MASSING　体量关系

GEOMETRY　几何关系

PARTI　基本构图

RICHARD MEIER　　　　　理查德·迈耶

THE ATHENEUM　　　　　　"雅典娜"游客中心
NEW HARMONY, INDIANA, USA　　新哈莫尼，印第安纳州，美国
1975-1979　　　　　　　　　1975－1979 年

SECTION A　剖面 A

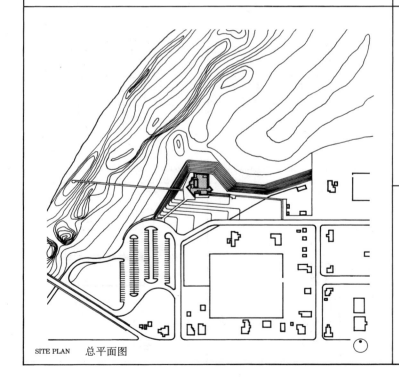

SITE PLAN　总平面图

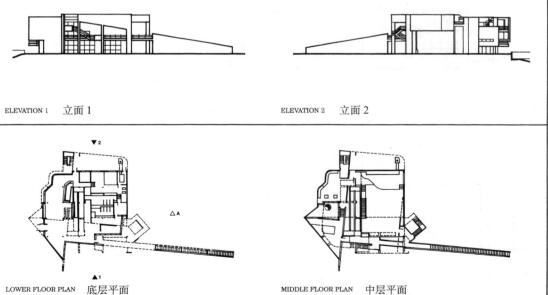

ELEVATION 1　立面 1　　　　　ELEVATION 2　立面 2

LOWER FLOOR PLAN　底层平面　　　MIDDLE FLOOR PLAN　中层平面

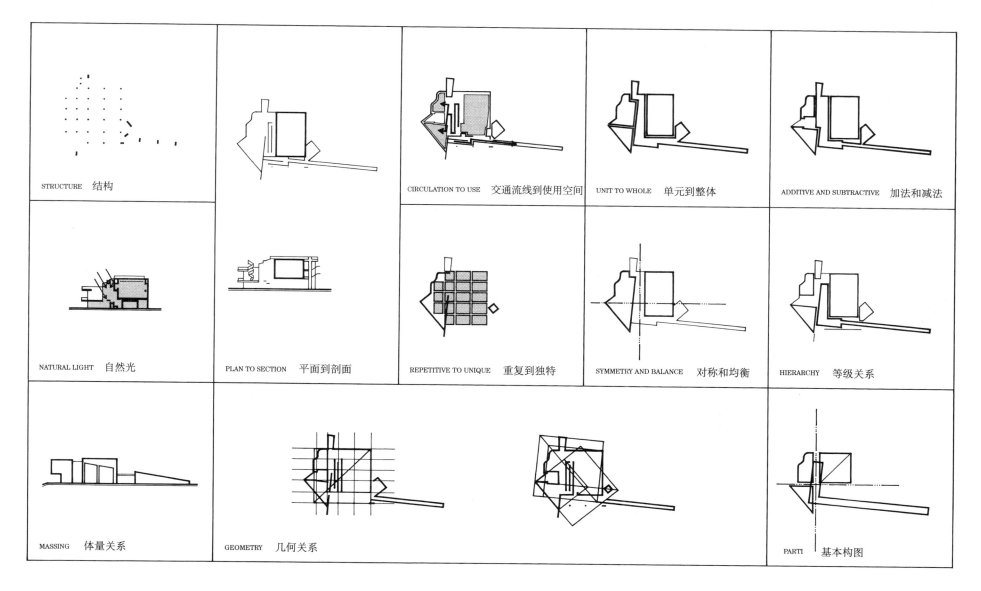

STRUCTURE 结构

CIRCULATION TO USE 交通流线到使用空间

UNIT TO WHOLE 单元到整体

ADDITIVE AND SUBTRACTIVE 加法和减法

NATURAL LIGHT 自然光

PLAN TO SECTION 平面到剖面

REPETITIVE TO UNIQUE 重复到独特

SYMMETRY AND BALANCE 对称和均衡

HIERARCHY 等级关系

MASSING 体量关系

GEOMETRY 几何关系

PARTI 基本构图

RICHARD MEIER

理查德·迈耶

ULM EXHIBITION AND ASSEMBLY BUILDING
ULM, GERMANY
1986-1992

乌尔姆展览与会议大厦
乌尔姆，德国
1986－1992 年

SECTION A　剖面 A

SECTION B　剖面 B

ELEVATION 1　立面 1

ELEVATION 2　立面 2

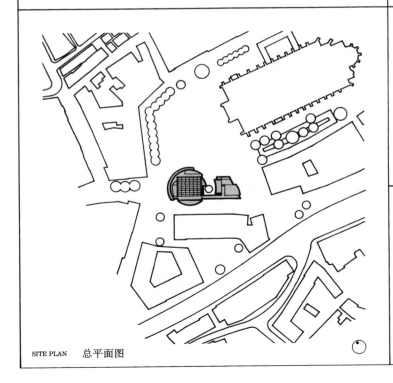

SITE PLAN　总平面图

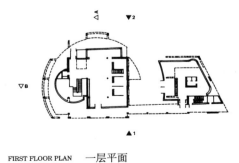

FIRST FLOOR PLAN　一层平面

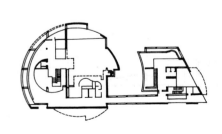

SECOND FLOOR PLAN　二层平面

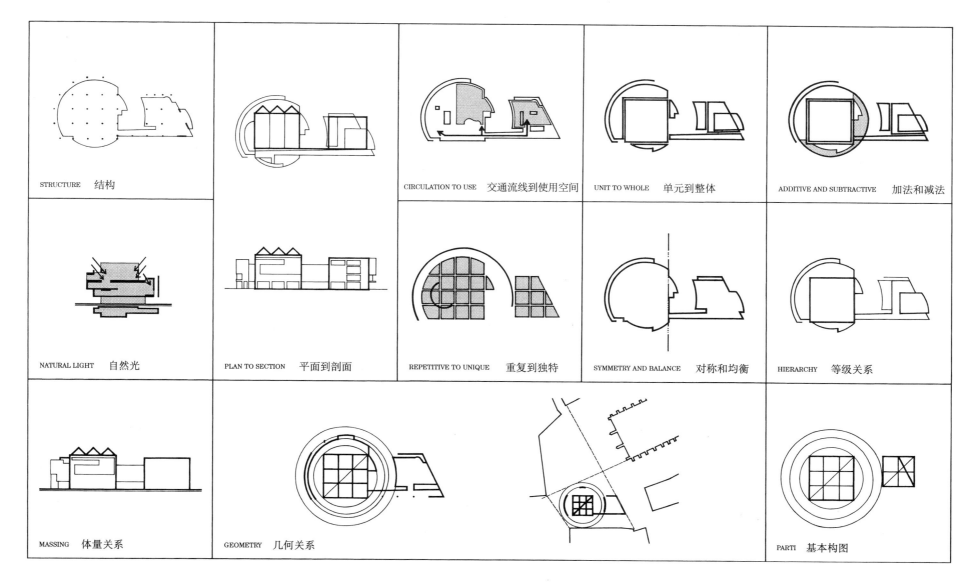

STRUCTURE　结构

CIRCULATION TO USE　交通流线到使用空间

UNIT TO WHOLE　单元到整体

ADDITIVE AND SUBTRACTIVE　加法和减法

NATURAL LIGHT　自然光

PLAN TO SECTION　平面到剖面

REPETITIVE TO UNIQUE　重复到独特

SYMMETRY AND BALANCE　对称和均衡

HIERARCHY　等级关系

MASSING　体量关系

GEOMETRY　几何关系

PARTI　基本构图

147

RICHARD MEIER　　理查德·迈耶

WEISHAUPT FORUM　　威斯豪普特中心
SCHWENDI, GERMANY　　施文迪，德国
1987-1992　　1987—1992 年

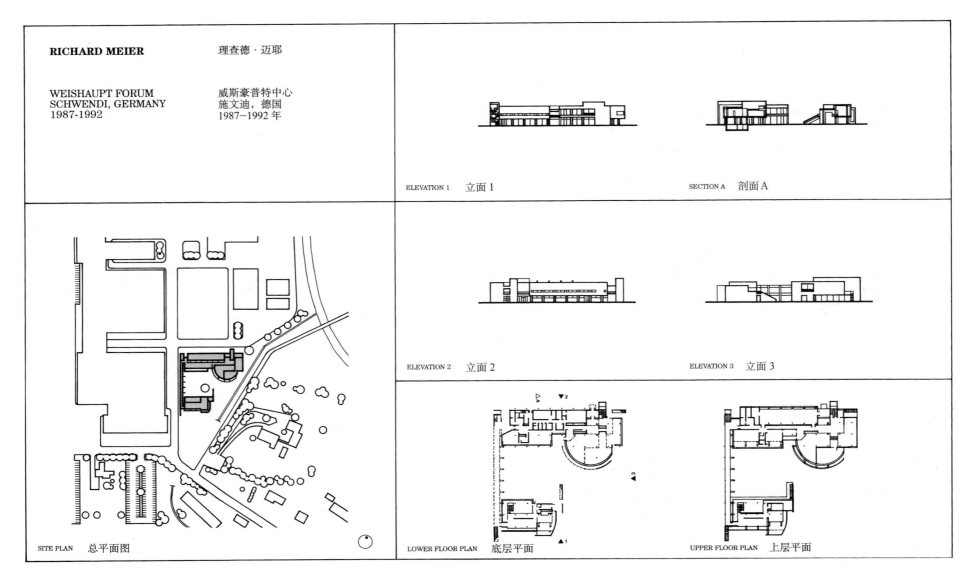

ELEVATION 1　立面 1

SECTION A　剖面 A

ELEVATION 2　立面 2

ELEVATION 3　立面 3

SITE PLAN　总平面图

LOWER FLOOR PLAN　底层平面

UPPER FLOOR PLAN　上层平面

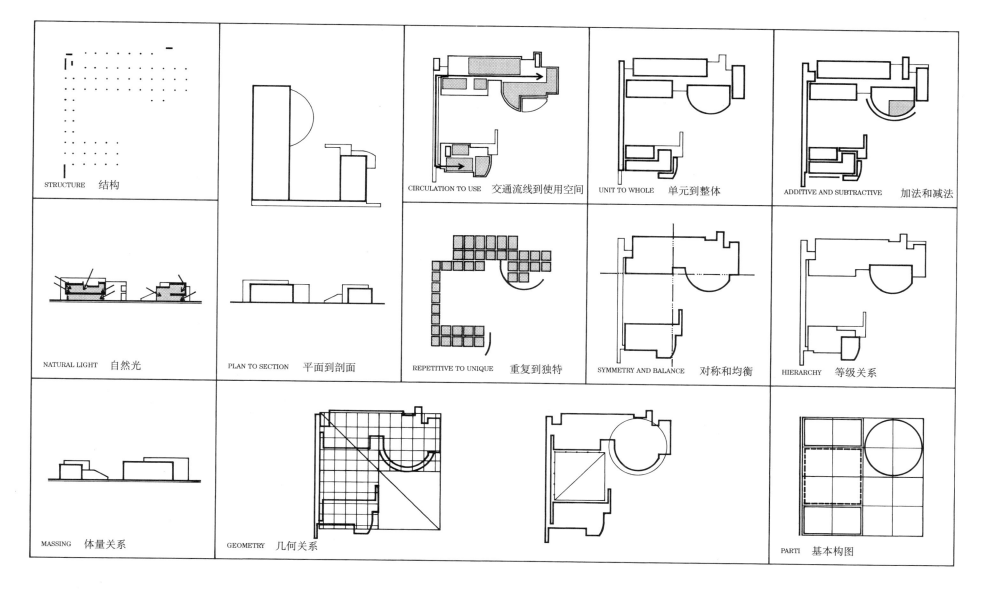

STRUCTURE　结构

CIRCULATION TO USE　交通流线到使用空间

UNIT TO WHOLE　单元到整体

ADDITIVE AND SUBTRACTIVE　加法和减法

NATURAL LIGHT　自然光

PLAN TO SECTION　平面到剖面

REPETITIVE TO UNIQUE　重复到独特

SYMMETRY AND BALANCE　对称和均衡

HIERARCHY　等级关系

MASSING　体量关系

GEOMETRY　几何关系

PARTI　基本构图

拉斐尔·莫内奥
RAFAEL MONEO

RAFAEL MONEO　　　　　拉斐尔·莫内奥

DON BENITO CULTURAL CENTER　　唐贝尼托文化中心
BADAJOZ, SPAIN　　　　　　　巴达霍斯，西班牙
1991–1997　　　　　　　　　　1991–1997 年

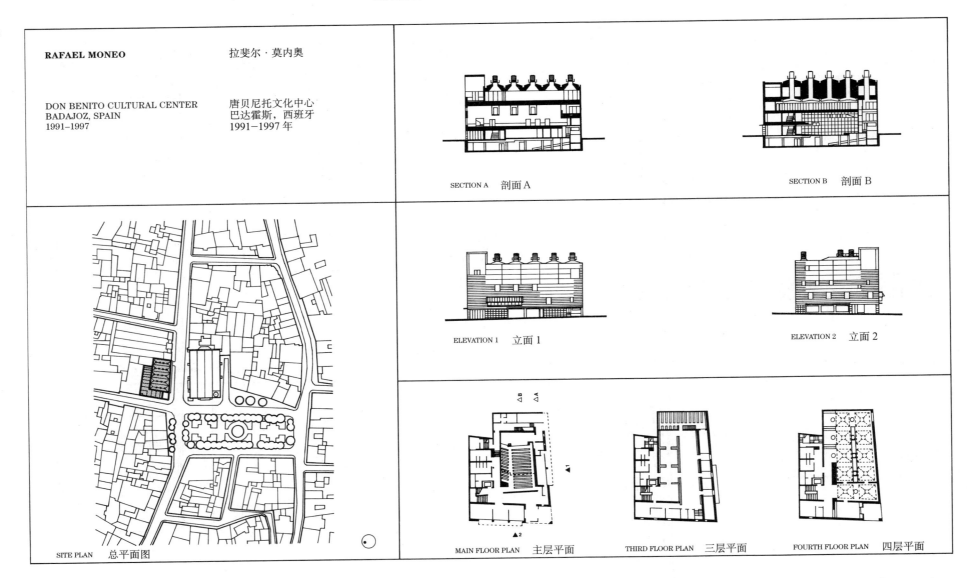

SECTION A　剖面A

SECTION B　剖面B

ELEVATION 1　立面 1

ELEVATION 2　立面 2

SITE PLAN　总平面图

MAIN FLOOR PLAN　主层平面

THIRD FLOOR PLAN　三层平面

FOURTH FLOOR PLAN　四层平面

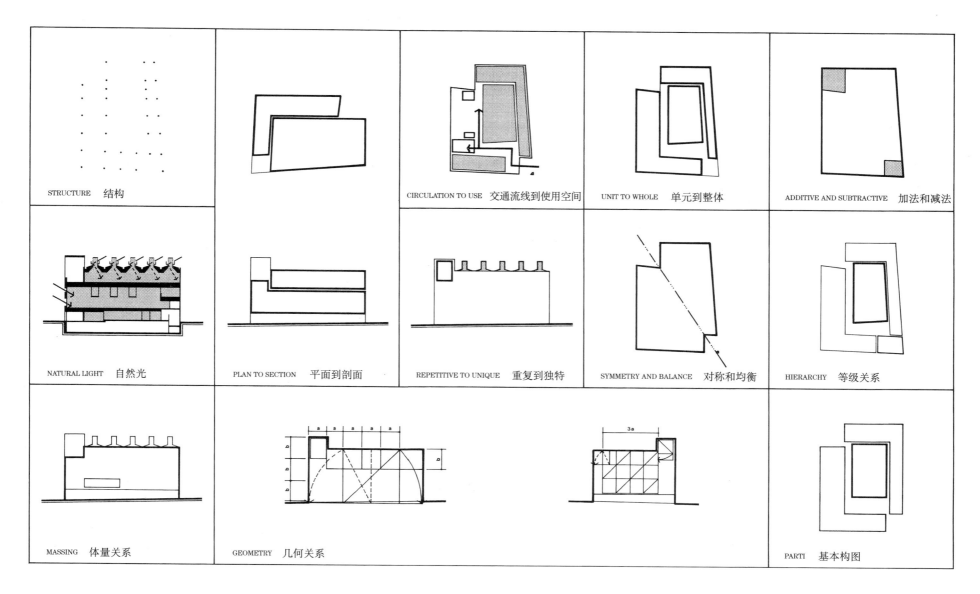

STRUCTURE　结构		CIRCULATION TO USE 交通流线到使用空间	UNIT TO WHOLE　单元到整体	ADDITIVE AND SUBTRACTIVE 加法和减法
NATURAL LIGHT　自然光	PLAN TO SECTION　平面到剖面	REPETITIVE TO UNIQUE　重复到独特	SYMMETRY AND BALANCE　对称和均衡	HIERARCHY　等级关系
MASSING　体量关系	GEOMETRY　几何关系		SYMMETRY AND BALANCE	PARTI　基本构图

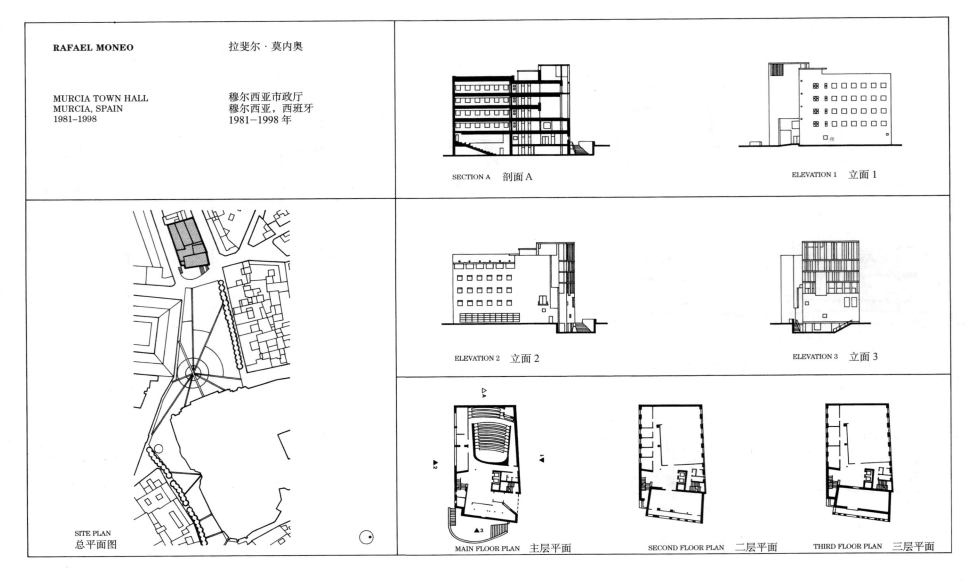

RAFAEL MONEO　　　　　　拉斐尔·莫内奥

MURCIA TOWN HALL　　　　　穆尔西亚市政厅
MURCIA, SPAIN　　　　　　　穆尔西亚，西班牙
1981–1998　　　　　　　　　　1981–1998 年

SECTION A　剖面 A

ELEVATION 1　立面 1

ELEVATION 2　立面 2

ELEVATION 3　立面 3

SITE PLAN
总平面图

MAIN FLOOR PLAN　主层平面

SECOND FLOOR PLAN　二层平面

THIRD FLOOR PLAN　三层平面

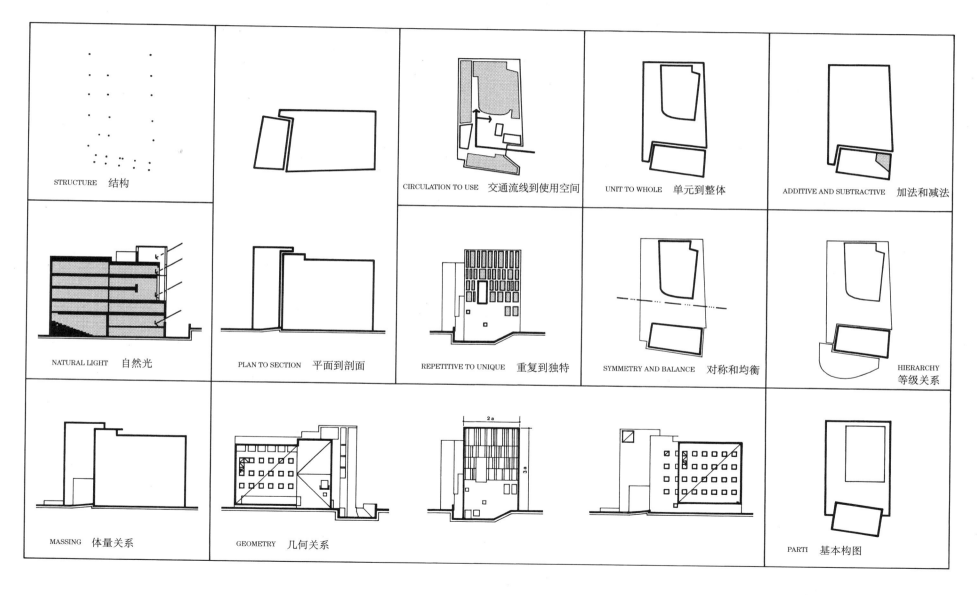

STRUCTURE　结构

CIRCULATION TO USE　交通流线到使用空间

UNIT TO WHOLE　单元到整体

ADDITIVE AND SUBTRACTIVE　加法和减法

NATURAL LIGHT　自然光

PLAN TO SECTION　平面到剖面

REPETITIVE TO UNIQUE　重复到独特

SYMMETRY AND BALANCE　对称和均衡

HIERARCHY
等级关系

MASSING　体量关系

GEOMETRY　几何关系

PARTI　基本构图

查尔斯·穆尔
CHARLES MOORE

CHARLES MOORE　　　　　查尔斯·穆尔

MOORE HOUSE　　　　　穆尔住宅
ORINDA, CALIFORNIA, USA　　奥林达，加利福尼亚州，美国
1962　　　　　　　　　　　　1962 年

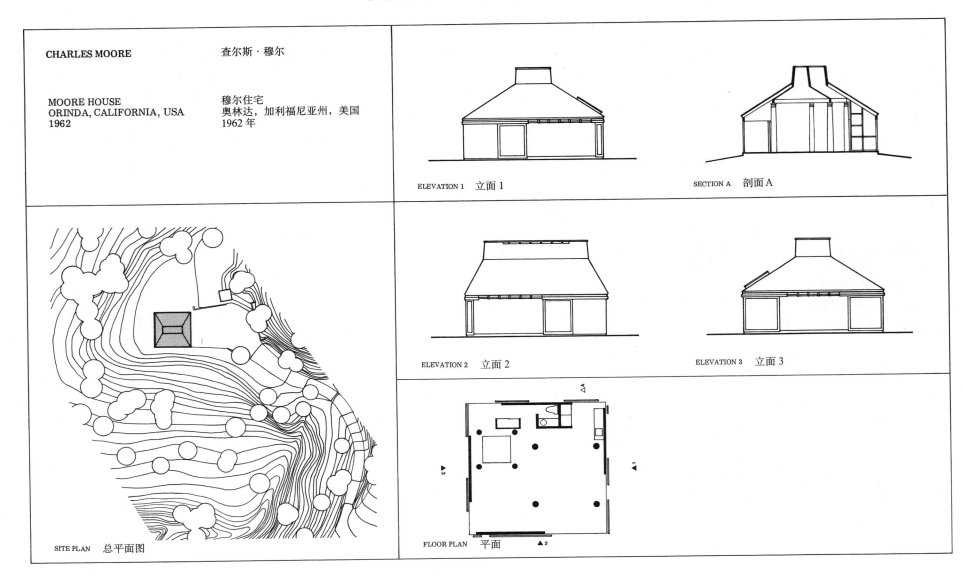

ELEVATION 1　立面 1

SECTION A　剖面 A

ELEVATION 2　立面 2

ELEVATION 3　立面 3

SITE PLAN　总平面图

FLOOR PLAN　平面

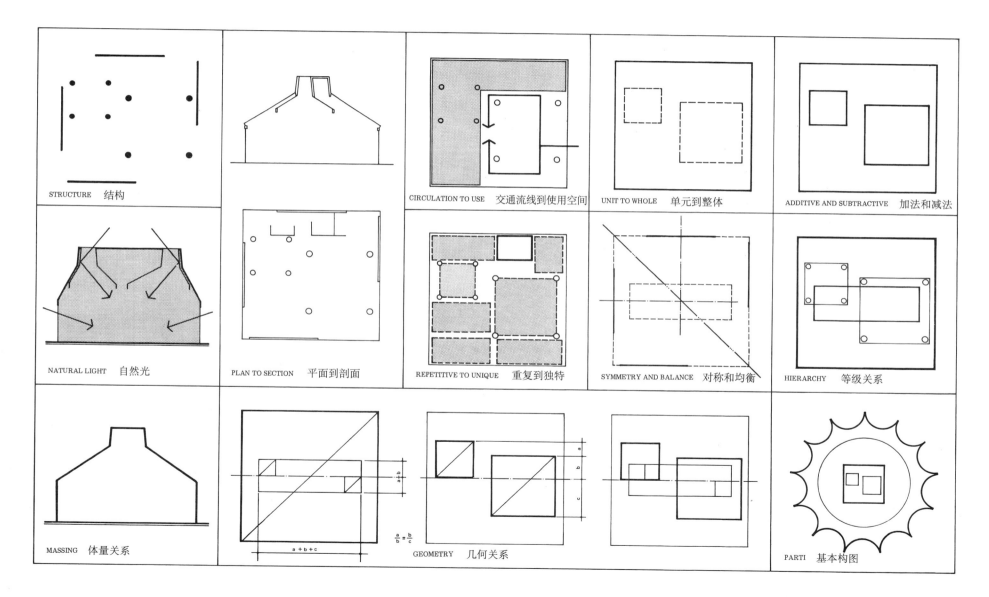

STRUCTURE　结构

CIRCULATION TO USE　交通流线到使用空间

UNIT TO WHOLE　单元到整体

ADDITIVE AND SUBTRACTIVE　加法和减法

NATURAL LIGHT　自然光

PLAN TO SECTION　平面到剖面

REPETITIVE TO UNIQUE　重复到独特

SYMMETRY AND BALANCE　对称和均衡

HIERARCHY　等级关系

MASSING　体量关系

GEOMETRY　几何关系

$\frac{a}{b} = \frac{b}{c}$

a + b + c

PARTI　基本构图

155

CHARLES MOORE 查尔斯·穆尔

CONDOMINIUM I
SEA RANCH, CALIFORNIA, USA
1964-1965

一期共管住宅
西兰奇，加利福尼亚州，美国
1964—1965 年

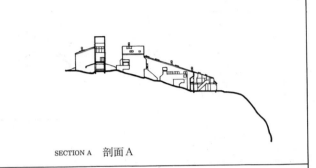

SECTION A　剖面 A

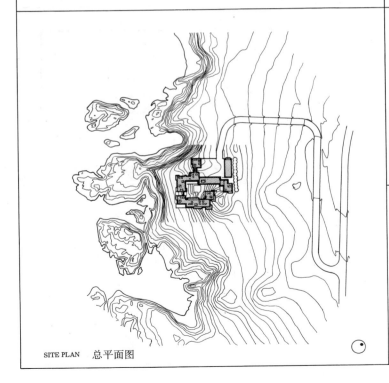

SITE PLAN　总平面图

ELEVATION 1　立面 1

ELEVATION 2　立面 2

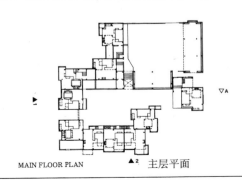

MAIN FLOOR PLAN　　▲2　主层平面

UPPER FLOOR PLAN　上层平面

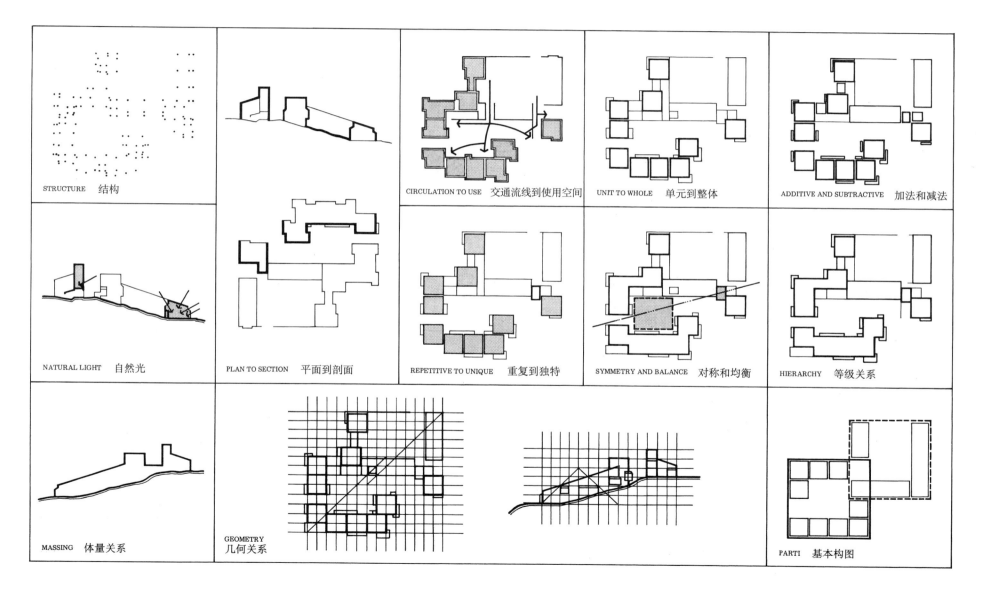

STRUCTURE　结构

CIRCULATION TO USE　交通流线到使用空间

UNIT TO WHOLE　单元到整体

ADDITIVE AND SUBTRACTIVE　加法和减法

NATURAL LIGHT　自然光

PLAN TO SECTION　平面到剖面

REPETITIVE TO UNIQUE　重复到独特

SYMMETRY AND BALANCE　对称和均衡

HIERARCHY　等级关系

MASSING　体量关系

GEOMETRY　几何关系

PARTI　基本构图

CHARLES MOORE　　　查尔斯·穆尔

HINES HOUSE　　　海因斯住宅
SEA RANCH, CALIFORNIA, USA　　西兰奇，加利福尼亚州，美国
1967　　　　　1967 年

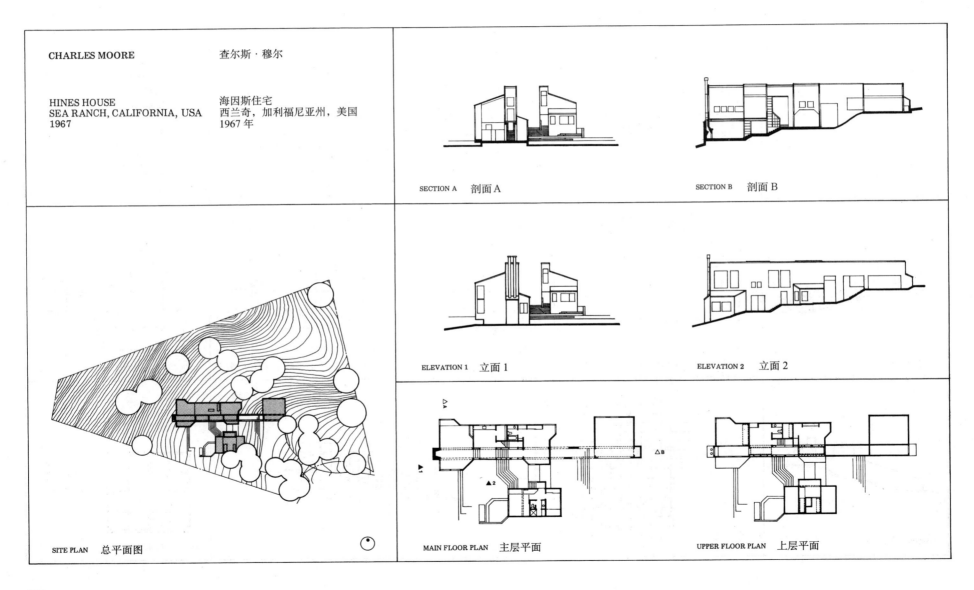

SECTION A　剖面 A

SECTION B　剖面 B

ELEVATION 1　立面 1

ELEVATION 2　立面 2

SITE PLAN　总平面图

MAIN FLOOR PLAN　主层平面

UPPER FLOOR PLAN　上层平面

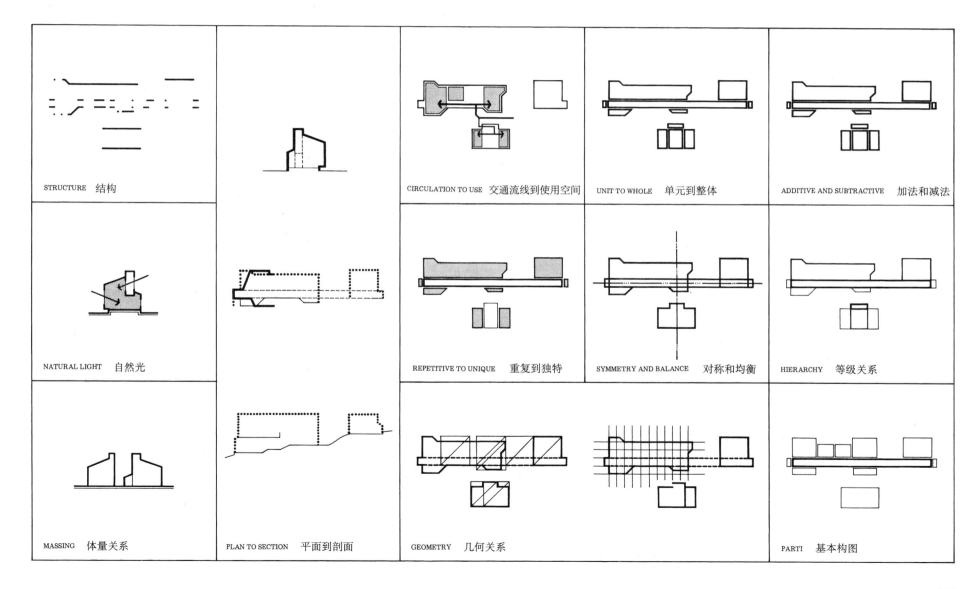

STRUCTURE 结构

CIRCULATION TO USE 交通流线到使用空间

UNIT TO WHOLE 单元到整体

ADDITIVE AND SUBTRACTIVE 加法和减法

NATURAL LIGHT 自然光

REPETITIVE TO UNIQUE 重复到独特

SYMMETRY AND BALANCE 对称和均衡

HIERARCHY 等级关系

MASSING 体量关系

PLAN TO SECTION 平面到剖面

GEOMETRY 几何关系

PARTI 基本构图

CHARLES MOORE
查尔斯·穆尔

BURNS HOUSE
SANTA MONICA CANYON, CALIFORNIA, USA
1974

伯恩斯住宅
圣莫尼卡峡谷，加利福尼亚州，美国
1974 年

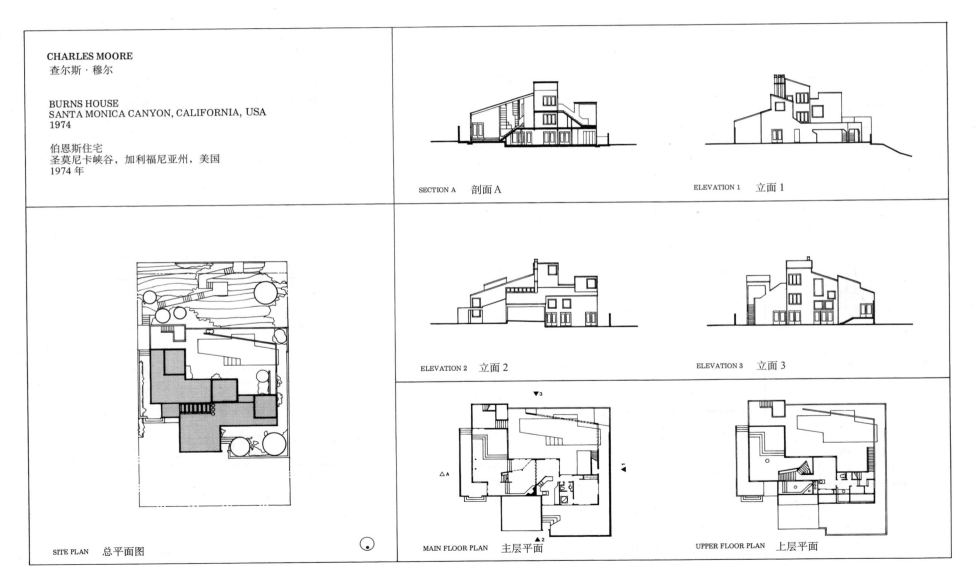

SECTION A   剖面 A

ELEVATION 1   立面 1

ELEVATION 2   立面 2

ELEVATION 3   立面 3

SITE PLAN   总平面图

MAIN FLOOR PLAN   主层平面

UPPER FLOOR PLAN   上层平面

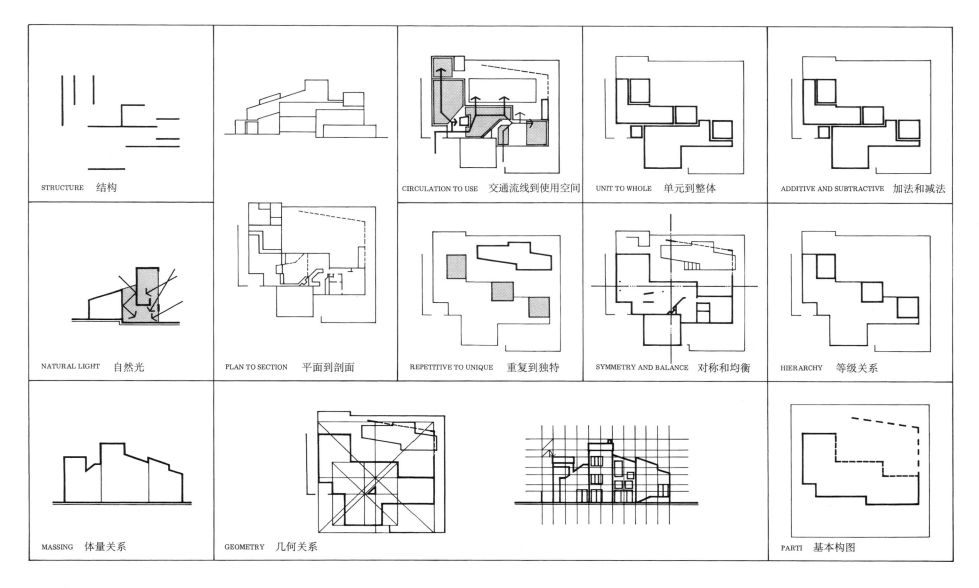

STRUCTURE　结构

CIRCULATION TO USE　交通流线到使用空间

UNIT TO WHOLE　单元到整体

ADDITIVE AND SUBTRACTIVE　加法和减法

NATURAL LIGHT　自然光

PLAN TO SECTION　平面到剖面

REPETITIVE TO UNIQUE　重复到独特

SYMMETRY AND BALANCE　对称和均衡

HIERARCHY　等级关系

MASSING　体量关系

GEOMETRY　几何关系

PARTI　基本构图

156

GLENN MURCUTT

格伦·穆科特

MAGNEY HOUSE
BINGIE POINT, MORUYA, AUSTRALIA
1982–1984

马格尼住宅
宾奇屿，莫鲁亚，澳大利亚
1982–1984 年

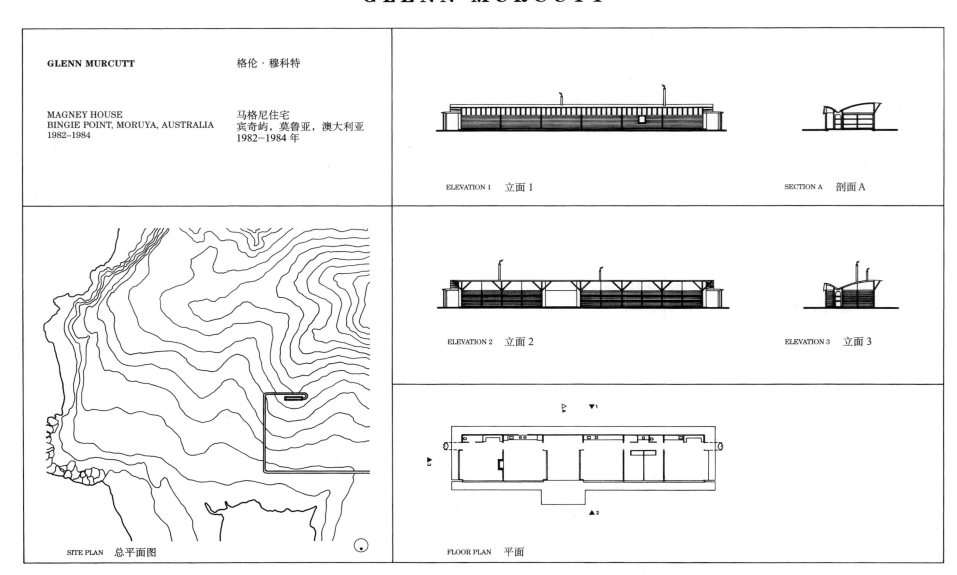

ELEVATION 1　立面 1

SECTION A　剖面 A

ELEVATION 2　立面 2

ELEVATION 3　立面 3

SITE PLAN　总平面图

FLOOR PLAN　平面

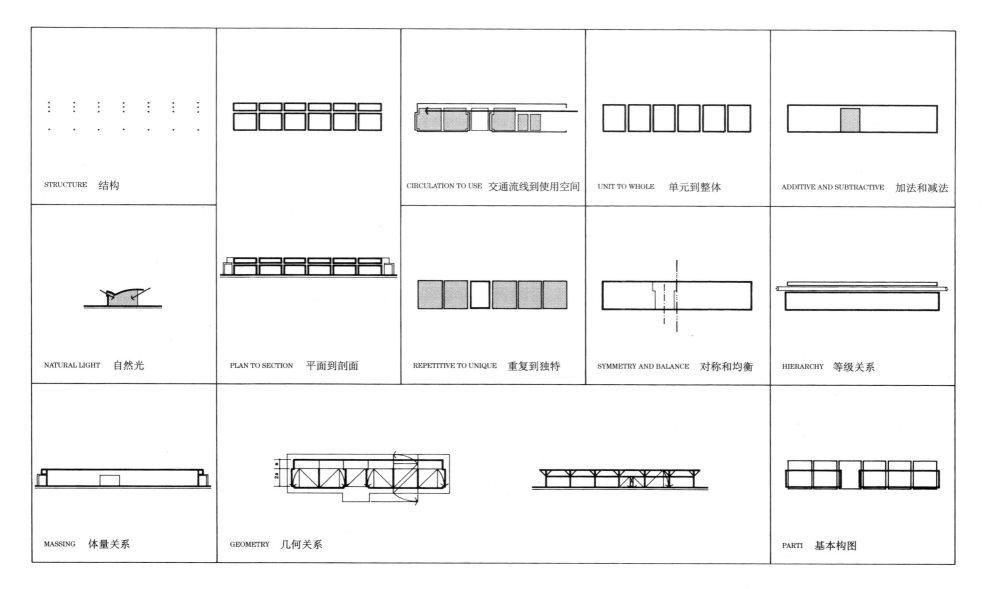

STRUCTURE 结构

CIRCULATION TO USE 交通流线到使用空间

UNIT TO WHOLE 单元到整体

ADDITIVE AND SUBTRACTIVE 加法和减法

NATURAL LIGHT 自然光

PLAN TO SECTION 平面到剖面

REPETITIVE TO UNIQUE 重复到独特

SYMMETRY AND BALANCE 对称和均衡

HIERARCHY 等级关系

MASSING 体量关系

GEOMETRY 几何关系

PARTI 基本构图

GLENN MURCUTT
格伦·穆科特

SIMPSON-LEE HOUSE
MT. WILSON, NEW SOUTH WALES, AUSTRALIA
1989–1994

辛普森－李住宅
威尔逊山地，新南威尔士州，澳大利亚
1989–1994 年

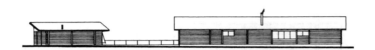

ELEVATION 1　立面 1

SECTION A　剖面 A

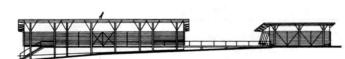

ELEVATION 2　立面 2

ELEVATION 3　立面 3

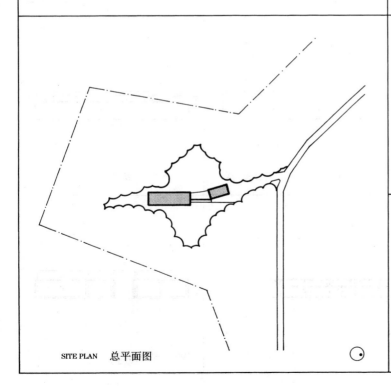

SITE PLAN　总平面图

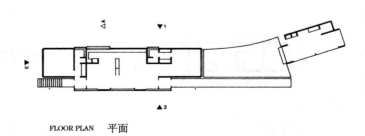

FLOOR PLAN　平面

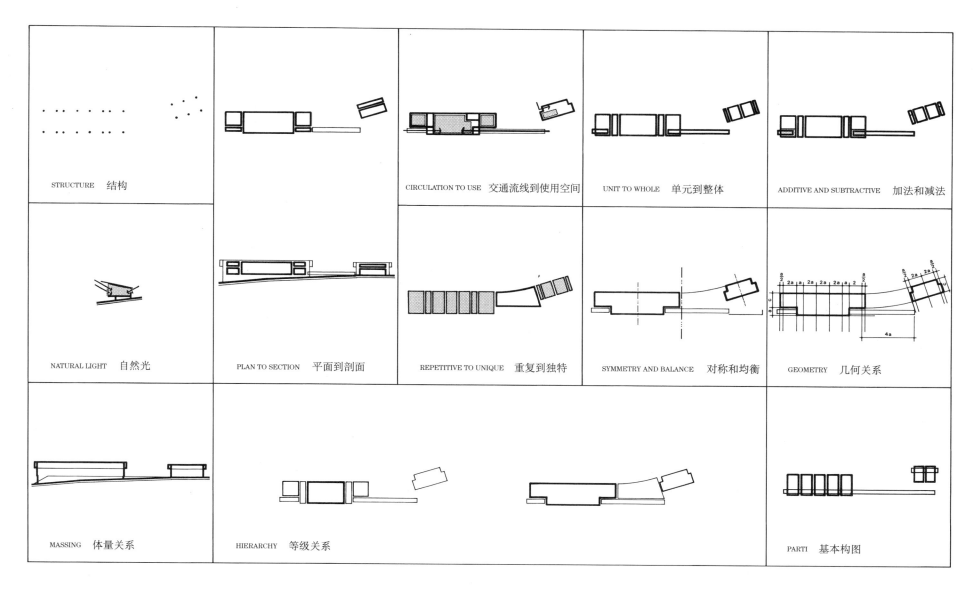

STRUCTURE 结构

CIRCULATION TO USE 交通流线到使用空间

UNIT TO WHOLE 单元到整体

ADDITIVE AND SUBTRACTIVE 加法和减法

NATURAL LIGHT 自然光

PLAN TO SECTION 平面到剖面

REPETITIVE TO UNIQUE 重复到独特

SYMMETRY AND BALANCE 对称和均衡

GEOMETRY 几何关系

MASSING 体量关系

HIERARCHY 等级关系

PARTI 基本构图

让·努韦尔
JEAN NOUVEL

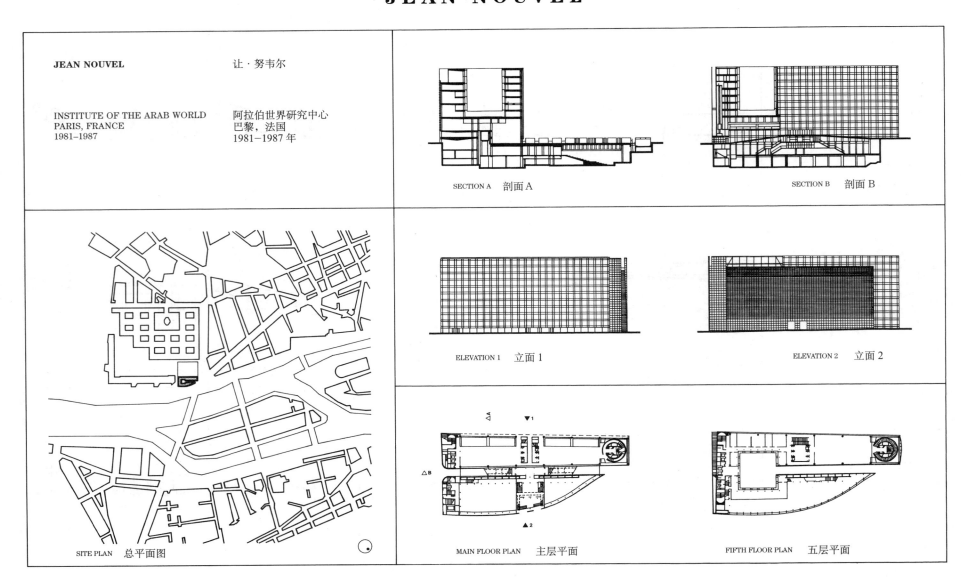

JEAN NOUVEL　　　　　　　让·努韦尔

INSTITUTE OF THE ARAB WORLD　　阿拉伯世界研究中心
PARIS, FRANCE　　　　　　　　　巴黎，法国
1981–1987　　　　　　　　　　　1981–1987 年

SECTION A　剖面 A

SECTION B　剖面 B

ELEVATION 1　立面 1

ELEVATION 2　立面 2

SITE PLAN　总平面图

MAIN FLOOR PLAN　主层平面

FIFTH FLOOR PLAN　五层平面

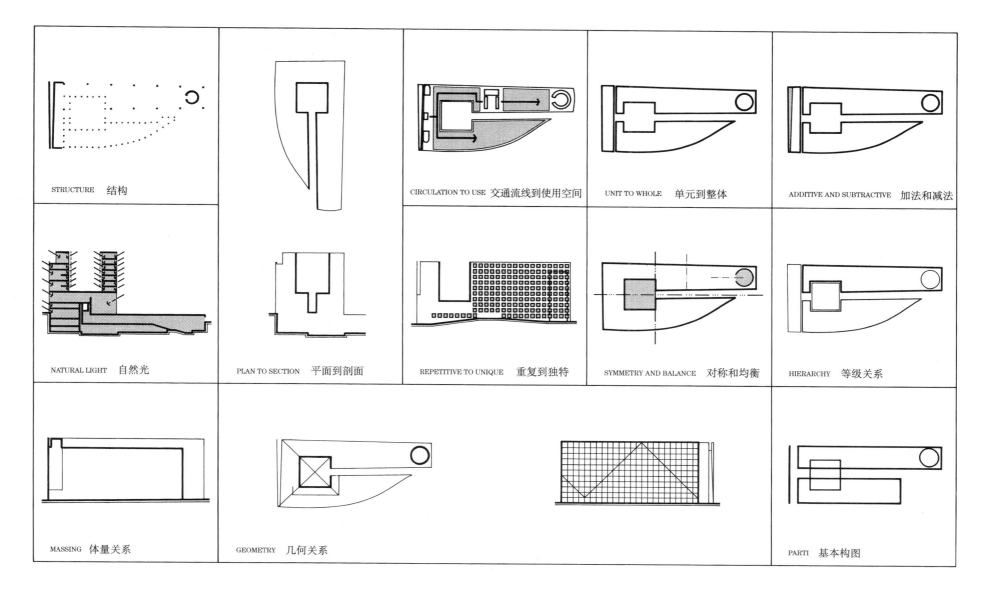

STRUCTURE　结构

CIRCULATION TO USE　交通流线到使用空间

UNIT TO WHOLE　单元到整体

ADDITIVE AND SUBTRACTIVE　加法和减法

NATURAL LIGHT　自然光

PLAN TO SECTION　平面到剖面

REPETITIVE TO UNIQUE　重复到独特

SYMMETRY AND BALANCE　对称和均衡

HIERARCHY　等级关系

MASSING　体量关系

GEOMETRY　几何关系

PARTI　基本构图

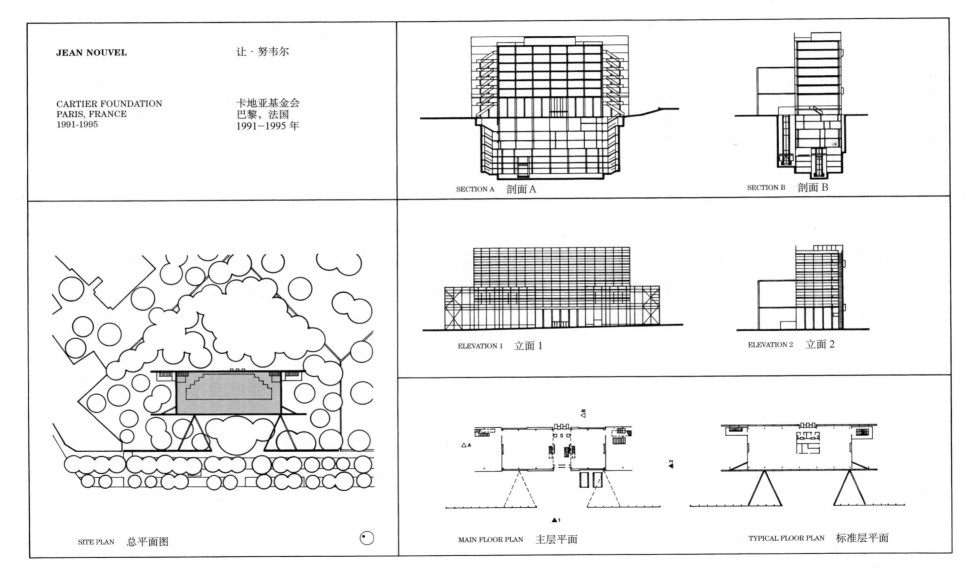

JEAN NOUVEL 让 · 努韦尔

CARTIER FOUNDATION 卡地亚基金会
PARIS, FRANCE 巴黎，法国
1991-1995 1991-1995 年

SECTION A 剖面 A

SECTION B 剖面 B

ELEVATION 1 立面 1

ELEVATION 2 立面 2

SITE PLAN 总平面图

MAIN FLOOR PLAN 主层平面

TYPICAL FLOOR PLAN 标准层平面

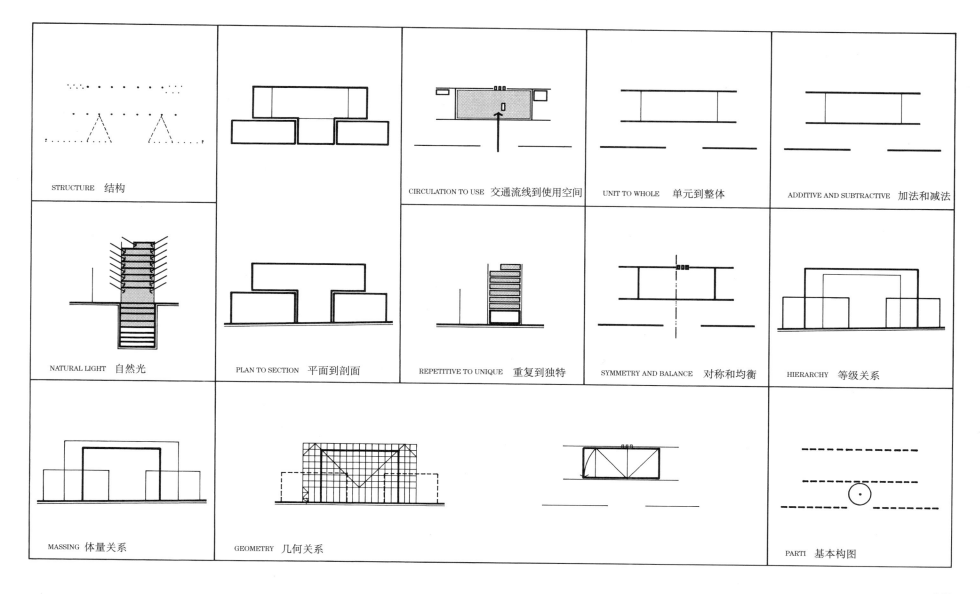

STRUCTURE 结构

CIRCULATION TO USE 交通流线到使用空间

UNIT TO WHOLE 单元到整体

ADDITIVE AND SUBTRACTIVE 加法和减法

NATURAL LIGHT 自然光

PLAN TO SECTION 平面到剖面

REPETITIVE TO UNIQUE 重复到独特

SYMMETRY AND BALANCE 对称和均衡

HIERARCHY 等级关系

MASSING 体量关系

GEOMETRY 几何关系

PARTI 基本构图

安德烈亚·帕拉第奥
ANDREA PALLADIO

ANDREA PALLADIO

安德烈亚·帕拉第奥

VILLA FOSCARI
MALCONTENTA, ITALY
c. 1549-1563

福斯卡里别墅
马尔孔滕塔，意大利
约 1549–1563 年

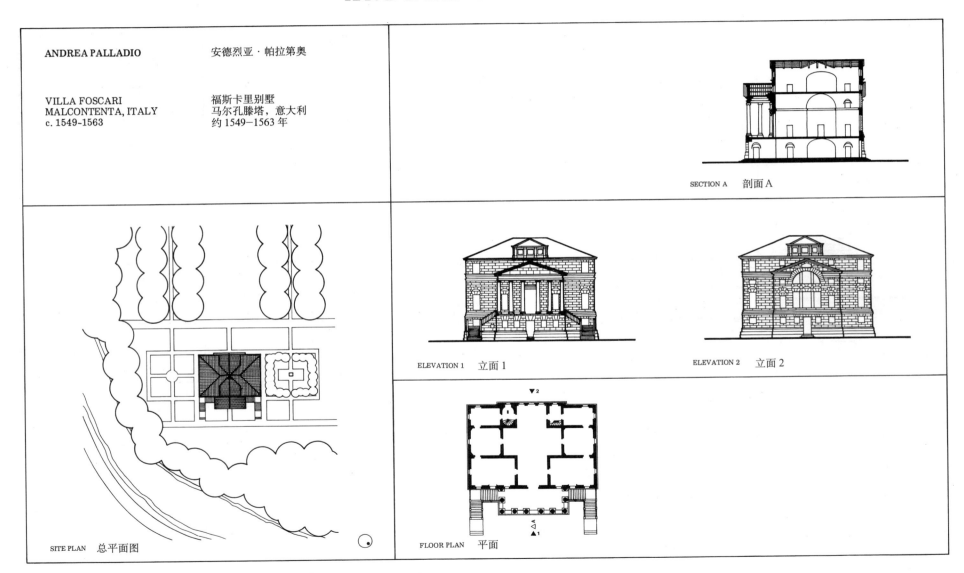

SECTION A   剖面A

ELEVATION 1   立面1

ELEVATION 2   立面2

SITE PLAN   总平面图

FLOOR PLAN   平面

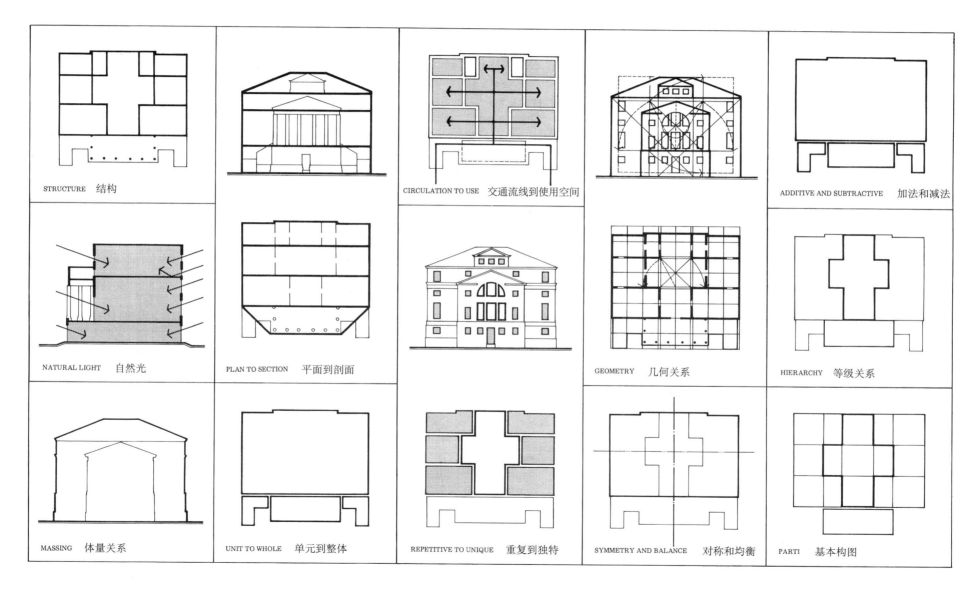

STRUCTURE　结构

CIRCULATION TO USE　交通流线到使用空间

ADDITIVE AND SUBTRACTIVE　加法和减法

NATURAL LIGHT　自然光

PLAN TO SECTION　平面到剖面

GEOMETRY　几何关系

HIERARCHY　等级关系

MASSING　体量关系

UNIT TO WHOLE　单元到整体

REPETITIVE TO UNIQUE　重复到独特

SYMMETRY AND BALANCE　对称和均衡

PARTI　基本构图

ANDREA PALLADIO　　　安德烈亚·帕拉第奥

CHURCH OF SAN GIORGIO MAGGIORE　圣乔治亚·马焦雷教堂
VENICE, ITALY　　　　　　　　威尼斯，意大利
1560–1580　　　　　　　　　　1560–1580 年

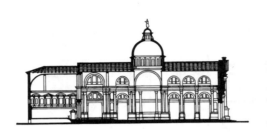

SECTION A　剖面 A

SECTION B　剖面 B

ELEVATION 1　立面 1

SITE PLAN　总平面图

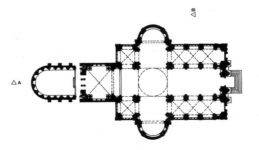

FLOOR PLAN　平面

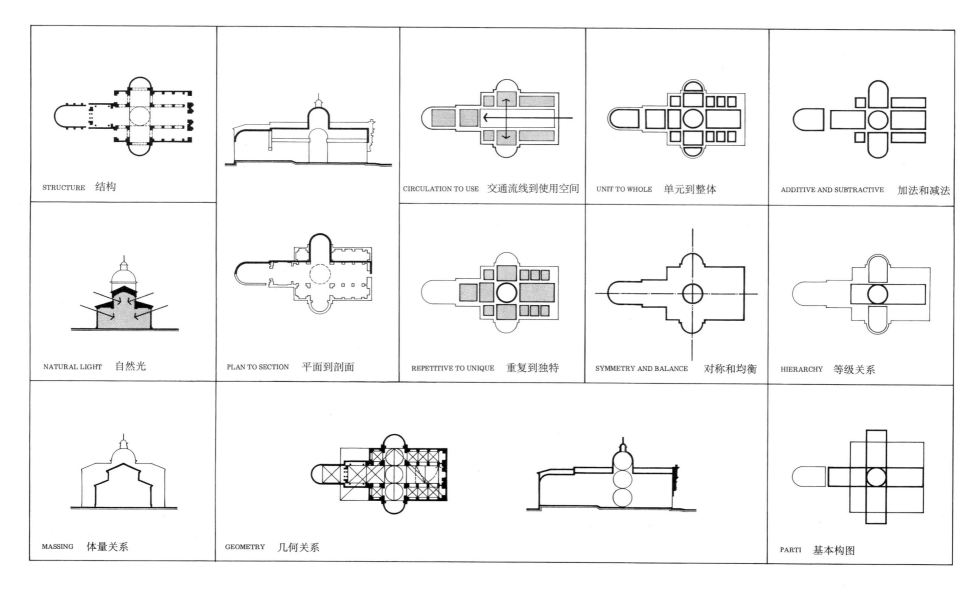

STRUCTURE　结构

NATURAL LIGHT　自然光

MASSING　体量关系

CIRCULATION TO USE　交通流线到使用空间

PLAN TO SECTION　平面到剖面

GEOMETRY　几何关系

UNIT TO WHOLE　单元到整体

REPETITIVE TO UNIQUE　重复到独特

SYMMETRY AND BALANCE　对称和均衡

ADDITIVE AND SUBTRACTIVE　加法和减法

HIERARCHY　等级关系

PARTI　基本构图

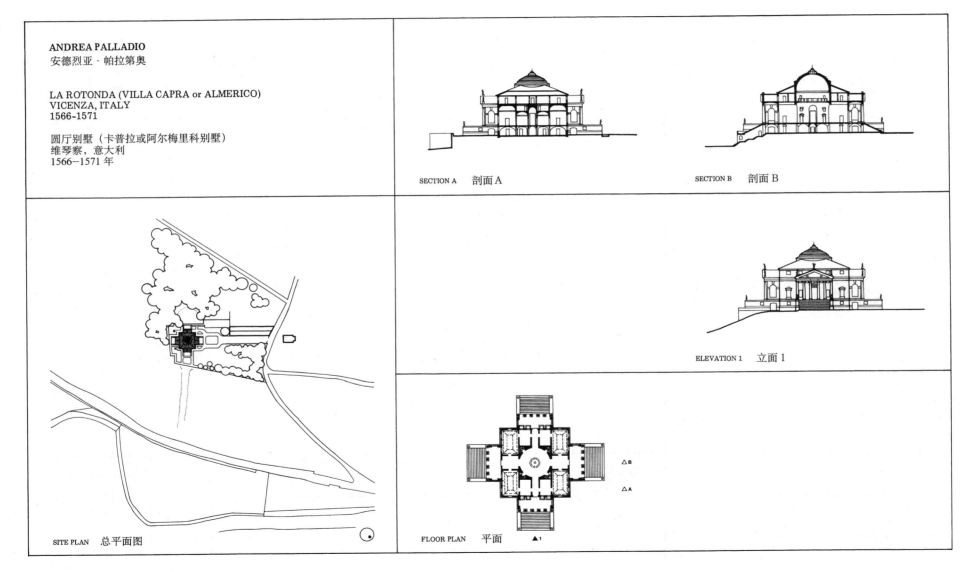

ANDREA PALLADIO
安德烈亚·帕拉第奥

LA ROTONDA (VILLA CAPRA or ALMERICO)
VICENZA, ITALY
1566-1571

圆厅别墅（卡普拉或阿尔梅里科别墅）
维琴察，意大利
1566-1571 年

SECTION A　剖面 A

SECTION B　剖面 B

ELEVATION 1　立面 1

SITE PLAN　总平面图

FLOOR PLAN　平面　▲1

△B

△A

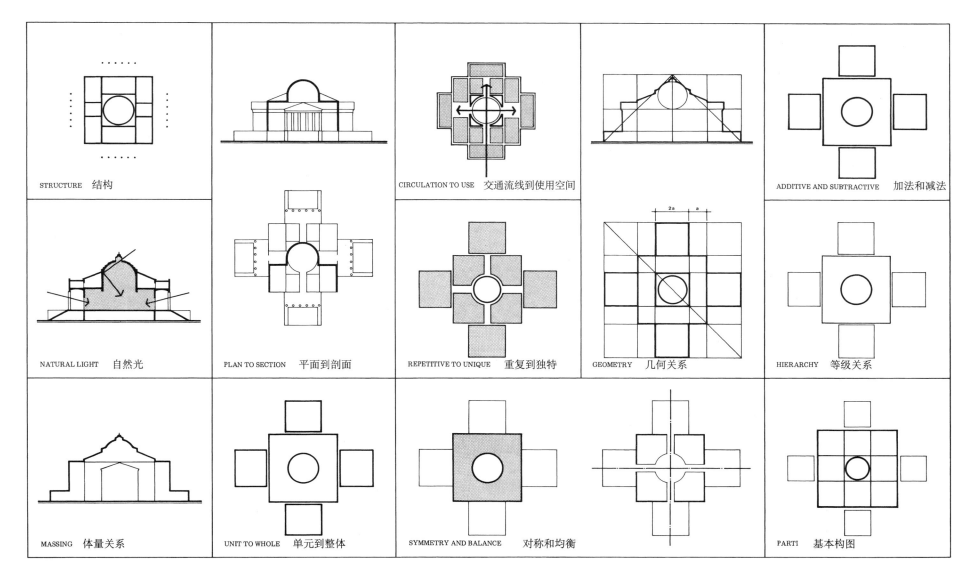

STRUCTURE 结构

CIRCULATION TO USE 交通流线到使用空间

ADDITIVE AND SUBTRACTIVE 加法和减法

NATURAL LIGHT 自然光

PLAN TO SECTION 平面到剖面

REPETITIVE TO UNIQUE 重复到独特

GEOMETRY 几何关系

HIERARCHY 等级关系

MASSING 体量关系

UNIT TO WHOLE 单元到整体

SYMMETRY AND BALANCE 对称和均衡

PARTI 基本构图

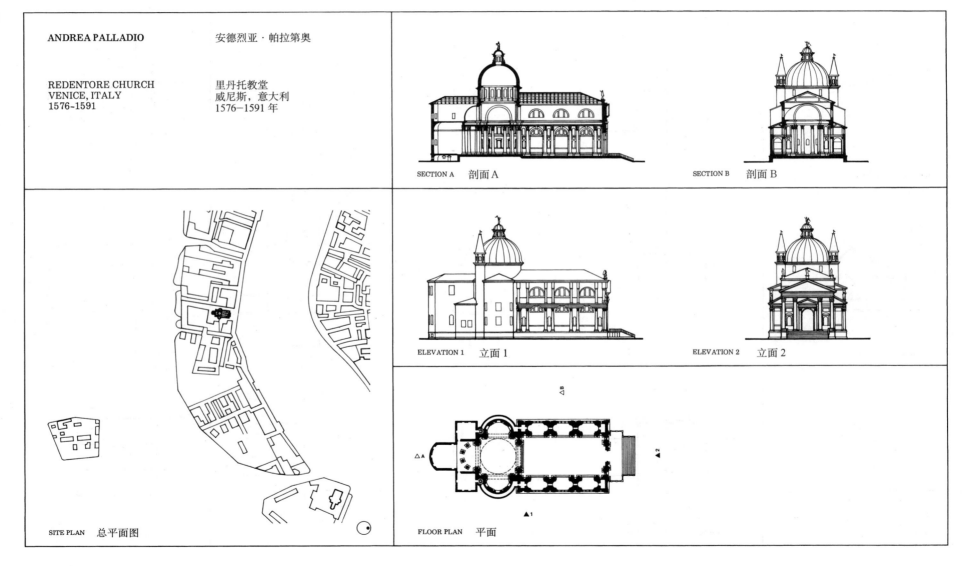

ANDREA PALLADIO　　　安德烈亚·帕拉第奥

REDENTORE CHURCH　　里丹托教堂
VENICE, ITALY　　　　威尼斯，意大利
1576–1591　　　　　　1576–1591 年

SECTION A　剖面 A

SECTION B　剖面 B

ELEVATION 1　立面 1

ELEVATION 2　立面 2

SITE PLAN　总平面图

FLOOR PLAN　平面

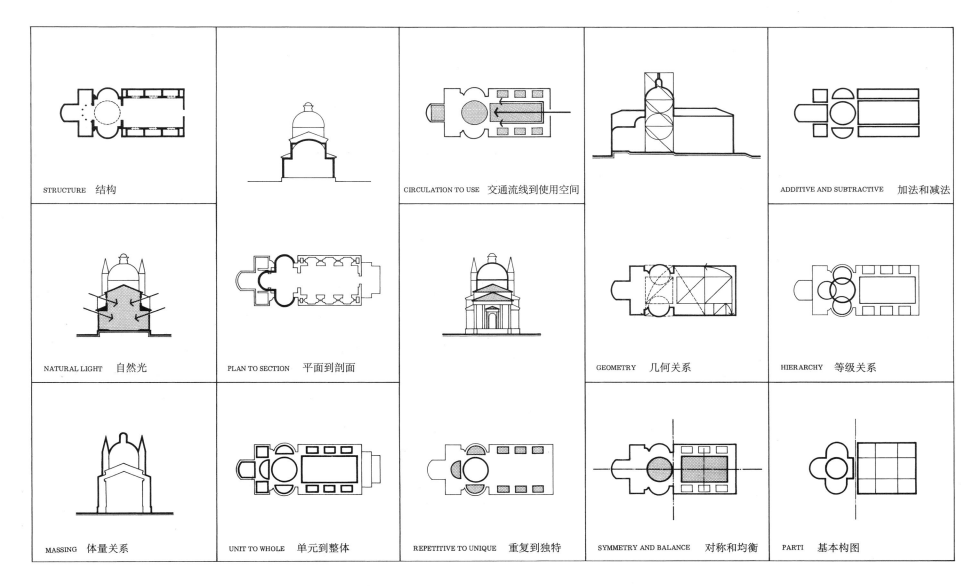

STRUCTURE　结构

CIRCULATION TO USE　交通流线到使用空间

ADDITIVE AND SUBTRACTIVE　加法和减法

NATURAL LIGHT　自然光

PLAN TO SECTION　平面到剖面

GEOMETRY　几何关系

HIERARCHY　等级关系

MASSING　体量关系

UNIT TO WHOLE　单元到整体

REPETITIVE TO UNIQUE　重复到独特

SYMMETRY AND BALANCE　对称和均衡

PARTI　基本构图

托马斯·菲弗
THOMAS PHIFER

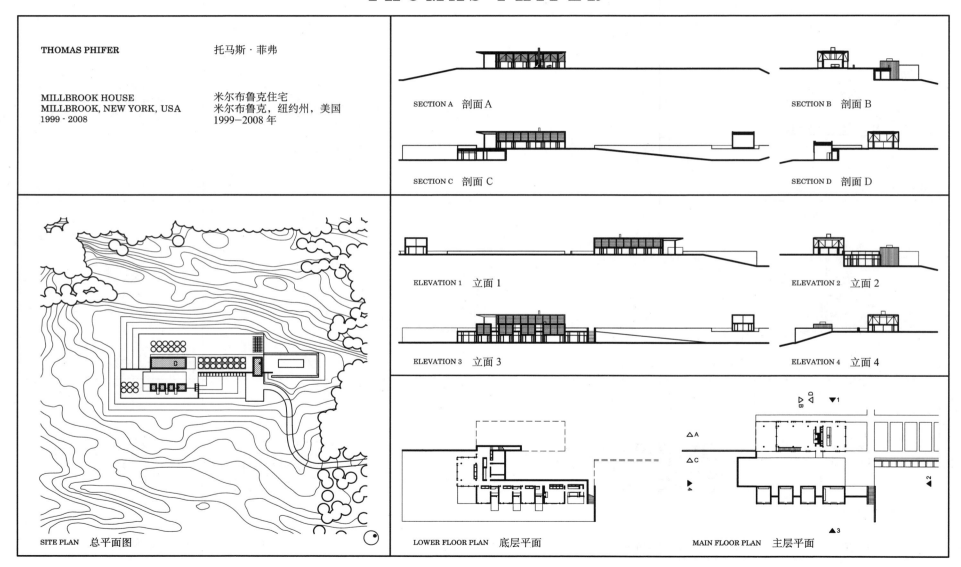

THOMAS PHIFER　　托马斯·菲弗

MILLBROOK HOUSE　　米尔布鲁克住宅
MILLBROOK, NEW YORK, USA　　米尔布鲁克，纽约州，美国
1999 · 2008　　1999—2008 年

SECTION A　剖面 A　　　　SECTION B　剖面 B

SECTION C　剖面 C　　　　SECTION D　剖面 D

ELEVATION 1　立面 1　　　　ELEVATION 2　立面 2

ELEVATION 3　立面 3　　　　ELEVATION 4　立面 4

SITE PLAN　总平面图

LOWER FLOOR PLAN　底层平面　　　　MAIN FLOOR PLAN　主层平面

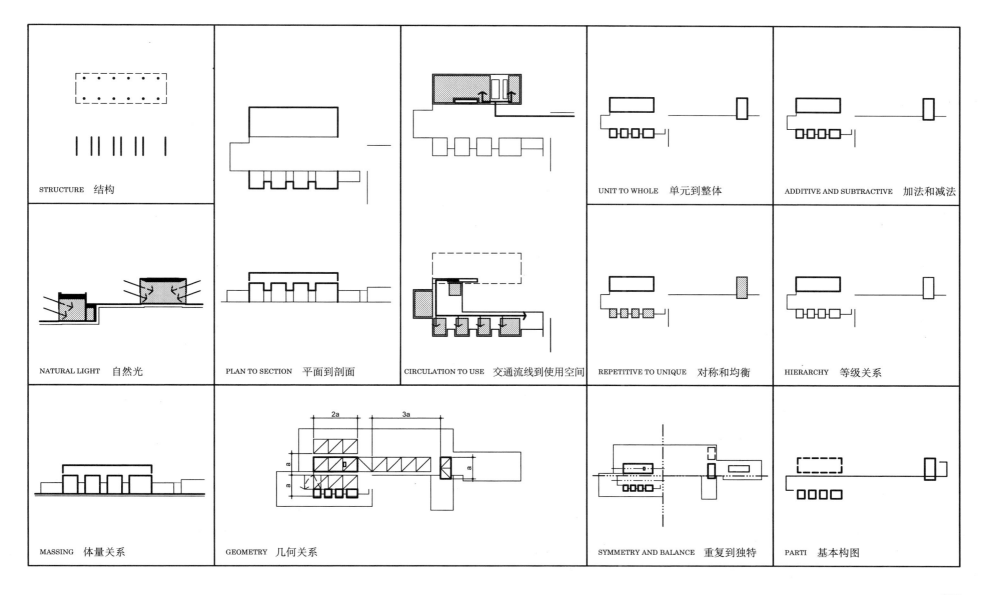

STRUCTURE　结构

NATURAL LIGHT　自然光

MASSING　体量关系

PLAN TO SECTION　平面到剖面

GEOMETRY　几何关系

CIRCULATION TO USE　交通流线到使用空间

UNIT TO WHOLE　单元到整体

REPETITIVE TO UNIQUE　对称和均衡

SYMMETRY AND BALANCE　重复到独特

ADDITIVE AND SUBTRACTIVE　加法和减法

HIERARCHY　等级关系

PARTI　基本构图

THOMAS PHIFER　　托马斯·菲弗

SALT POINT HOUSE　　盐点住宅
SALT POINT, NEW YORK, USA　　盐点，纽约州，美国
2004 · 2007　　2004—2007 年

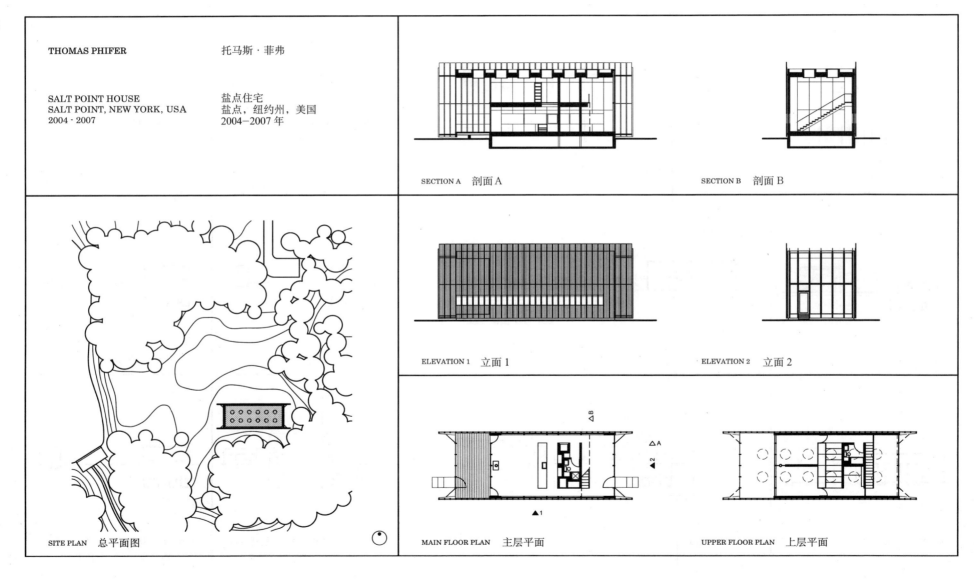

SECTION A　剖面 A

SECTION B　剖面 B

ELEVATION 1　立面 1

ELEVATION 2　立面 2

SITE PLAN　总平面图

MAIN FLOOR PLAN　主层平面

UPPER FLOOR PLAN　上层平面

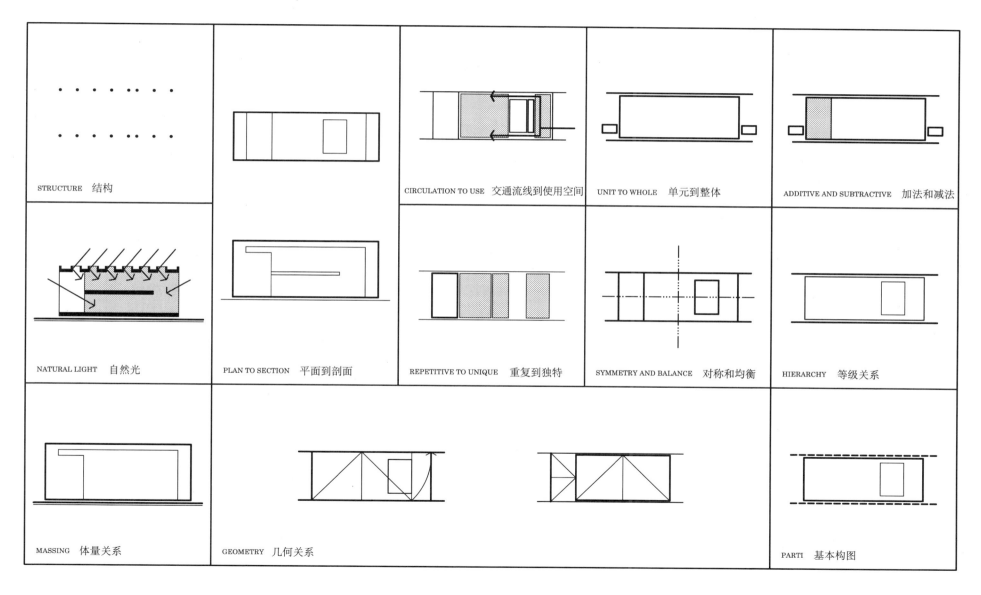

STRUCTURE　结构

CIRCULATION TO USE　交通流线到使用空间

UNIT TO WHOLE　单元到整体

ADDITIVE AND SUBTRACTIVE　加法和减法

NATURAL LIGHT　自然光

PLAN TO SECTION　平面到剖面

REPETITIVE TO UNIQUE　重复到独特

SYMMETRY AND BALANCE　对称和均衡

HIERARCHY　等级关系

MASSING　体量关系

GEOMETRY　几何关系

PARTI　基本构图

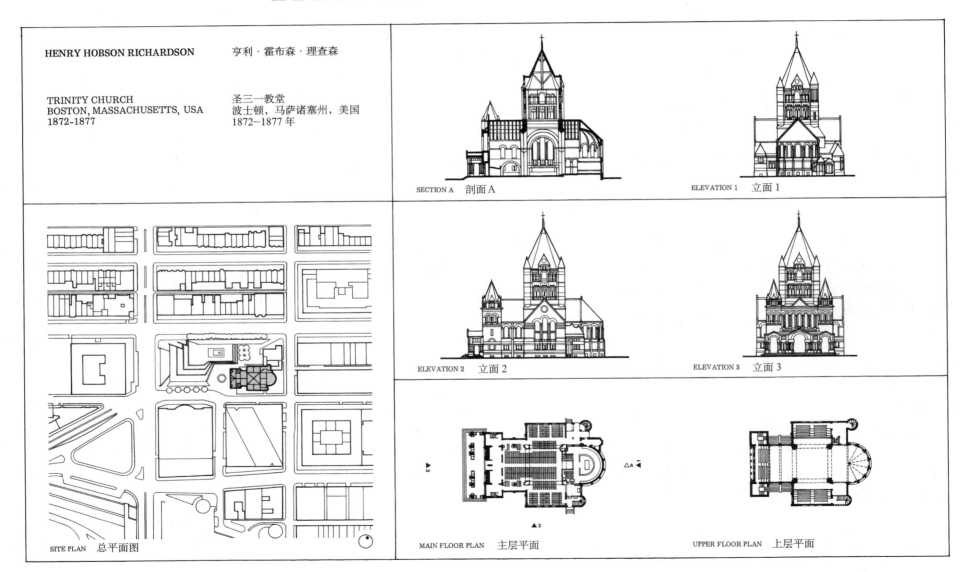

HENRY HOBSON RICHARDSON　　　　亨利·霍布森·理查森

TRINITY CHURCH　　　　圣三一教堂
BOSTON, MASSACHUSETTS, USA　　　　波士顿，马萨诸塞州，美国
1872–1877　　　　1872–1877 年

SECTION A　剖面 A

ELEVATION 1　立面 1

ELEVATION 2　立面 2

ELEVATION 3　立面 3

SITE PLAN　总平面图

MAIN FLOOR PLAN　主层平面

UPPER FLOOR PLAN　上层平面

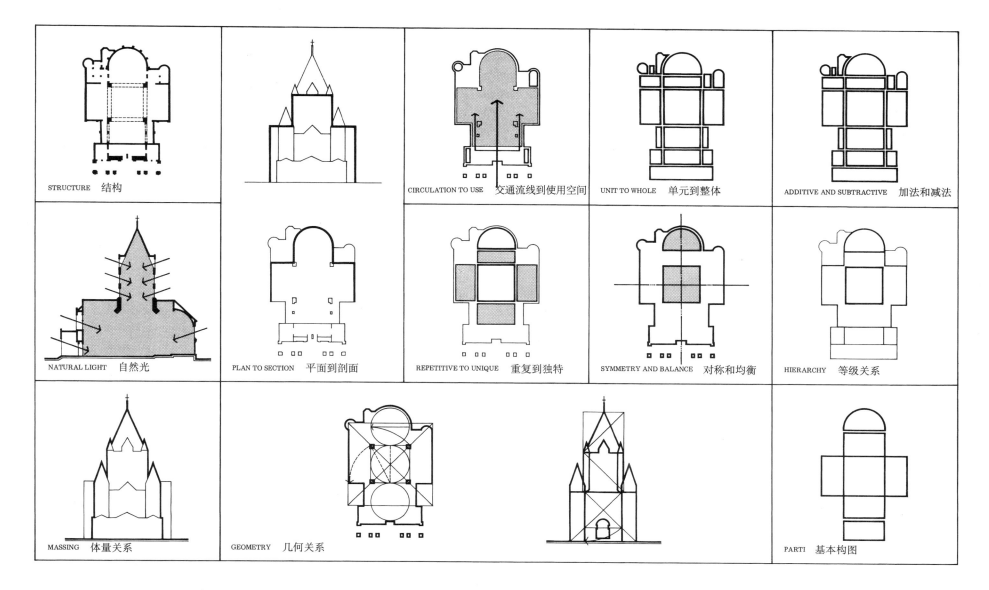

STRUCTURE　结构

CIRCULATION TO USE　交通流线到使用空间

UNIT TO WHOLE　单元到整体

ADDITIVE AND SUBTRACTIVE　加法和减法

NATURAL LIGHT　自然光

PLAN TO SECTION　平面到剖面

REPETITIVE TO UNIQUE　重复到独特

SYMMETRY AND BALANCE　对称和均衡

HIERARCHY　等级关系

MASSING　体量关系

GEOMETRY　几何关系

PARTI　基本构图

HENRY HOBSON RICHARDSON　　　亨利·霍布森·理查森

SEVER HALL　　　　　　　　　　塞弗楼
HARVARD UNIVERSITY　　　　　　哈佛大学
CAMBRIDGE, MASSACHUSETTS, USA　剑桥，马萨诸塞州，美国
1878–1880　　　　　　　　　　　1878－1880 年

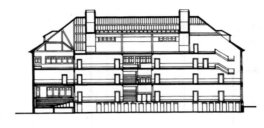

SECTION A　剖面 A

SECTION B　剖面 B

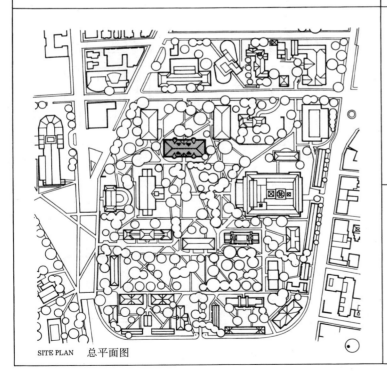

SITE PLAN　总平面图

ELEVATION 1　立面 1

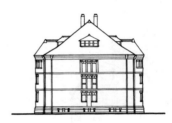

ELEVATION 2　立面 2

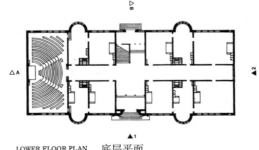

LOWER FLOOR PLAN　底层平面

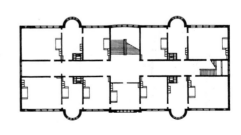

UPPER FLOOR PLAN　上层平面

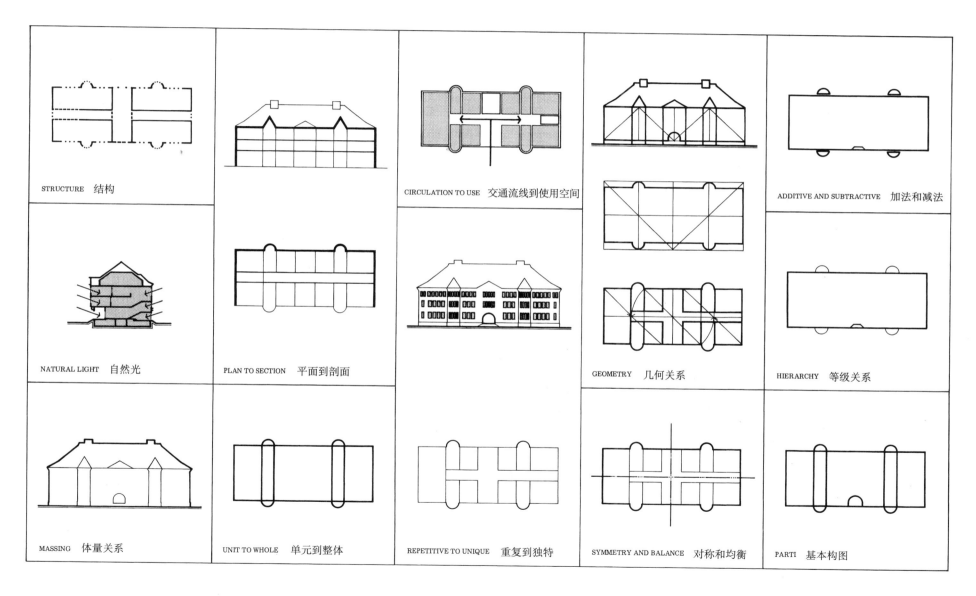

STRUCTURE 结构

CIRCULATION TO USE 交通流线到使用空间

ADDITIVE AND SUBTRACTIVE 加法和减法

NATURAL LIGHT 自然光

PLAN TO SECTION 平面到剖面

GEOMETRY 几何关系

HIERARCHY 等级关系

MASSING 体量关系

UNIT TO WHOLE 单元到整体

REPETITIVE TO UNIQUE 重复到独特

SYMMETRY AND BALANCE 对称和均衡

PARTI 基本构图

HENRY HOBSON RICHARDSON　　亨利·霍布森·理查森

ALLEGHENY COUNTY COURTHOUSE　阿勒格尼县法院
PITTSBURGH, PENNSYLVANIA, USA　匹兹堡，宾夕法尼亚州，美国
1883-1888　　　　　　　　　　1883-1888 年

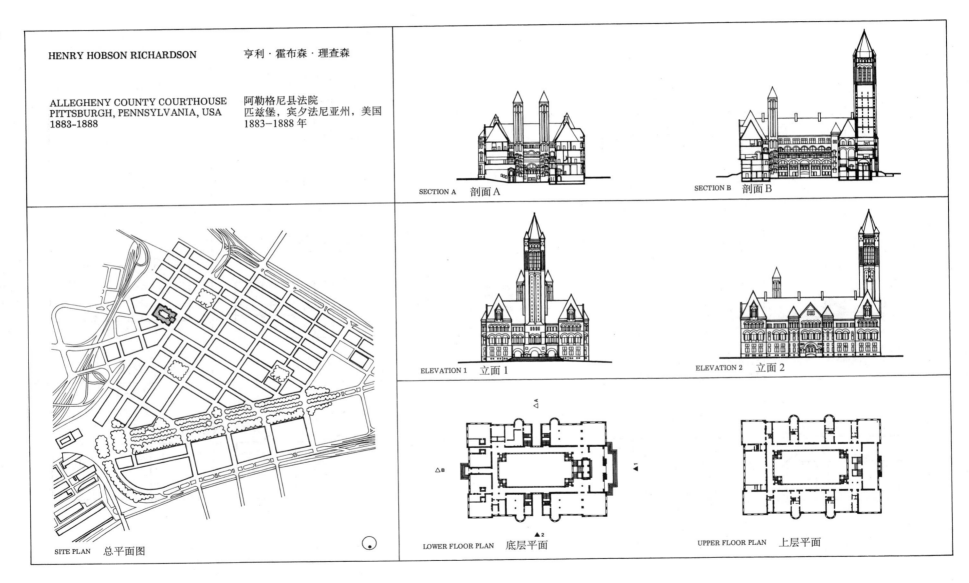

SECTION A　剖面A

SECTION B　剖面B

ELEVATION 1　立面 1

ELEVATION 2　立面 2

SITE PLAN　总平面图

LOWER FLOOR PLAN　底层平面

UPPER FLOOR PLAN　上层平面

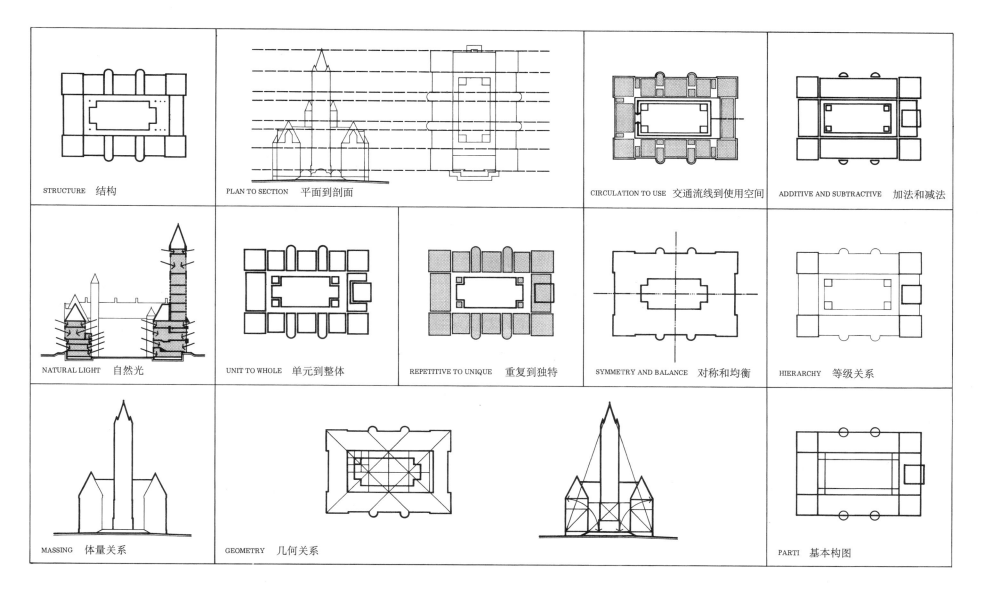

STRUCTURE　结构

PLAN TO SECTION　平面到剖面

CIRCULATION TO USE　交通流线到使用空间

ADDITIVE AND SUBTRACTIVE　加法和减法

NATURAL LIGHT　自然光

UNIT TO WHOLE　单元到整体

REPETITIVE TO UNIQUE　重复到独特

SYMMETRY AND BALANCE　对称和均衡

HIERARCHY　等级关系

MASSING　体量关系

GEOMETRY　几何关系

PARTI　基本构图

HENRY HOBSON RICHARDSON   亨利·霍布森·理查森

J. J. GLESSNER HOUSE   J·J·格莱斯纳住宅
CHICAGO, ILLINOIS, USA   芝加哥，伊利诺伊州，美国
1885-1887   1885-1887 年

SECTION A   剖面 A

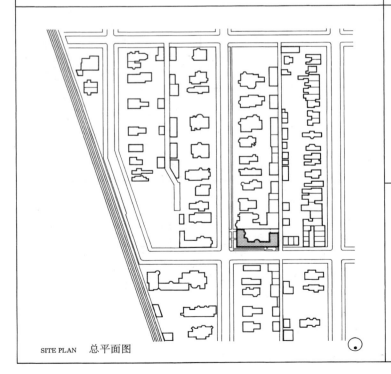

SITE PLAN   总平面图

ELEVATION 1   立面 1

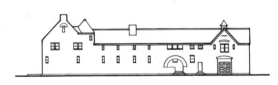

ELEVATION 2   立面 2

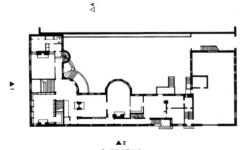

MAIN FLOOR PLAN   主层平面

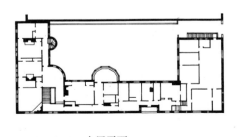

UPPER FLOOR PLAN   上层平面

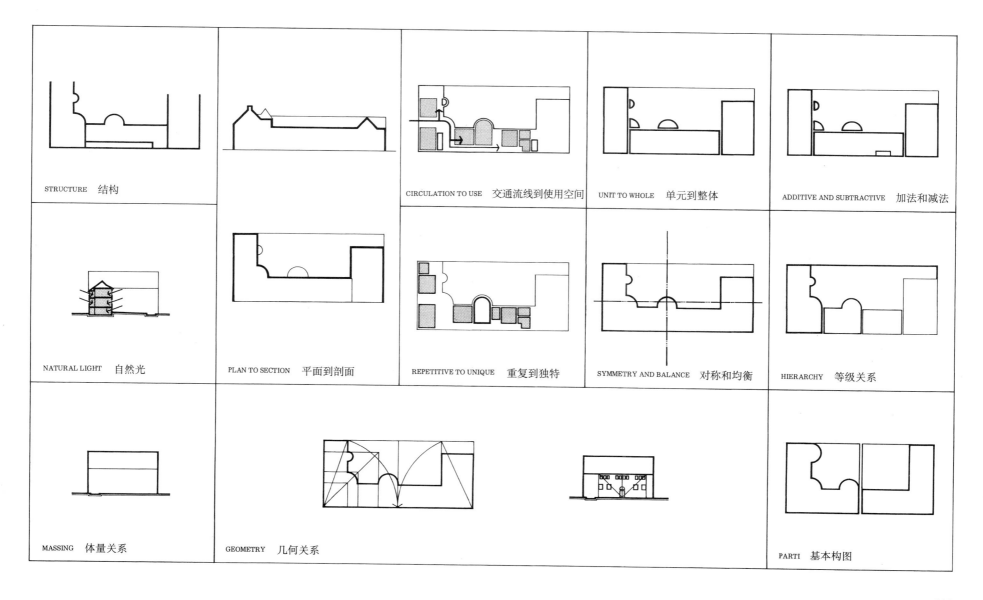

STRUCTURE　结构

CIRCULATION TO USE　交通流线到使用空间

UNIT TO WHOLE　单元到整体

ADDITIVE AND SUBTRACTIVE　加法和减法

NATURAL LIGHT　自然光

PLAN TO SECTION　平面到剖面

REPETITIVE TO UNIQUE　重复到独特

SYMMETRY AND BALANCE　对称和均衡

HIERARCHY　等级关系

MASSING　体量关系

GEOMETRY　几何关系

PARTI　基本构图

阿尔瓦罗·西扎
ALVARO SIZA

ALVARO SIZA　　　　　　　　阿尔瓦罗·西扎

SANTA MARIA CHURCH　　　　圣塔玛丽亚教堂
MARCO DE CANAVESES, PORTUGAL　马尔库·迪卡纳韦塞什，葡萄牙
1990 · 1996　　　　　　　　　1990—1996 年

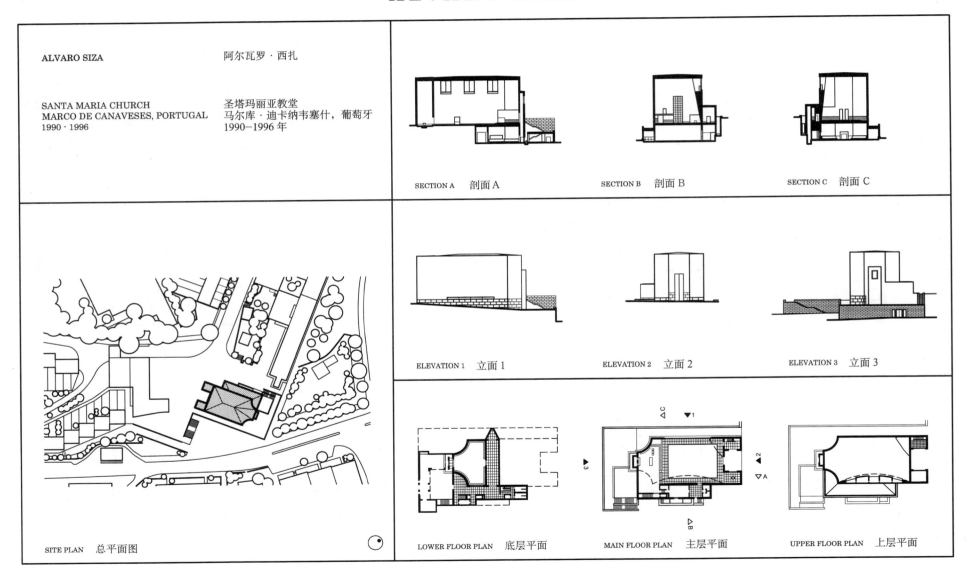

SECTION A　剖面 A　　　　　SECTION B　剖面 B　　　　　SECTION C　剖面 C

ELEVATION 1　立面 1　　　　ELEVATION 2　立面 2　　　　ELEVATION 3　立面 3

SITE PLAN　总平面图

LOWER FLOOR PLAN　底层平面　　MAIN FLOOR PLAN　主层平面　　UPPER FLOOR PLAN　上层平面

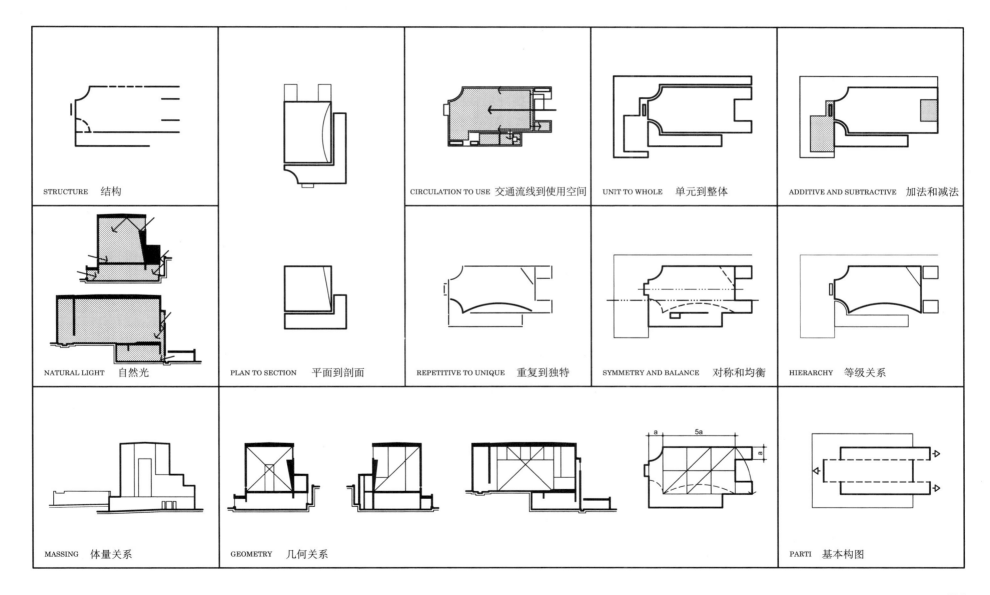

STRUCTURE　结构

CIRCULATION TO USE 交通流线到使用空间

UNIT TO WHOLE　单元到整体

ADDITIVE AND SUBTRACTIVE　加法和减法

NATURAL LIGHT　自然光

PLAN TO SECTION　平面到剖面

REPETITIVE TO UNIQUE　重复到独特

SYMMETRY AND BALANCE　对称和均衡

HIERARCHY　等级关系

MASSING　体量关系

GEOMETRY　几何关系

PARTI　基本构图

ALVARO SIZA　　　　　　阿尔瓦罗·西扎

VAN MIDDELEM · DUPONT HOUSE　　VAN MIDDELEM—DUPONT 住宅
OUDENBURG, BELGIUM　　奥登布尔格，比利时
1994 · 1997　　1994—1997 年

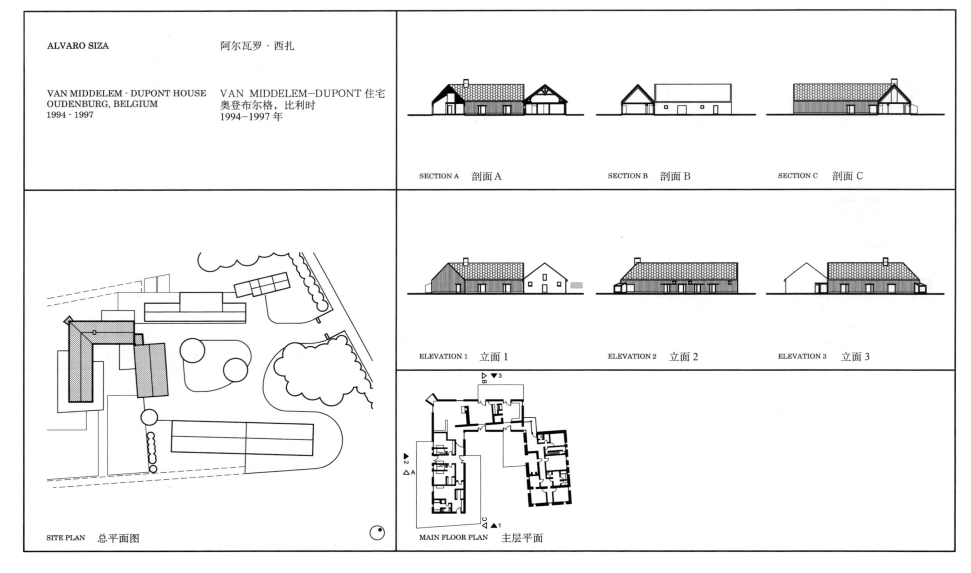

SECTION A　剖面 A　　　　SECTION B　剖面 B　　　　SECTION C　剖面 C

ELEVATION 1　立面 1　　　　ELEVATION 2　立面 2　　　　ELEVATION 3　立面 3

SITE PLAN　总平面图

MAIN FLOOR PLAN　主层平面

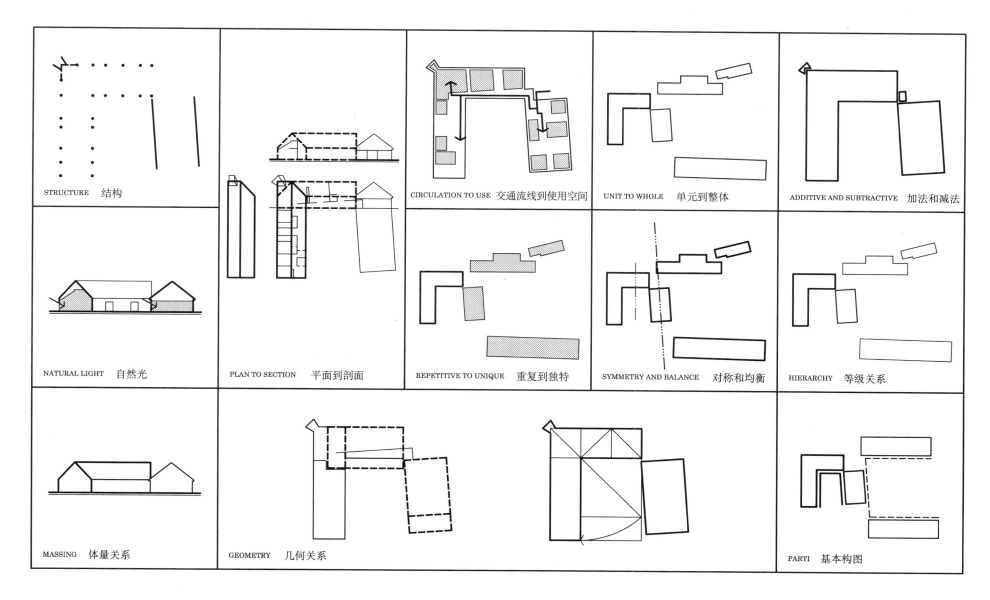

STRUCTURE　结构

CIRCULATION TO USE　交通流线到使用空间

UNIT TO WHOLE　单元到整体

ADDITIVE AND SUBTRACTIVE　加法和减法

NATURAL LIGHT　自然光

PLAN TO SECTION　平面到剖面

REPETITIVE TO UNIQUE　重复到独特

SYMMETRY AND BALANCE　对称和均衡

HIERARCHY　等级关系

MASSING　体量关系

GEOMETRY　几何关系

PARTI　基本构图

詹姆斯·斯特林
JAMES STIRLING

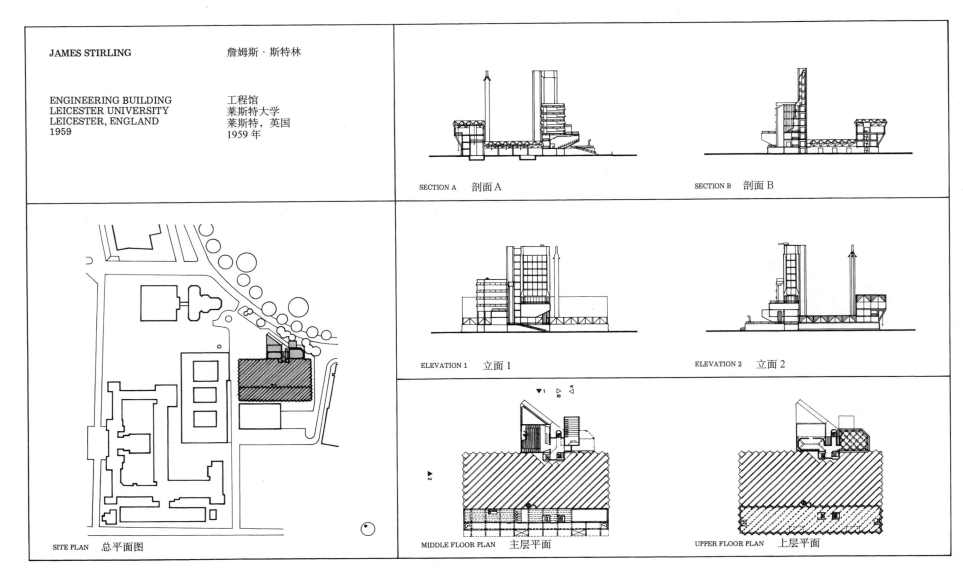

JAMES STIRLING   詹姆斯·斯特林

ENGINEERING BUILDING   工程馆
LEICESTER UNIVERSITY   莱斯特大学
LEICESTER, ENGLAND   莱斯特，英国
1959   1959 年

SECTION A   剖面 A

SECTION B   剖面 B

ELEVATION 1   立面 1

ELEVATION 2   立面 2

SITE PLAN   总平面图

MIDDLE FLOOR PLAN   主层平面

UPPER FLOOR PLAN   上层平面

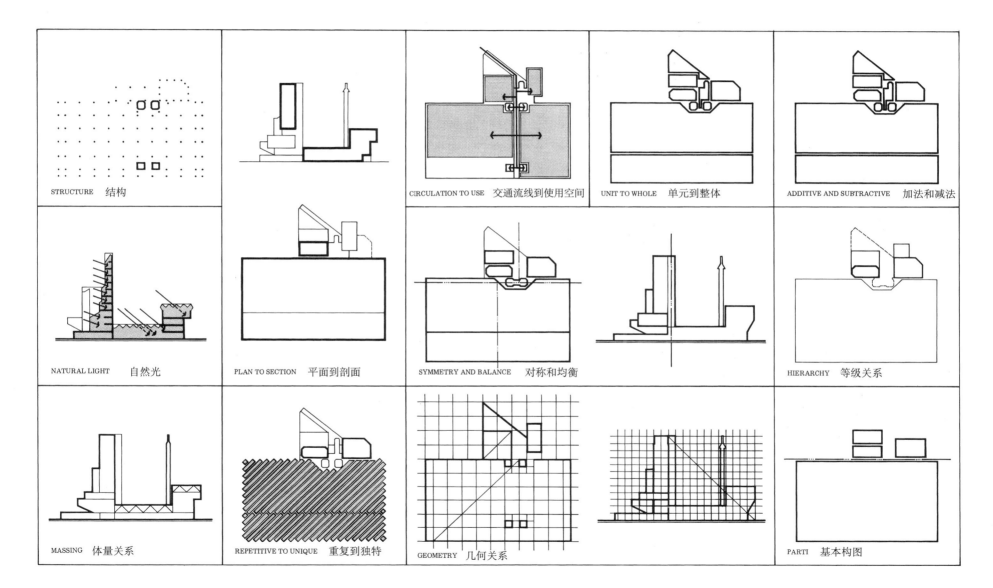

STRUCTURE　结构

CIRCULATION TO USE　交通流线到使用空间

UNIT TO WHOLE　单元到整体

ADDITIVE AND SUBTRACTIVE　加法和减法

NATURAL LIGHT　自然光

PLAN TO SECTION　平面到剖面

SYMMETRY AND BALANCE　对称和均衡

HIERARCHY　等级关系

MASSING　体量关系

REPETITIVE TO UNIQUE　重复到独特

GEOMETRY　几何关系

PARTI　基本构图

JAMES STIRLING 詹姆斯·斯特林

HISTORY FACULTY BUILDING 历史系图书馆
CAMBRIDGE UNIVERSITY 剑桥大学
CAMBRIDGE, ENGLAND 剑桥，英国
1964 1964 年

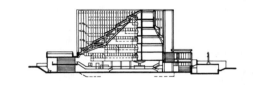

SECTION A 剖面 A   SECTION B 剖面 B

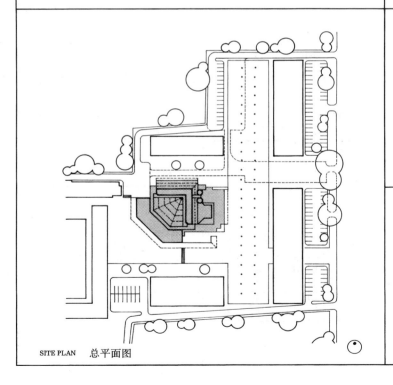

SITE PLAN 总平面图

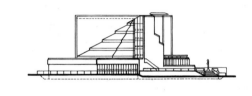

ELEVATION 1 立面 1   ELEVATION 2 立面 2

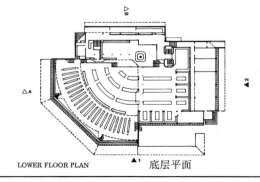

LOWER FLOOR PLAN 底层平面

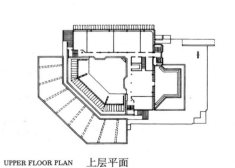

UPPER FLOOR PLAN 上层平面

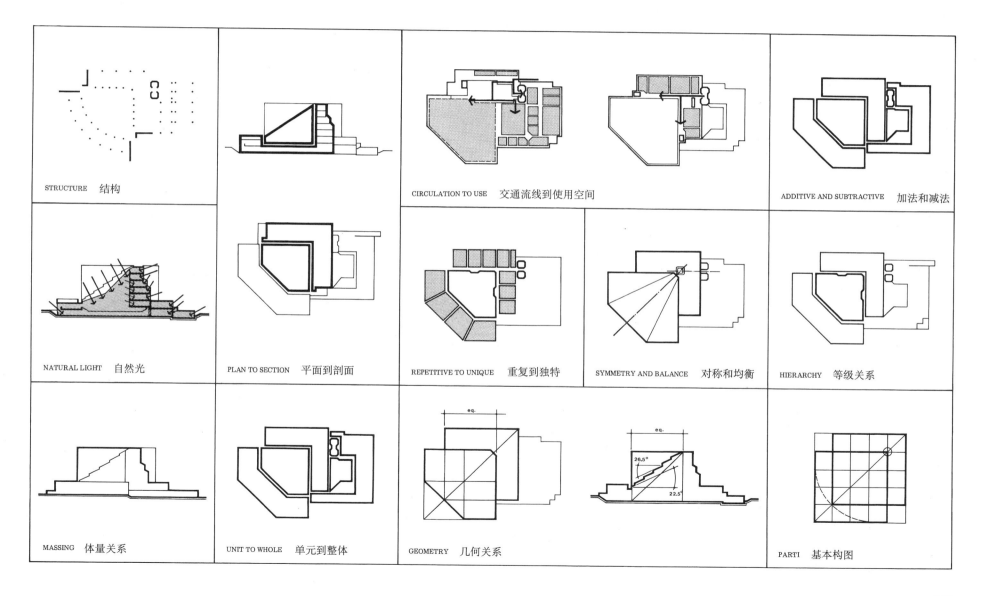

STRUCTURE　结构

CIRCULATION TO USE　交通流线到使用空间

ADDITIVE AND SUBTRACTIVE　加法和减法

NATURAL LIGHT　自然光

PLAN TO SECTION　平面到剖面

REPETITIVE TO UNIQUE　重复到独特

SYMMETRY AND BALANCE　对称和均衡

HIERARCHY　等级关系

MASSING　体量关系

UNIT TO WHOLE　单元到整体

GEOMETRY　几何关系

PARTI　基本构图

JAMES STIRLING　　　詹姆斯·斯特林

FLOREY BUILDING　　弗洛雷大楼
QUEENS COLLEGE　　皇后学院
OXFORD, ENGLAND　　牛津，英国
1966　　　　　　　　1966 年

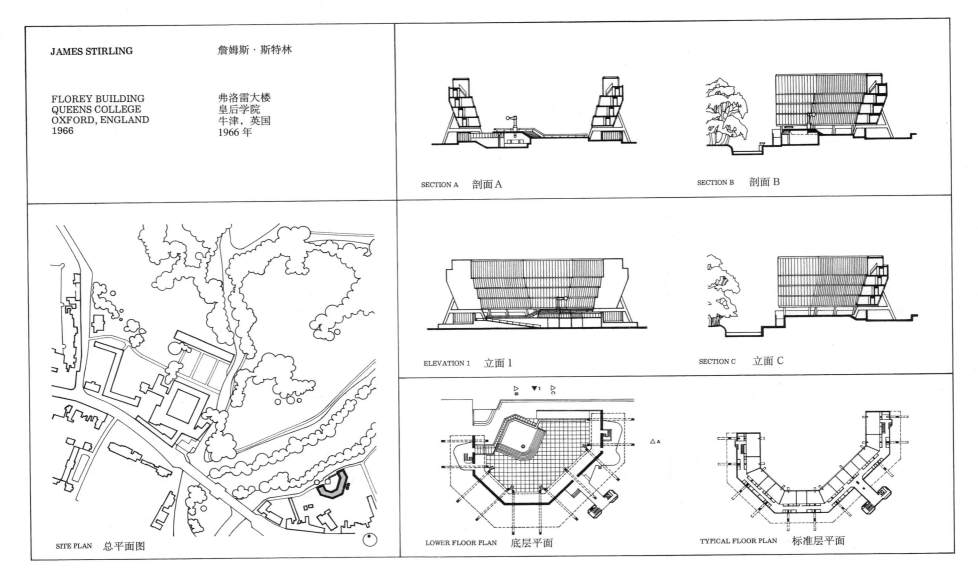

SECTION A　剖面 A

SECTION B　剖面 B

ELEVATION 1　立面 1

SECTION C　立面 C

SITE PLAN　总平面图

LOWER FLOOR PLAN　底层平面

TYPICAL FLOOR PLAN　标准层平面

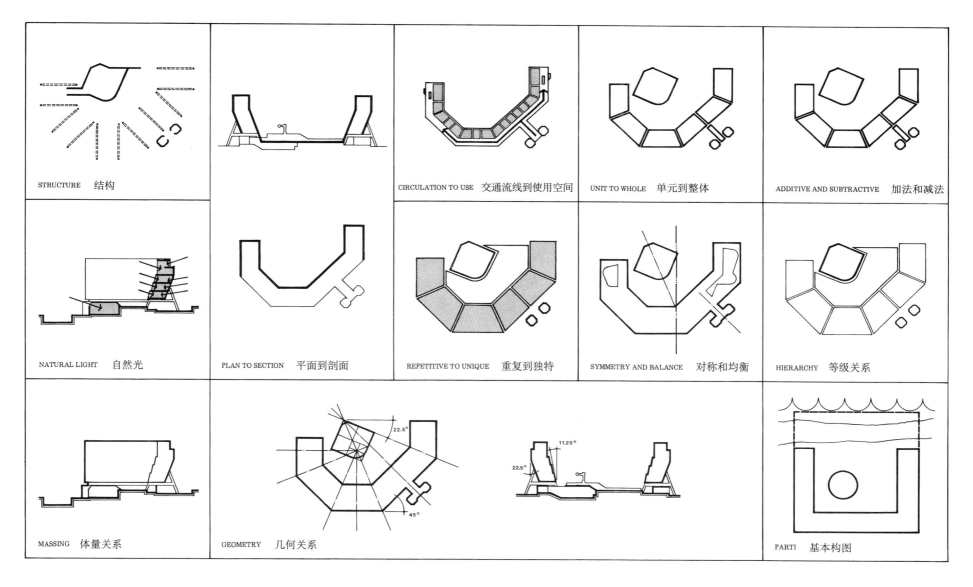

STRUCTURE　结构		CIRCULATION TO USE　交通流线到使用空间	UNIT TO WHOLE　单元到整体	ADDITIVE AND SUBTRACTIVE　加法和减法
NATURAL LIGHT　自然光	PLAN TO SECTION　平面到剖面	REPETITIVE TO UNIQUE　重复到独特	SYMMETRY AND BALANCE　对称和均衡	HIERARCHY　等级关系
MASSING　体量关系	GEOMETRY　几何关系			PARTI　基本构图

JAMES STIRLING　　　　　　詹姆斯·斯特林

OLIVETTI TRAINING SCHOOL　　奥利韦蒂专科学校
HASLEMERE, SURREY, ENGLAND　黑斯尔米尔，萨里郡，英国
1969　　　　　　　　　　　　1969 年

SECTION A　剖面 A

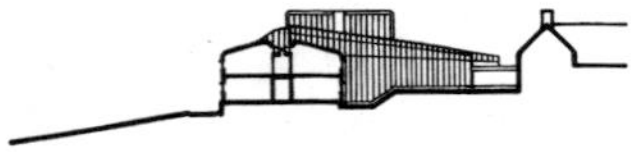

SECTION B　剖面 B

ELEVATION 1　立面 1

SITE PLAN　总平面图

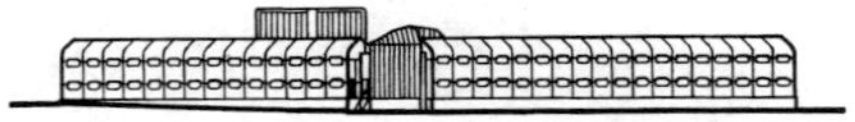

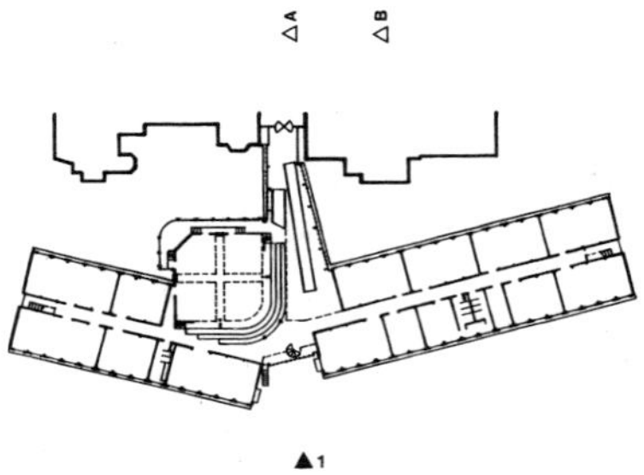

LOWER FLOOR PLAN　底层平面

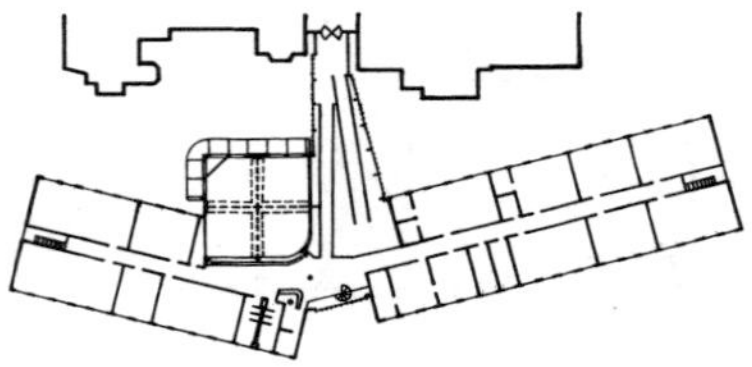

UPPER FLOOR PLAN　上层平面

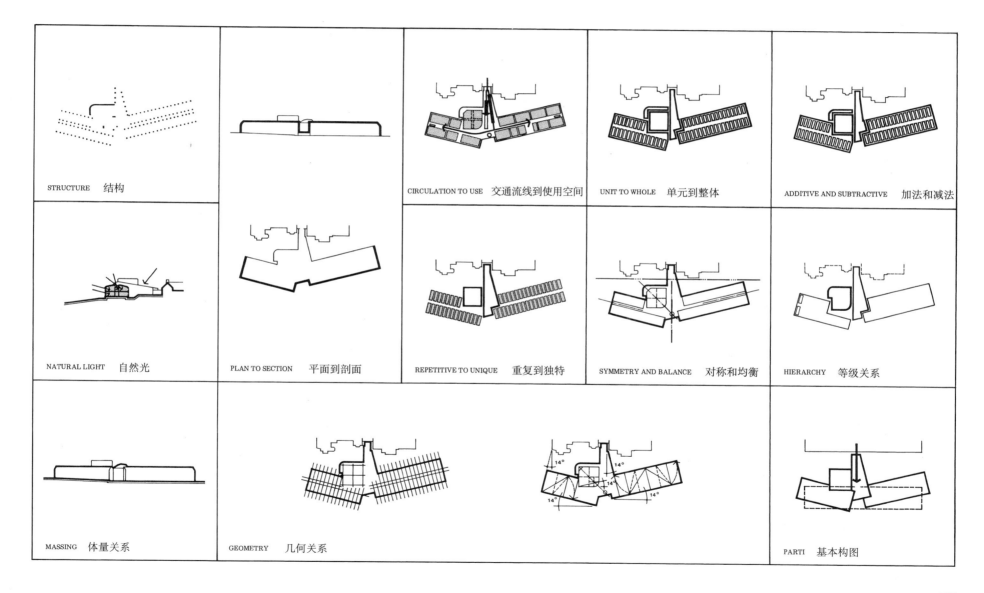

STRUCTURE　结构		CIRCULATION TO USE　交通流线到使用空间	UNIT TO WHOLE　单元到整体	ADDITIVE AND SUBTRACTIVE　加法和减法
NATURAL LIGHT　自然光	PLAN TO SECTION　平面到剖面	REPETITIVE TO UNIQUE　重复到独特	SYMMETRY AND BALANCE　对称和均衡	HIERARCHY　等级关系
MASSING　体量关系	GEOMETRY　几何关系			PARTI　基本构图

路易斯·沙利文
LOUIS SULLIVAN

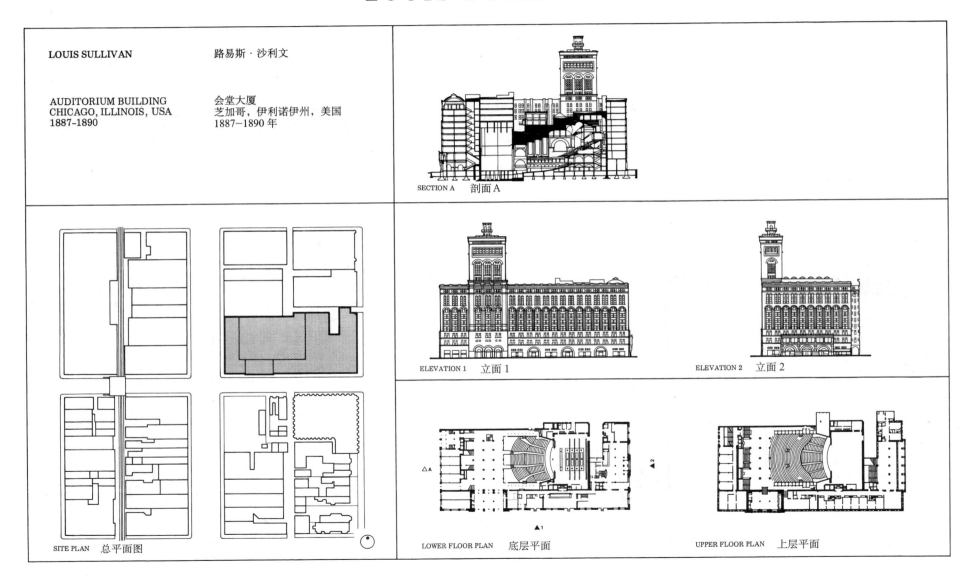

LOUIS SULLIVAN　　　　路易斯·沙利文

AUDITORIUM BUILDING　　会堂大厦
CHICAGO, ILLINOIS, USA　　芝加哥，伊利诺伊州，美国
1887-1890　　　　　　　　1887-1890 年

SECTION A　剖面 A

ELEVATION 1　立面 1　　　　ELEVATION 2　立面 2

SITE PLAN　总平面图

LOWER FLOOR PLAN　底层平面　　UPPER FLOOR PLAN　上层平面

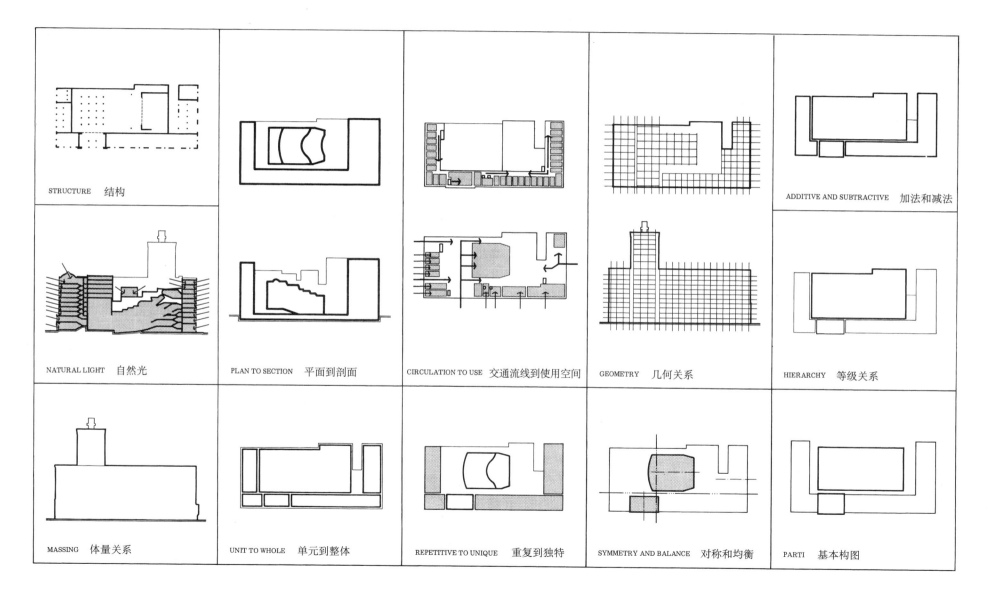

STRUCTURE　结构

NATURAL LIGHT　自然光

MASSING　体量关系

PLAN TO SECTION　平面到剖面

UNIT TO WHOLE　单元到整体

CIRCULATION TO USE　交通流线到使用空间

REPETITIVE TO UNIQUE　重复到独特

GEOMETRY　几何关系

SYMMETRY AND BALANCE　对称和均衡

ADDITIVE AND SUBTRACTIVE　加法和减法

HIERARCHY　等级关系

PARTI　基本构图

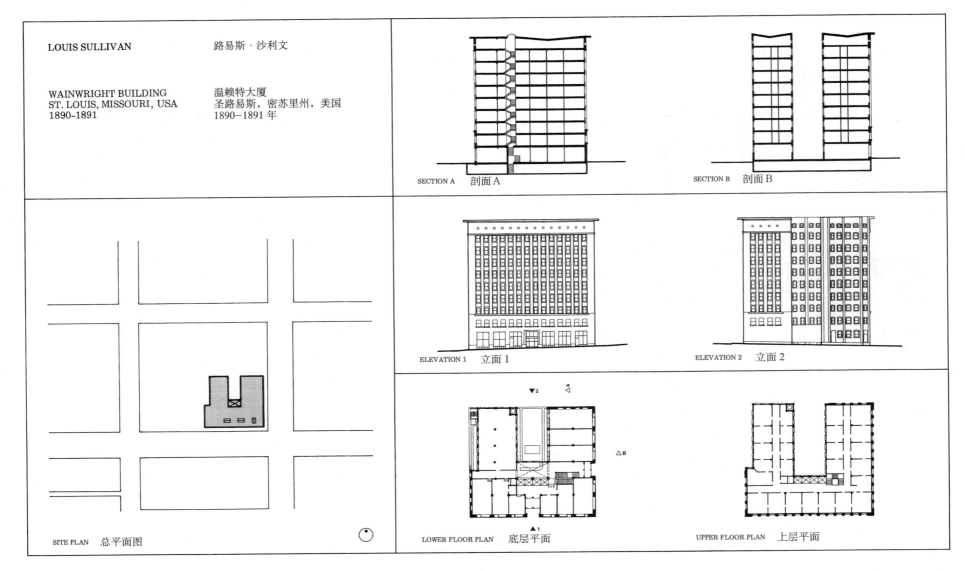

LOUIS SULLIVAN　　　　路易斯·沙利文

WAINWRIGHT BUILDING　　温赖特大厦
ST. LOUIS, MISSOURI, USA　圣路易斯，密苏里州，美国
1890-1891　　　　　　　1890—1891 年

SECTION A　剖面 A

SECTION B　剖面 B

ELEVATION 1　立面 1

ELEVATION 2　立面 2

SITE PLAN　总平面图

LOWER FLOOR PLAN　底层平面

UPPER FLOOR PLAN　上层平面

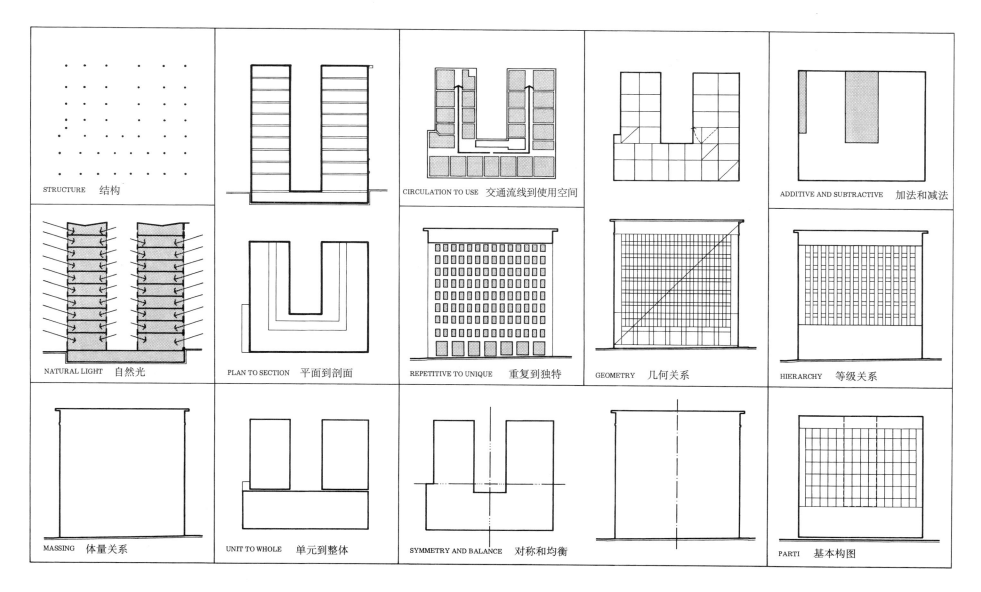

STRUCTURE　结构		CIRCULATION TO USE　交通流线到使用空间		ADDITIVE AND SUBTRACTIVE　加法和减法
NATURAL LIGHT　自然光	PLAN TO SECTION　平面到剖面	REPETITIVE TO UNIQUE　重复到独特	GEOMETRY　几何关系	HIERARCHY　等级关系
MASSING　体量关系	UNIT TO WHOLE　单元到整体	SYMMETRY AND BALANCE　对称和均衡		PARTI　基本构图

LOUIS SULLIVAN
路易斯·沙利文

CARSON PIRIE AND SCOTT STORE
(SCHLESINGER AND MAYER DEPARTMENT STORE)
CHICAGO, ILLINOIS, USA
1899-1903

卡森·皮里与斯科特百货大楼
（施莱辛格与梅耶百货大楼）
芝加哥，伊利诺伊州，美国
1899-1903 年

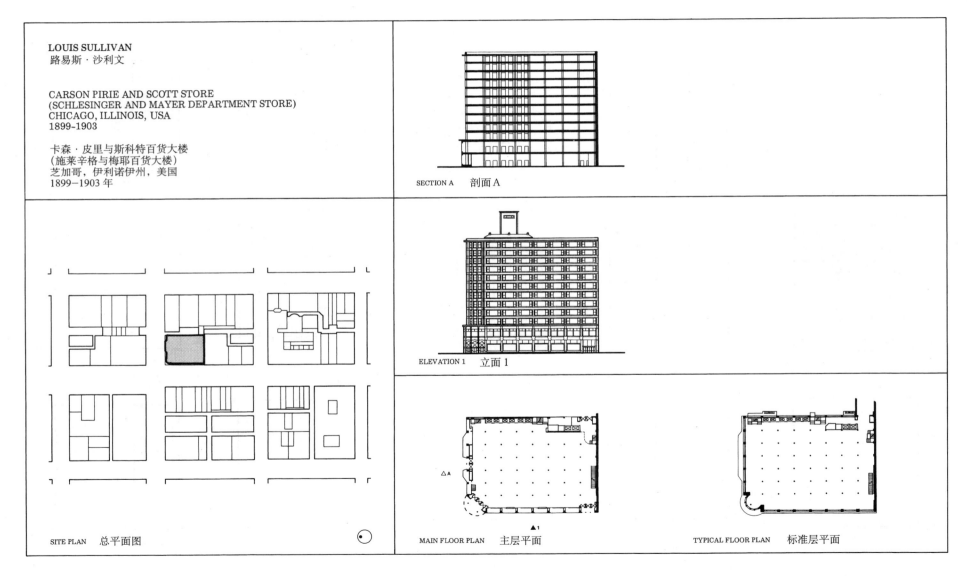

SECTION A   剖面 A

ELEVATION 1   立面 1

SITE PLAN   总平面图

MAIN FLOOR PLAN   主层平面

TYPICAL FLOOR PLAN   标准层平面

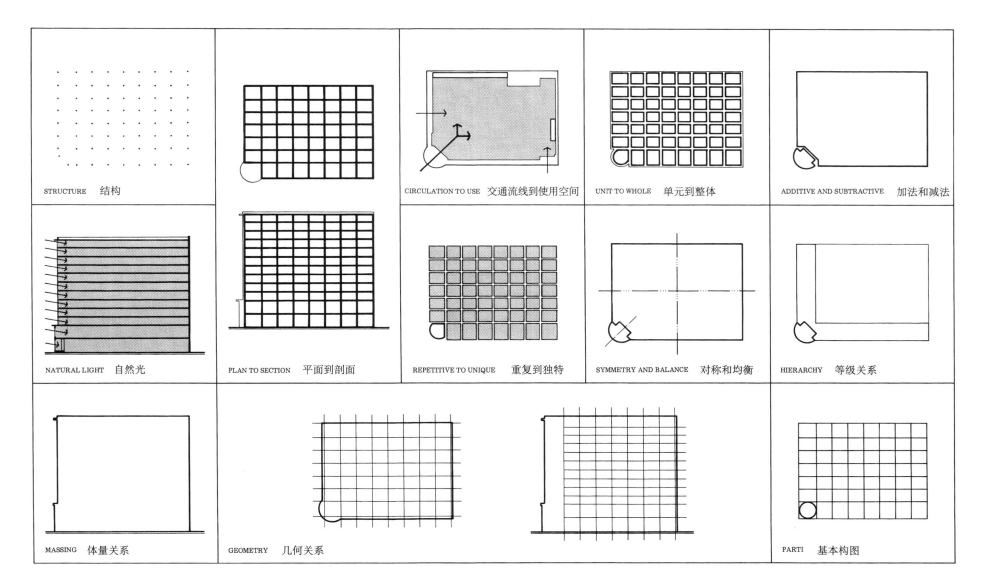

STRUCTURE　结构

CIRCULATION TO USE　交通流线到使用空间

UNIT TO WHOLE　单元到整体

ADDITIVE AND SUBTRACTIVE　加法和减法

NATURAL LIGHT　自然光

PLAN TO SECTION　平面到剖面

REPETITIVE TO UNIQUE　重复到独特

SYMMETRY AND BALANCE　对称和均衡

HIERARCHY　等级关系

MASSING　体量关系

GEOMETRY　几何关系

PARTI　基本构图

LOUIS SULLIVAN　　　　路易斯·沙利文

NATIONAL FARMERS' BANK　　国家农民银行
OWATONNA, MINNESOTA, USA　奥瓦通纳，明尼苏达州，美国
1907-1908　　　　　　　　　1907–1908 年

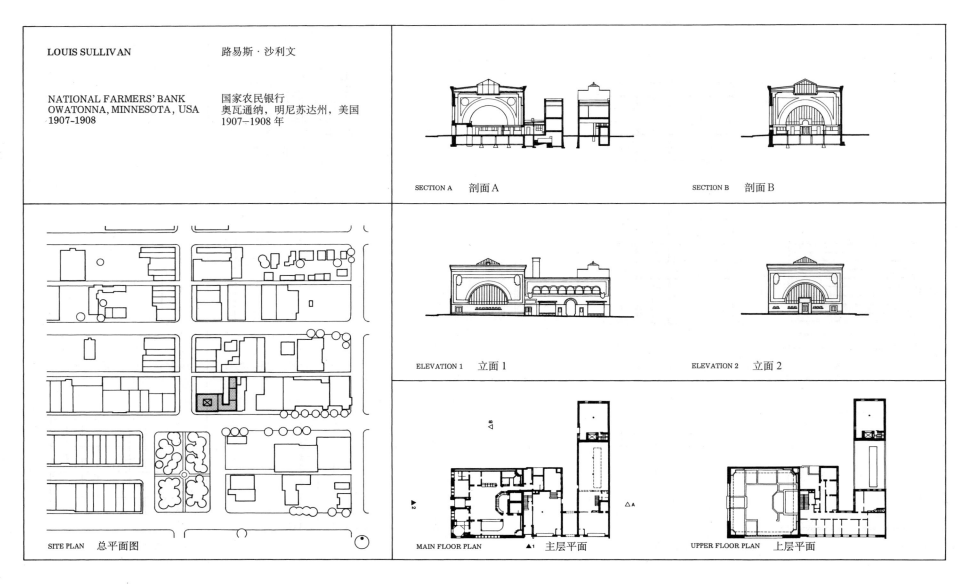

SECTION A　剖面 A　　　　　　　　　　SECTION B　剖面 B

ELEVATION 1　立面 1　　　　　　　　　ELEVATION 2　立面 2

SITE PLAN　总平面图

MAIN FLOOR PLAN　▲1　主层平面　　　　UPPER FLOOR PLAN　上层平面

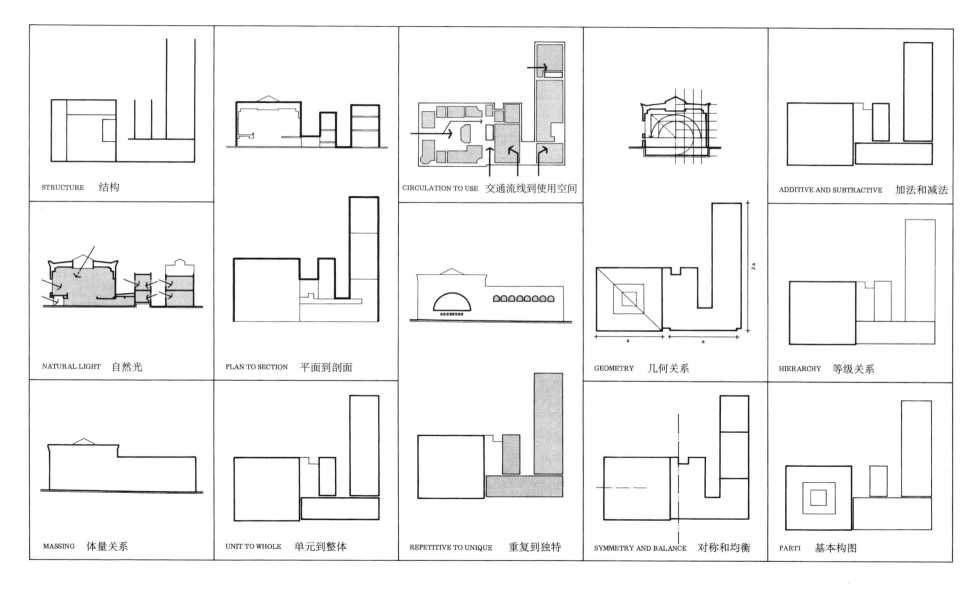

STRUCTURE 结构

CIRCULATION TO USE 交通流线到使用空间

ADDITIVE AND SUBTRACTIVE 加法和减法

NATURAL LIGHT 自然光

PLAN TO SECTION 平面到剖面

GEOMETRY 几何关系

HIERARCHY 等级关系

MASSING 体量关系

UNIT TO WHOLE 单元到整体

REPETITIVE TO UNIQUE 重复到独特

SYMMETRY AND BALANCE 对称和均衡

PARTI 基本构图

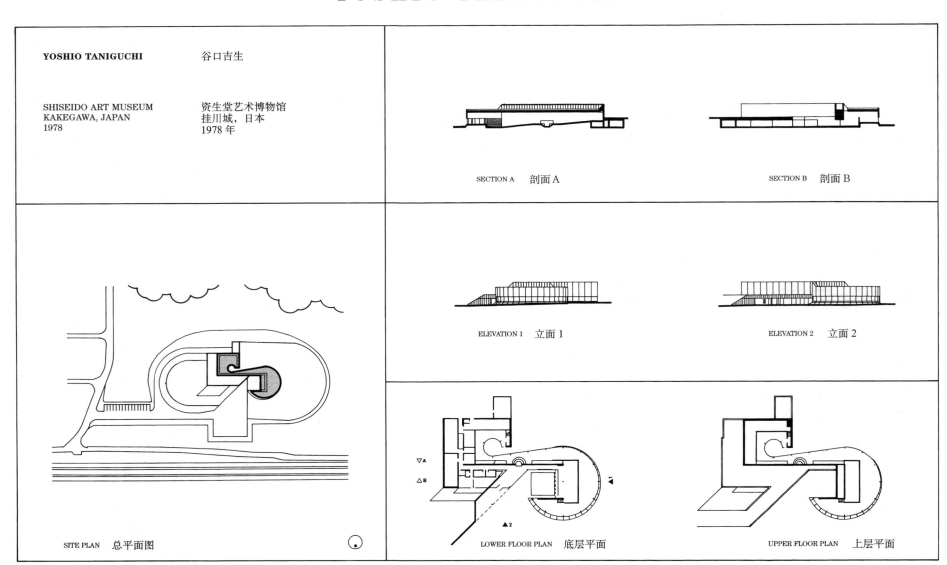

YOSHIO TANIGUCHI   谷口吉生

SHISEIDO ART MUSEUM   资生堂艺术博物馆
KAKEGAWA, JAPAN   挂川城，日本
1978   1978 年

SECTION A   剖面 A   SECTION B   剖面 B

ELEVATION 1   立面 1   ELEVATION 2   立面 2

SITE PLAN   总平面图

LOWER FLOOR PLAN   底层平面   UPPER FLOOR PLAN   上层平面

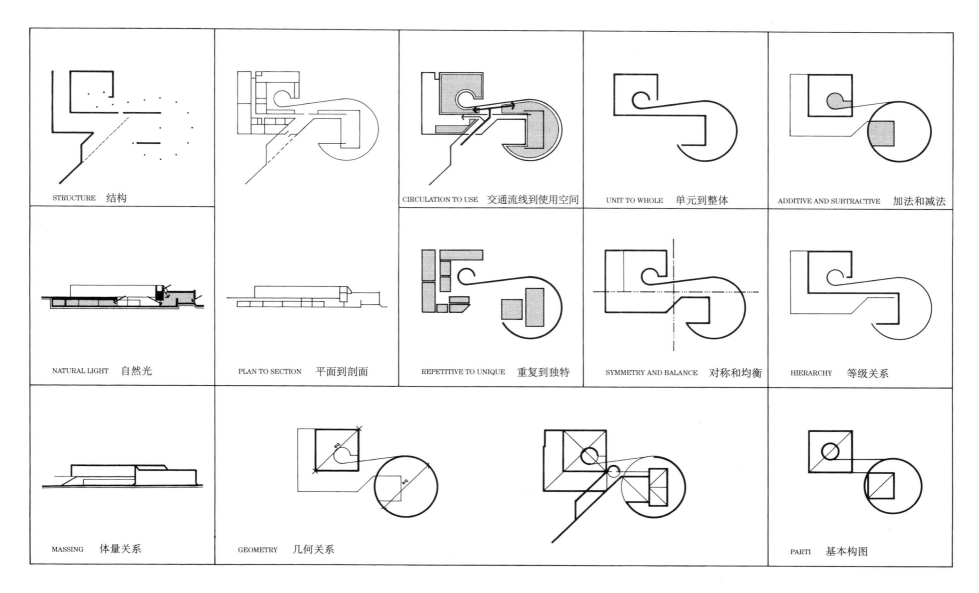

STRUCTURE　结构

CIRCULATION TO USE　交通流线到使用空间

UNIT TO WHOLE　单元到整体

ADDITIVE AND SUBTRACTIVE　加法和减法

NATURAL LIGHT　自然光

PLAN TO SECTION　平面到剖面

REPETITIVE TO UNIQUE　重复到独特

SYMMETRY AND BALANCE　对称和均衡

HIERARCHY　等级关系

MASSING　体量关系

GEOMETRY　几何关系

PARTI　基本构图

YOSHIO TANIGUCHI
谷口吉生

KASAI RINKAI PARK VIEW POINT VISITORS CENTER
TOKYO, JAPAN
1995

葛西临海公园游客中心
东京，日本
1995 年

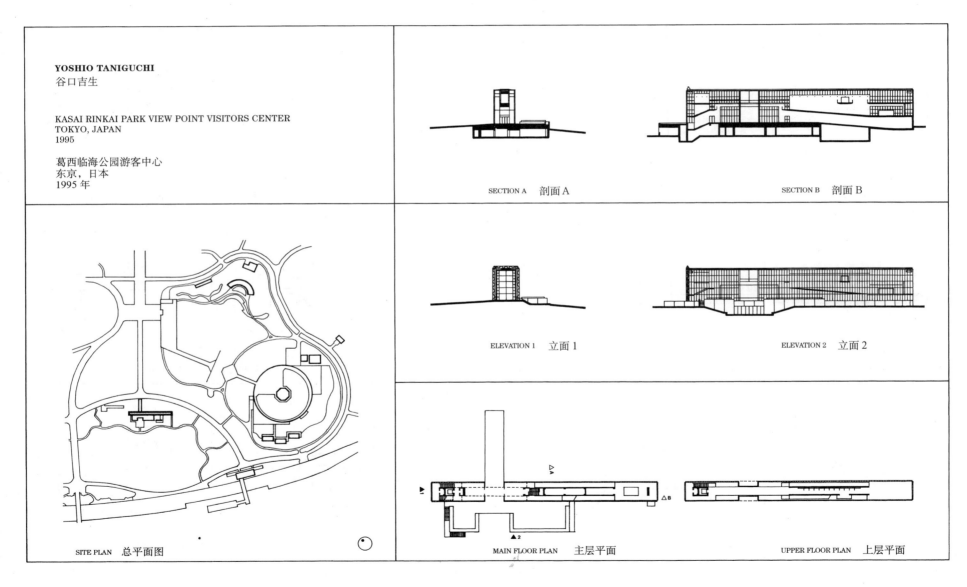

SECTION A 剖面 A

SECTION B 剖面 B

ELEVATION 1 立面 1

ELEVATION 2 立面 2

SITE PLAN 总平面图

MAIN FLOOR PLAN 主层平面

UPPER FLOOR PLAN 上层平面

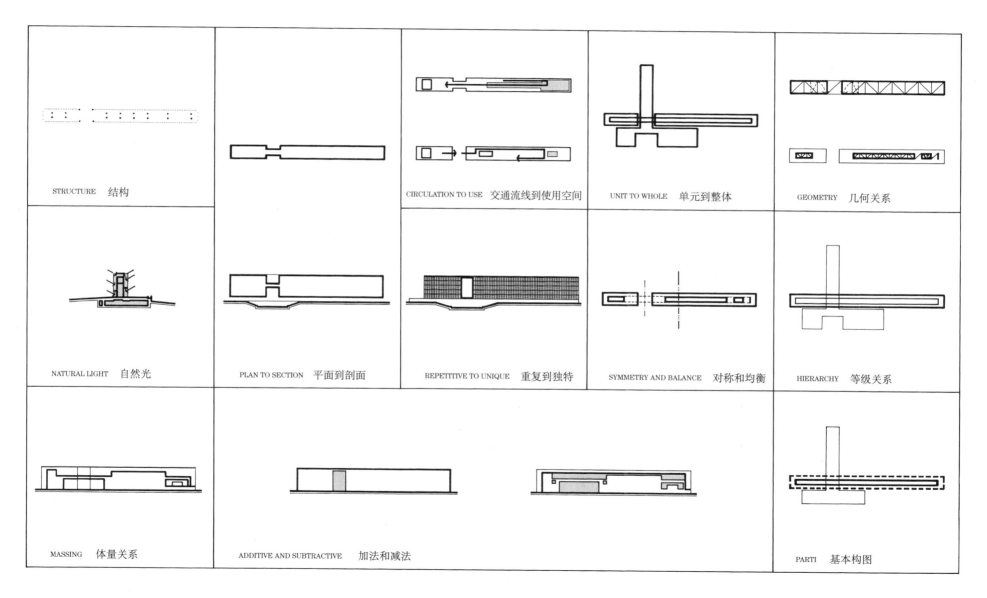

STRUCTURE　结构

CIRCULATION TO USE　交通流线到使用空间

UNIT TO WHOLE　单元到整体

GEOMETRY　几何关系

NATURAL LIGHT　自然光

PLAN TO SECTION　平面到剖面

REPETITIVE TO UNIQUE　重复到独特

SYMMETRY AND BALANCE　对称和均衡

HIERARCHY　等级关系

MASSING　体量关系

ADDITIVE AND SUBTRACTIVE　加法和减法

PARTI　基本构图

朱塞佩·泰拉尼
GIUSEPPE TERRAGNI

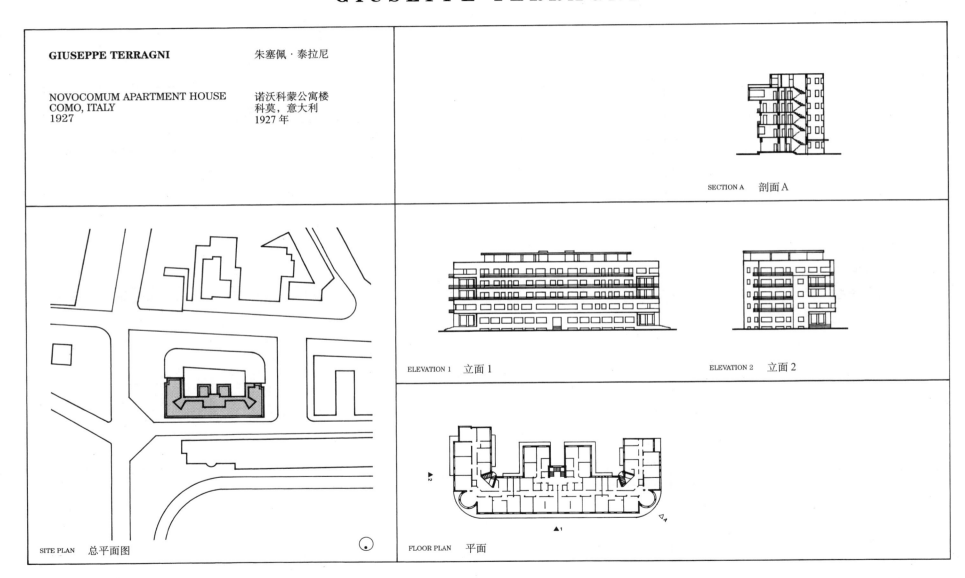

GIUSEPPE TERRAGNI

朱塞佩·泰拉尼

NOVOCOMUM APARTMENT HOUSE
COMO, ITALY
1927

诺沃科蒙公寓楼
科莫，意大利
1927 年

SECTION A   剖面 A

ELEVATION 1   立面 1

ELEVATION 2   立面 2

SITE PLAN   总平面图

FLOOR PLAN   平面

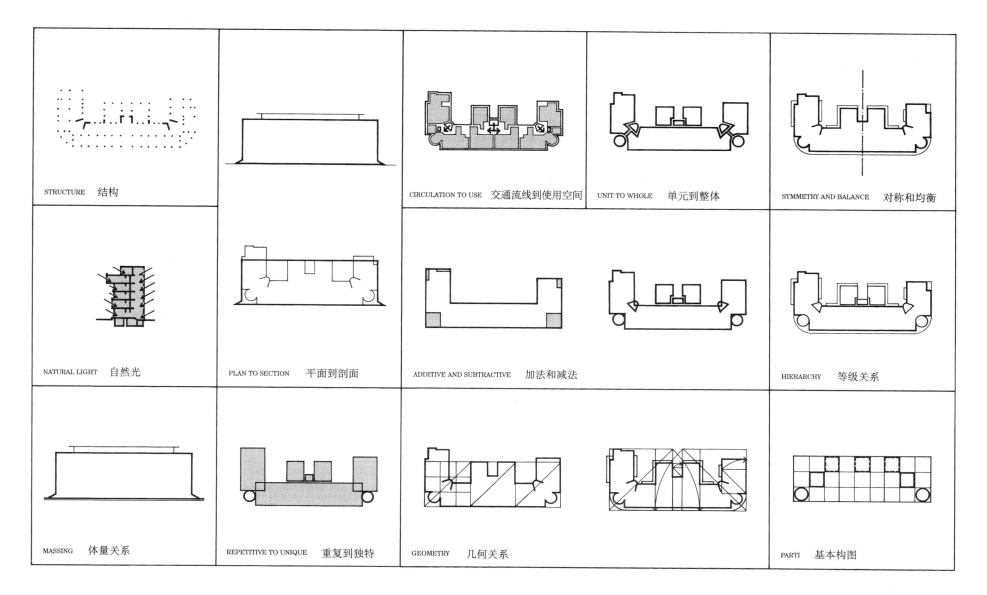

STRUCTURE　结构

CIRCULATION TO USE　交通流线到使用空间

UNIT TO WHOLE　单元到整体

SYMMETRY AND BALANCE　对称和均衡

NATURAL LIGHT　自然光

PLAN TO SECTION　平面到剖面

ADDITIVE AND SUBTRACTIVE　加法和减法

HIERARCHY　等级关系

MASSING　体量关系

REPETITIVE TO UNIQUE　重复到独特

GEOMETRY　几何关系

PARTI　基本构图

210

GIUSEPPE TERRAGNI　　　朱塞佩·泰拉尼

CASA DEL FASCIO　　　法肖公寓
COMO, ITALY　　　科莫，意大利
1932-1936　　　1932－1936 年

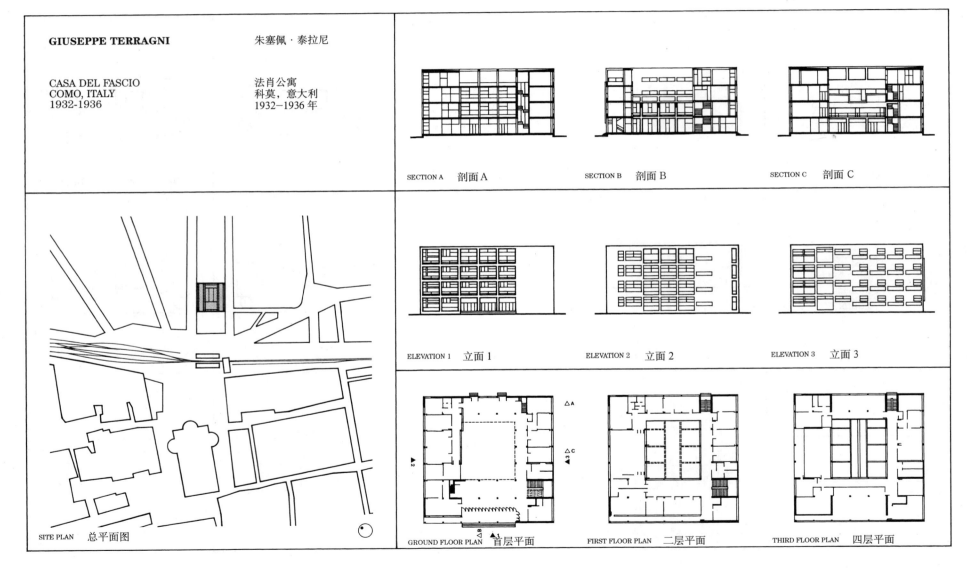

SECTION A　剖面 A　　　SECTION B　剖面 B　　　SECTION C　剖面 C

ELEVATION 1　立面 1　　　ELEVATION 2　立面 2　　　ELEVATION 3　立面 3

SITE PLAN　总平面图

GROUND FLOOR PLAN　首层平面　　　FIRST FLOOR PLAN　二层平面　　　THIRD FLOOR PLAN　四层平面

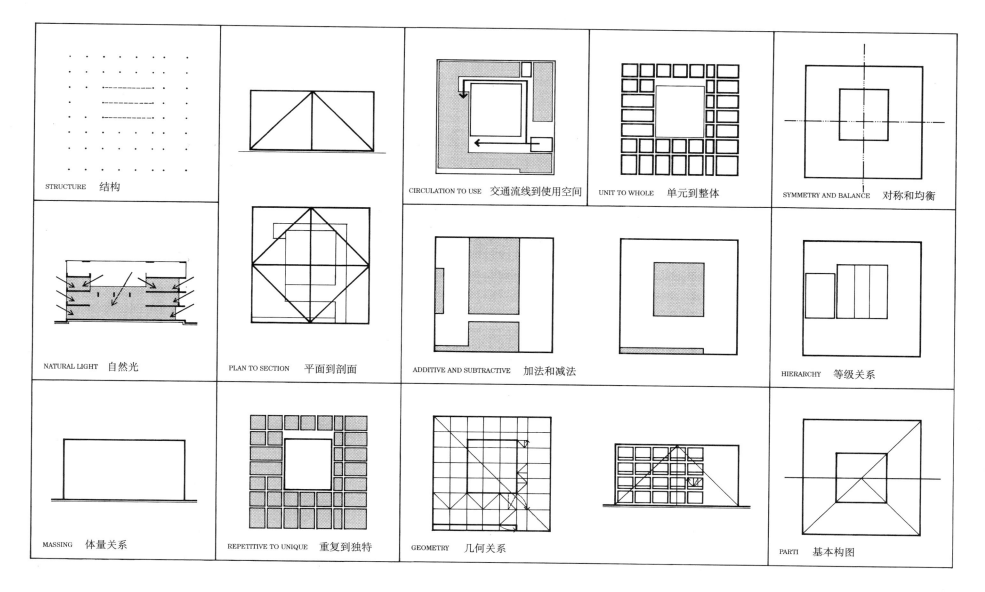

STRUCTURE　结构

CIRCULATION TO USE　交通流线到使用空间

UNIT TO WHOLE　单元到整体

SYMMETRY AND BALANCE　对称和均衡

NATURAL LIGHT　自然光

PLAN TO SECTION　平面到剖面

ADDITIVE AND SUBTRACTIVE　加法和减法

HIERARCHY　等级关系

MASSING　体量关系

REPETITIVE TO UNIQUE　重复到独特

GEOMETRY　几何关系

PARTI　基本构图

217

GIUSEPPE TERRAGNI　　朱塞佩·泰拉尼

SANT' ELIA NURSERY SCHOOL　　圣埃利娅幼儿园
COMO, ITALY　　科莫，意大利
1936-1937　　1936—1937 年

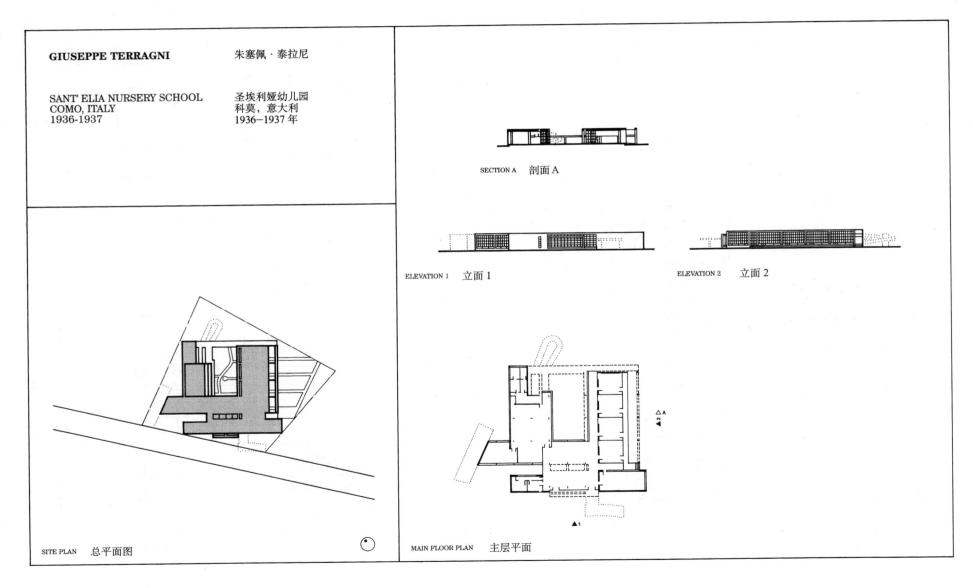

SITE PLAN　　总平面图

SECTION A　　剖面 A

ELEVATION 1　　立面 1

ELEVATION 2　　立面 2

MAIN FLOOR PLAN　　主层平面

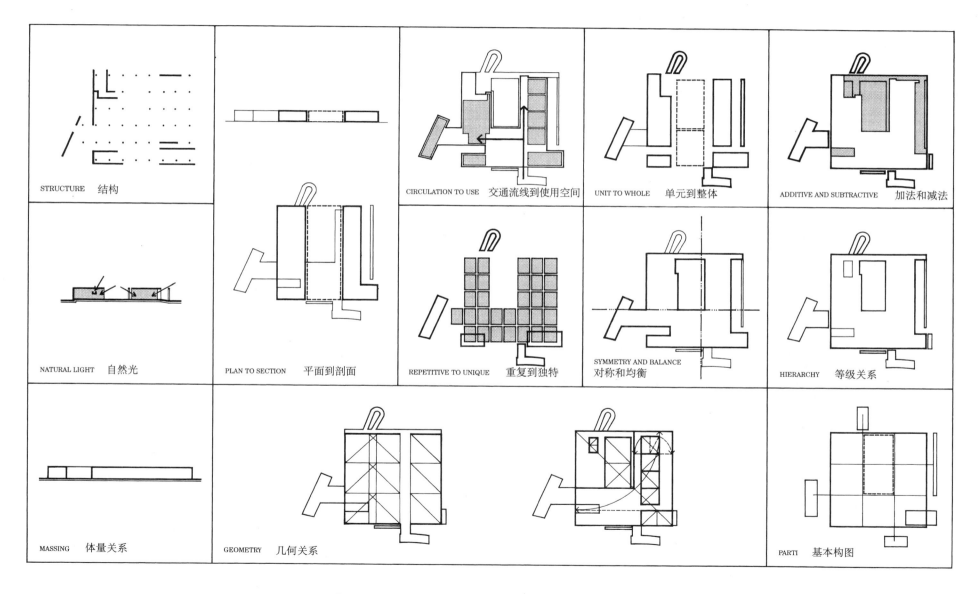

STRUCTURE　结构

NATURAL LIGHT　自然光

MASSING　体量关系

CIRCULATION TO USE　交通流线到使用空间

UNIT TO WHOLE　单元到整体

ADDITIVE AND SUBTRACTIVE　加法和减法

PLAN TO SECTION　平面到剖面

REPETITIVE TO UNIQUE　重复到独特

SYMMETRY AND BALANCE　对称和均衡

HIERARCHY　等级关系

GEOMETRY　几何关系

PARTI　基本构图

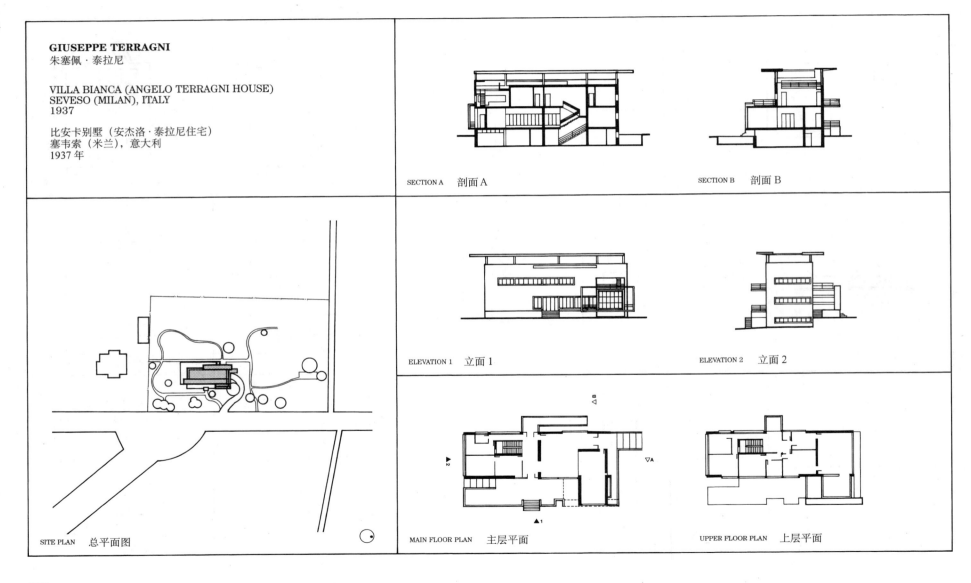

GIUSEPPE TERRAGNI
朱塞佩·泰拉尼

VILLA BIANCA (ANGELO TERRAGNI HOUSE)
SEVESO (MILAN), ITALY
1937

比安卡别墅（安杰洛·泰拉尼住宅）
塞韦索（米兰），意大利
1937 年

SECTION A　剖面 A

SECTION B　剖面 B

ELEVATION 1　立面 1

ELEVATION 2　立面 2

SITE PLAN　总平面图

MAIN FLOOR PLAN　主层平面

UPPER FLOOR PLAN　上层平面

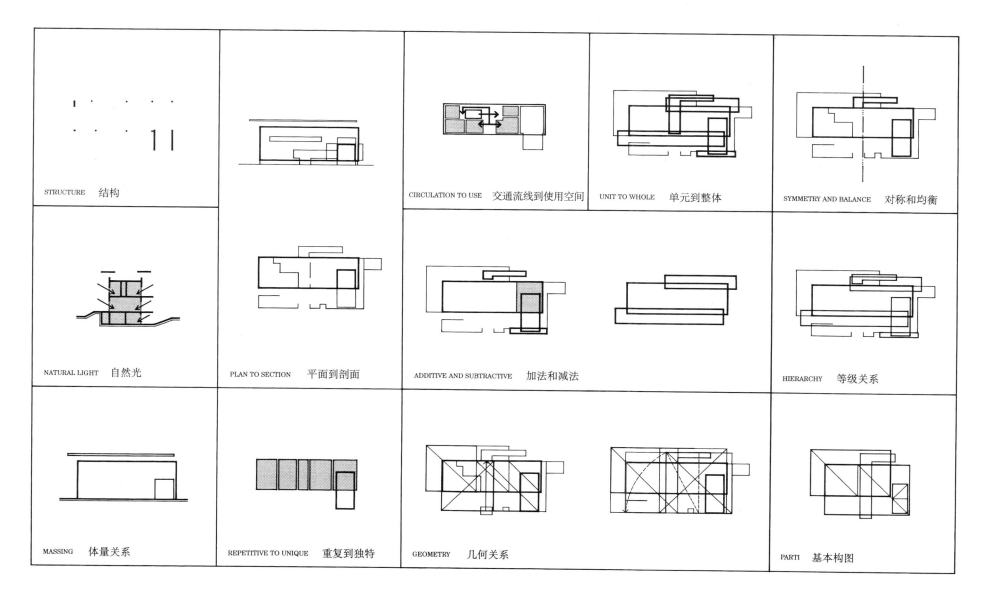

STRUCTURE　结构

CIRCULATION TO USE　交通流线到使用空间

UNIT TO WHOLE　单元到整体

SYMMETRY AND BALANCE　对称和均衡

NATURAL LIGHT　自然光

PLAN TO SECTION　平面到剖面

ADDITIVE AND SUBTRACTIVE　加法和减法

HIERARCHY　等级关系

MASSING　体量关系

REPETITIVE TO UNIQUE　重复到独特

GEOMETRY　几何关系

PARTI　基本构图

路德维希·密斯·凡·德·罗
LUDWIG MIES VAN DER ROHE

LUDWIG MIES VAN DER ROHE
路德维希·密斯·凡·德·罗

GERMAN PAVILION AT INTERNATIONAL EXHIBITION
BARCELONA, SPAIN
1928-1929

国际博览会德国馆
巴塞罗那，西班牙
1928-1929 年

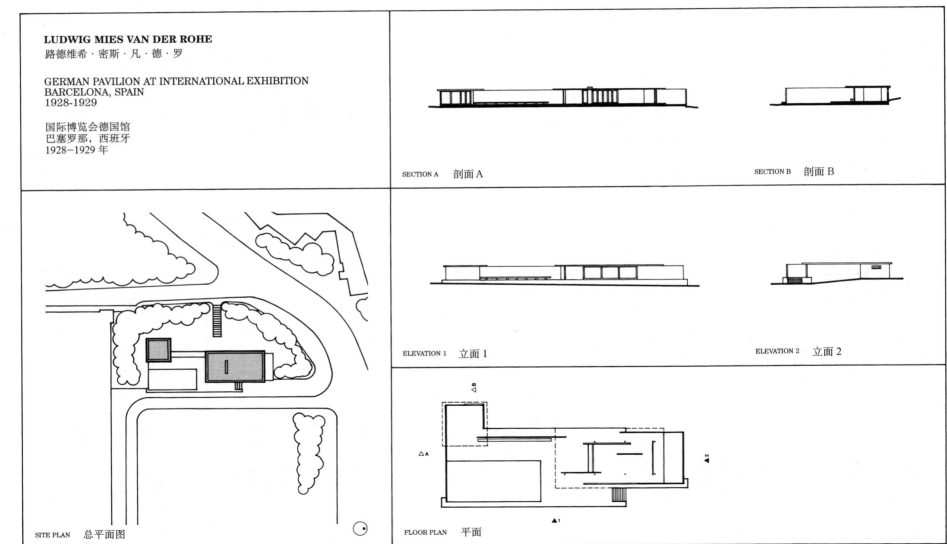

SECTION A　剖面 A

SECTION B　剖面 B

ELEVATION 1　立面 1

ELEVATION 2　立面 2

SITE PLAN　总平面图

FLOOR PLAN　平面

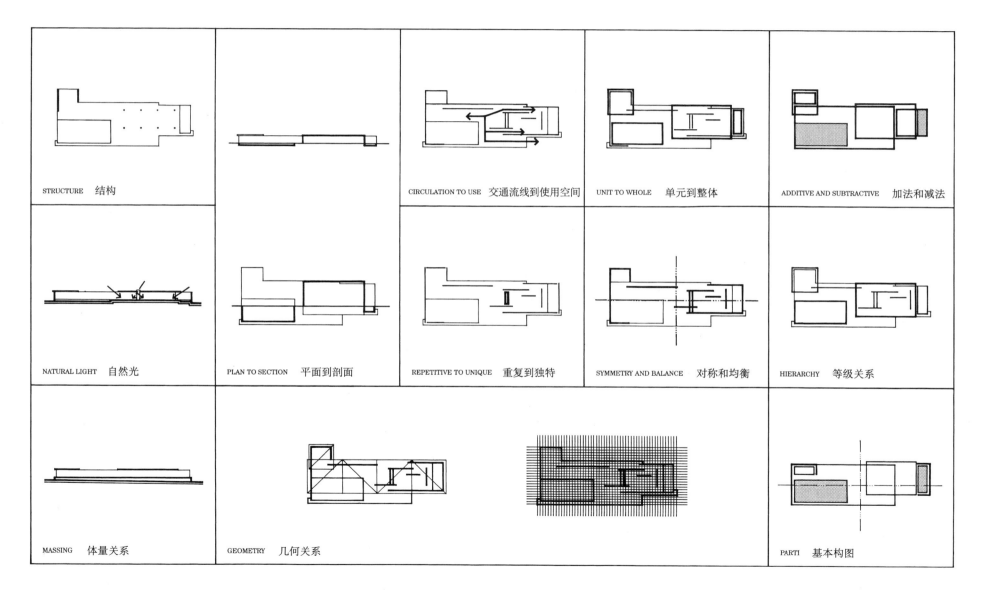

STRUCTURE　结构

CIRCULATION TO USE　交通流线到使用空间

UNIT TO WHOLE　单元到整体

ADDITIVE AND SUBTRACTIVE　加法和减法

NATURAL LIGHT　自然光

PLAN TO SECTION　平面到剖面

REPETITIVE TO UNIQUE　重复到独特

SYMMETRY AND BALANCE　对称和均衡

HIERARCHY　等级关系

MASSING　体量关系

GEOMETRY　几何关系

PARTI　基本构图

LUDWIG MIES VAN DER ROHE　路德维希·密斯·凡·德·罗

TUGENDHAT HOUSE
BRNO, CZECHOSLOVAKIA
1928-1930

吐根哈特住宅
布尔诺，捷克
1928—1930 年

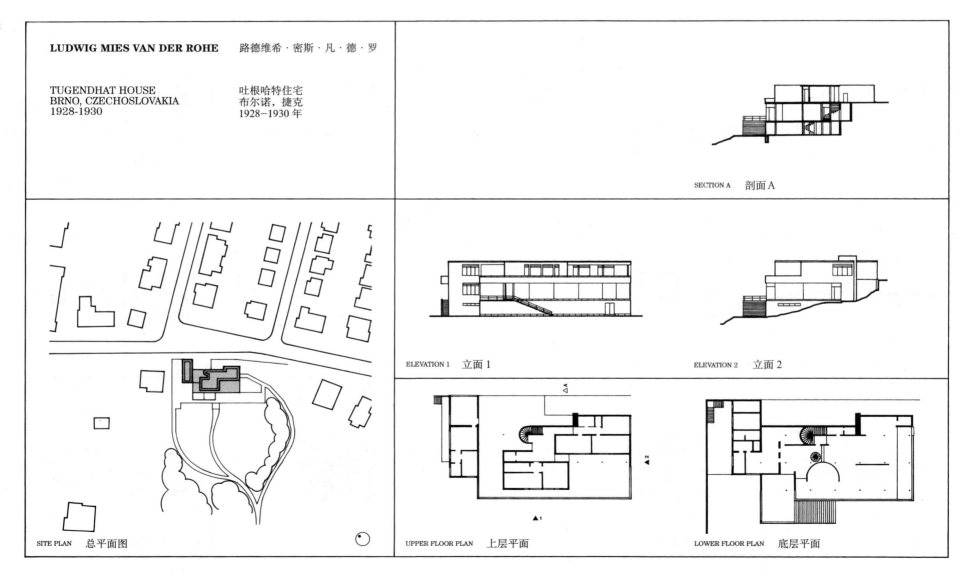

SECTION A　剖面 A

ELEVATION 1　立面 1

ELEVATION 2　立面 2

SITE PLAN　总平面图

UPPER FLOOR PLAN　上层平面

LOWER FLOOR PLAN　底层平面

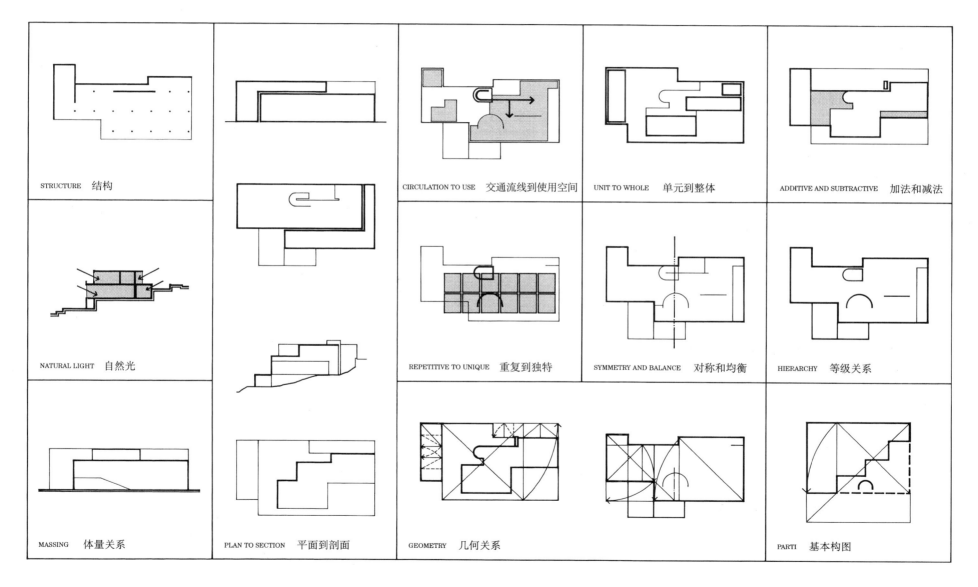

STRUCTURE　结构

NATURAL LIGHT　自然光

MASSING　体量关系

PLAN TO SECTION　平面到剖面

CIRCULATION TO USE　交通流线到使用空间

REPETITIVE TO UNIQUE　重复到独特

GEOMETRY　几何关系

UNIT TO WHOLE　单元到整体

SYMMETRY AND BALANCE　对称和均衡

ADDITIVE AND SUBTRACTIVE　加法和减法

HIERARCHY　等级关系

PARTI　基本构图

225

LUDWIG MIES VAN DER ROHE
路德维希·密斯·凡·德·罗

FARNSWORTH HOUSE
FOX RIVER VALLEY (NEAR PLANO), ILLINOIS, USA
1945-1951

范斯沃斯住宅
福克斯河谷（普莱诺附近），伊利诺伊州，美国
1945—1951 年

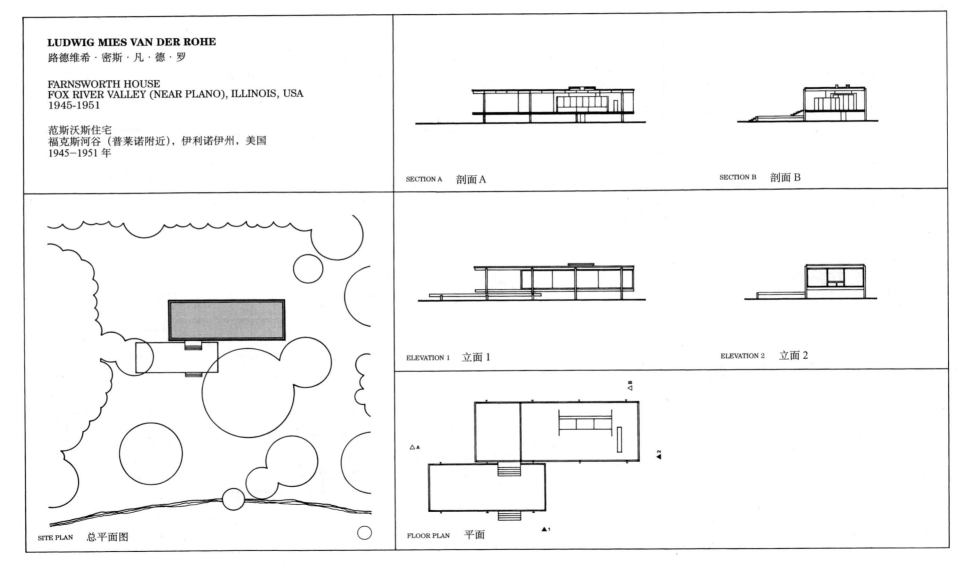

SECTION A　剖面 A

SECTION B　剖面 B

ELEVATION 1　立面 1

ELEVATION 2　立面 2

SITE PLAN　总平面图

FLOOR PLAN　平面

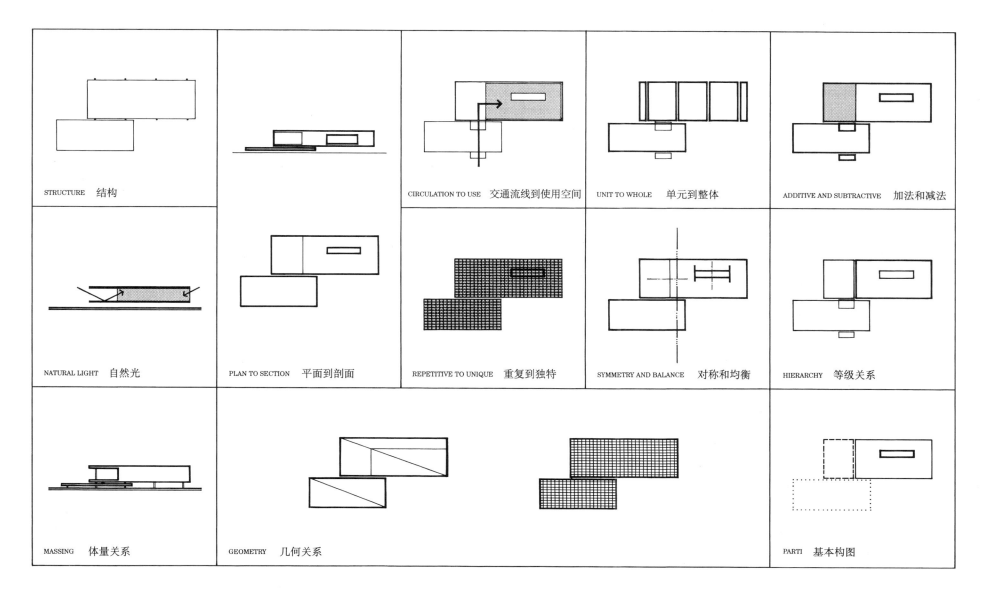

STRUCTURE　结构

CIRCULATION TO USE　交通流线到使用空间

UNIT TO WHOLE　单元到整体

ADDITIVE AND SUBTRACTIVE　加法和减法

NATURAL LIGHT　自然光

PLAN TO SECTION　平面到剖面

REPETITIVE TO UNIQUE　重复到独特

SYMMETRY AND BALANCE　对称和均衡

HIERARCHY　等级关系

MASSING　体量关系

GEOMETRY　几何关系

PARTI　基本构图

LUDWIG MIES VAN DER ROHE
路德维希·密斯·凡·德·罗

CROWN HALL (ARCHITECTURE, CITY
PLANNING AND DESIGN BUILDING)
ILLINOIS INSTITUTE OF TECHNOLOGY
CHICAGO, ILLINOIS, USA
1950-1956

克朗楼（建筑，城市规划与设计大楼）
伊利诺伊理工学院
芝加哥，伊利诺伊州，美国
1950－1956 年

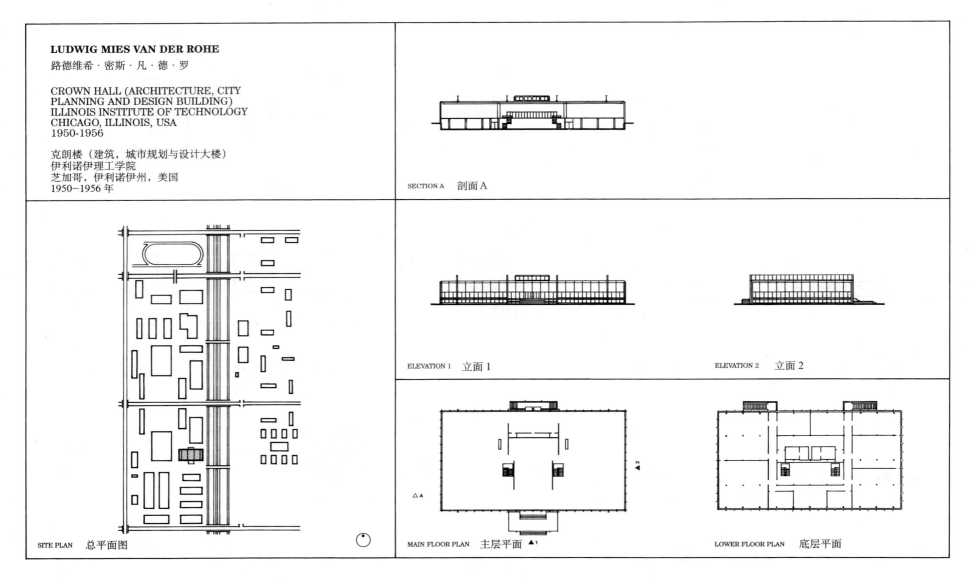

SECTION A 剖面A

ELEVATION 1 立面1

ELEVATION 2 立面2

SITE PLAN 总平面图

MAIN FLOOR PLAN 主层平面 ▲1

LOWER FLOOR PLAN 底层平面

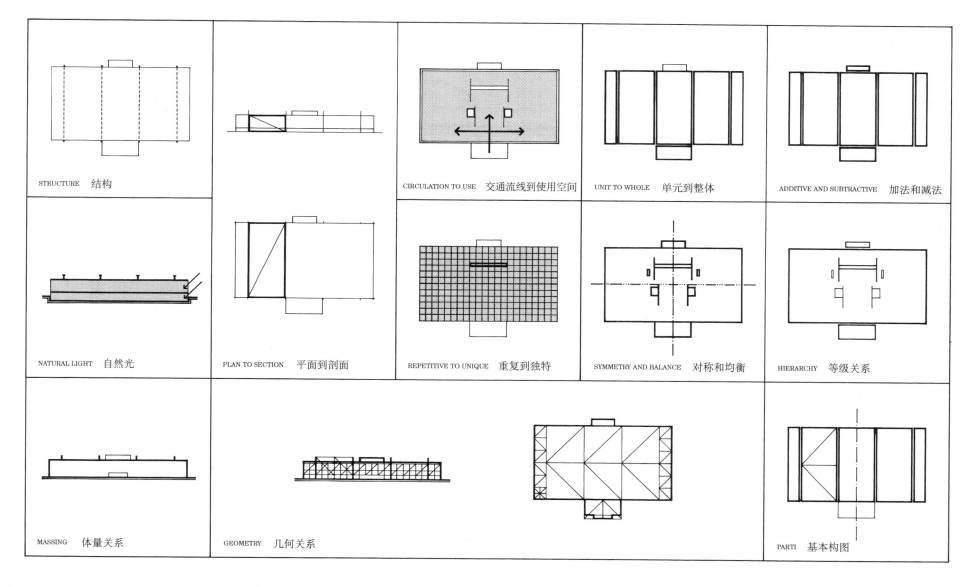

STRUCTURE　结构

CIRCULATION TO USE　交通流线到使用空间

UNIT TO WHOLE　单元到整体

ADDITIVE AND SUBTRACTIVE　加法和减法

NATURAL LIGHT　自然光

PLAN TO SECTION　平面到剖面

REPETITIVE TO UNIQUE　重复到独特

SYMMETRY AND BALANCE　对称和均衡

HIERARCHY　等级关系

MASSING　体量关系

GEOMETRY　几何关系

PARTI　基本构图

ROBERT VENTURI　　　　罗伯特·文丘里

VANNA VENTURI HOUSE　　万娜·文丘里住宅
PHILADELPHIA, PENNSYLVANIA, USA　　费城，宾夕法尼亚州，美国
1962　　1962 年

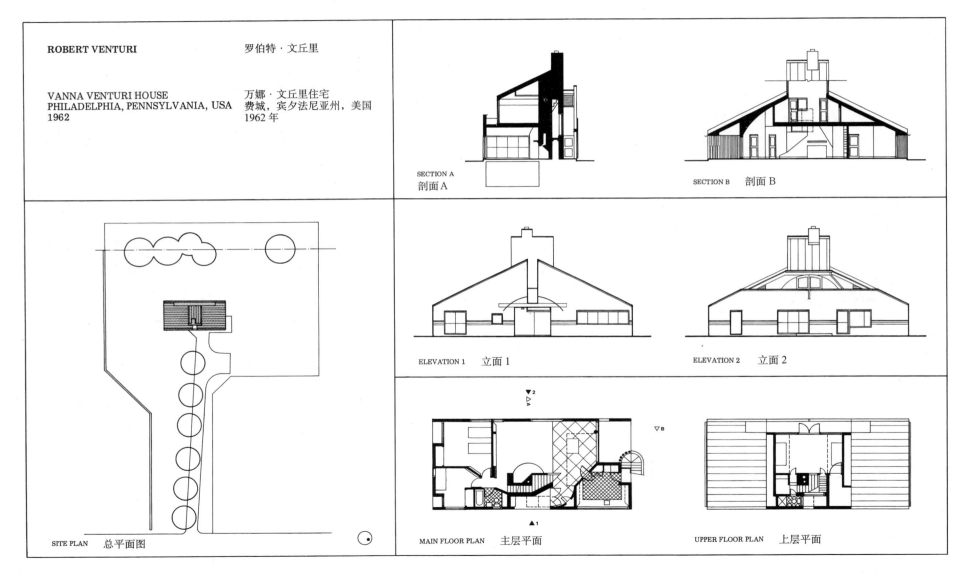

SECTION A　剖面 A

SECTION B　剖面 B

SITE PLAN　总平面图

ELEVATION 1　立面 1

ELEVATION 2　立面 2

MAIN FLOOR PLAN　主层平面

UPPER FLOOR PLAN　上层平面

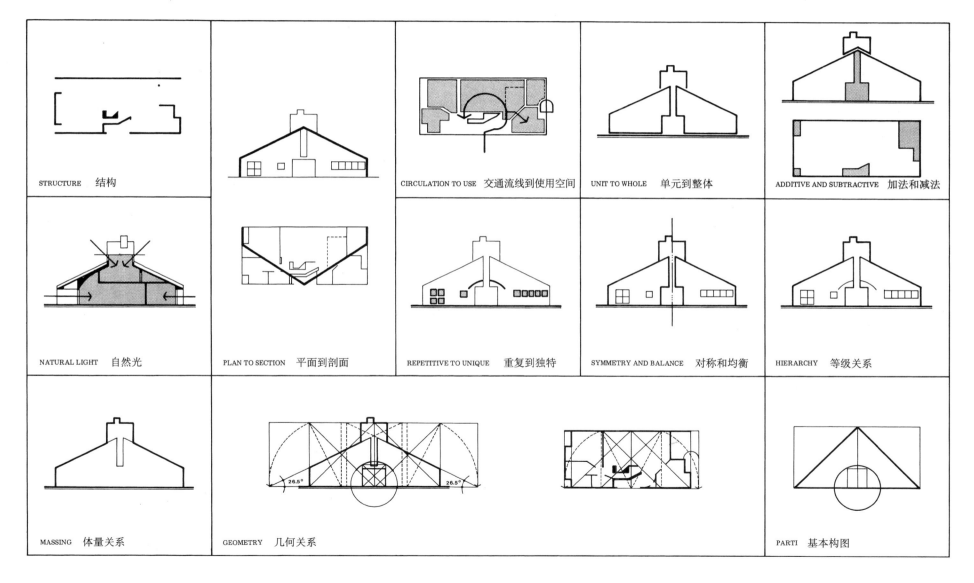

STRUCTURE　结构

CIRCULATION TO USE　交通流线到使用空间

UNIT TO WHOLE　单元到整体

ADDITIVE AND SUBTRACTIVE　加法和减法

NATURAL LIGHT　自然光

PLAN TO SECTION　平面到剖面

REPETITIVE TO UNIQUE　重复到独特

SYMMETRY AND BALANCE　对称和均衡

HIERARCHY　等级关系

MASSING　体量关系

GEOMETRY　几何关系

PARTI　基本构图

231

ROBERT VENTURI 罗伯特·文丘里

FIRE STATION NUMBER 4 消防站 4 号
COLUMBUS, INDIANA, USA 哥伦布，印第安纳州，美国
1966 1966 年

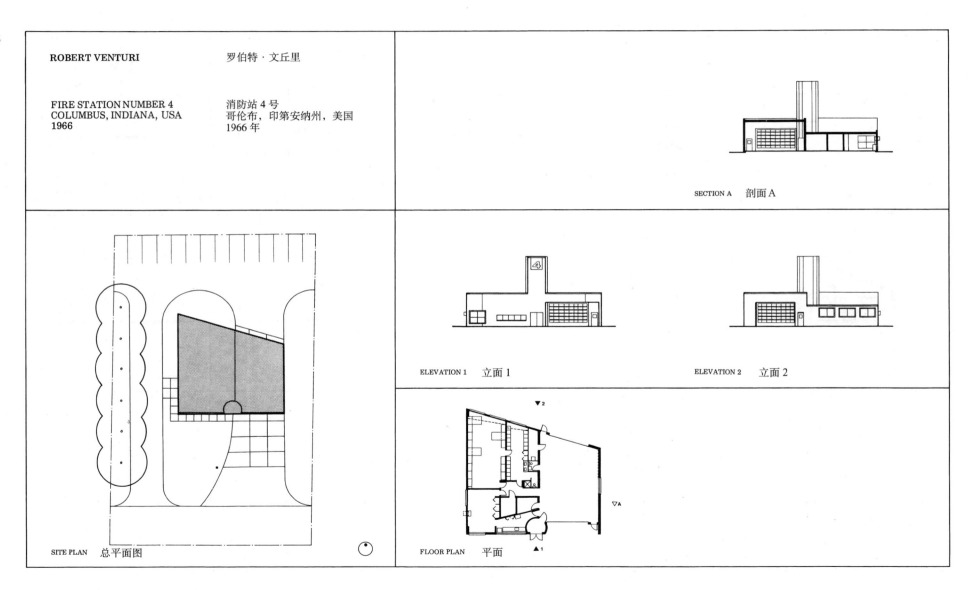

SECTION A 剖面 A

ELEVATION 1 立面 1

ELEVATION 2 立面 2

SITE PLAN 总平面图

FLOOR PLAN 平面

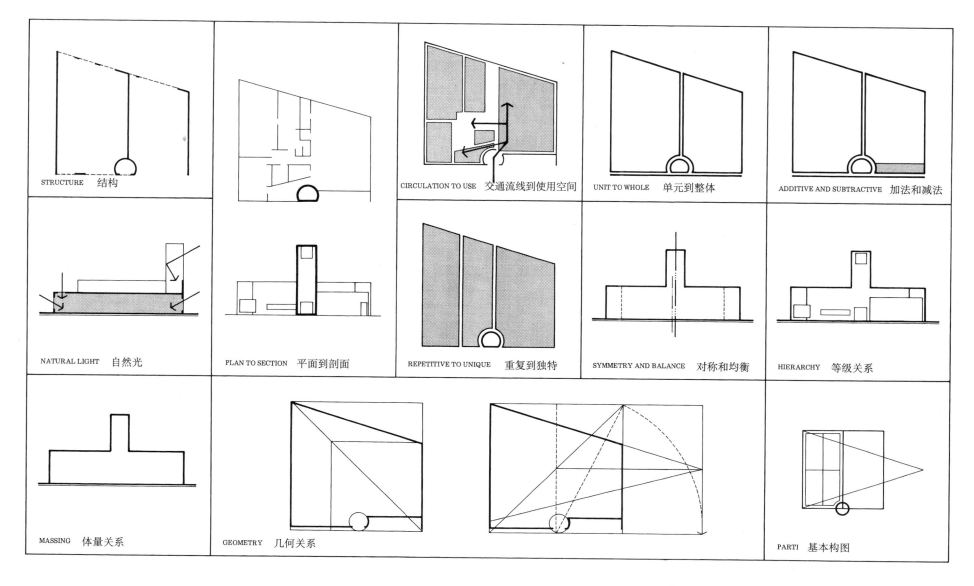

STRUCTURE　结构

CIRCULATION TO USE　交通流线到使用空间

UNIT TO WHOLE　单元到整体

ADDITIVE AND SUBTRACTIVE　加法和减法

NATURAL LIGHT　自然光

PLAN TO SECTION　平面到剖面

REPETITIVE TO UNIQUE　重复到独特

SYMMETRY AND BALANCE　对称和均衡

HIERARCHY　等级关系

MASSING　体量关系

GEOMETRY　几何关系

PARTI　基本构图

ROBERT VENTURI　　　罗伯特·文丘里

PETER BRANT HOUSE　　　彼得·布兰特住宅
GREENWICH, CONNECTICUT, USA　　格林威治，康涅狄格州，美国
1973　　　　　　　　　　1973 年

SECTION A　剖面 A　　　　　　　　SECTION B　剖面 B

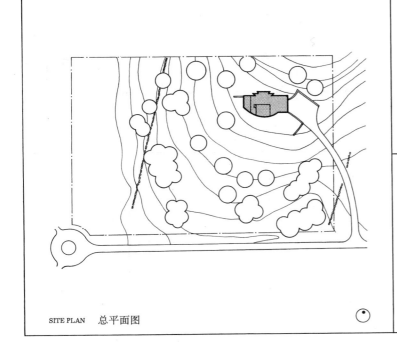

SITE PLAN　总平面图

ELEVATION 1　立面 1

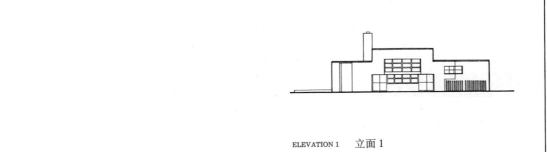

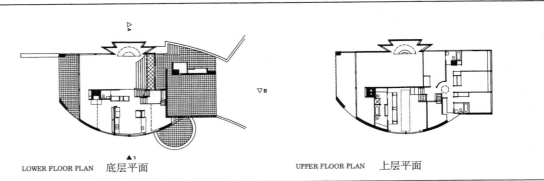

LOWER FLOOR PLAN　底层平面　　　UPPER FLOOR PLAN　上层平面

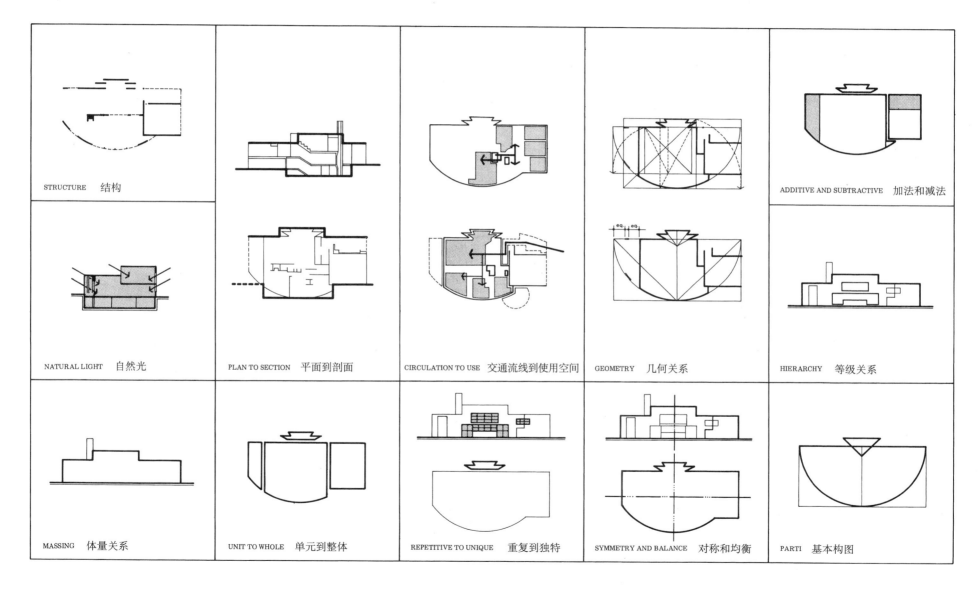

STRUCTURE　结构

ADDITIVE AND SUBTRACTIVE　加法和减法

NATURAL LIGHT　自然光

PLAN TO SECTION　平面到剖面

CIRCULATION TO USE　交通流线到使用空间

GEOMETRY　几何关系

HIERARCHY　等级关系

MASSING　体量关系

UNIT TO WHOLE　单元到整体

REPETITIVE TO UNIQUE　重复到独特

SYMMETRY AND BALANCE　对称和均衡

PARTI　基本构图

235

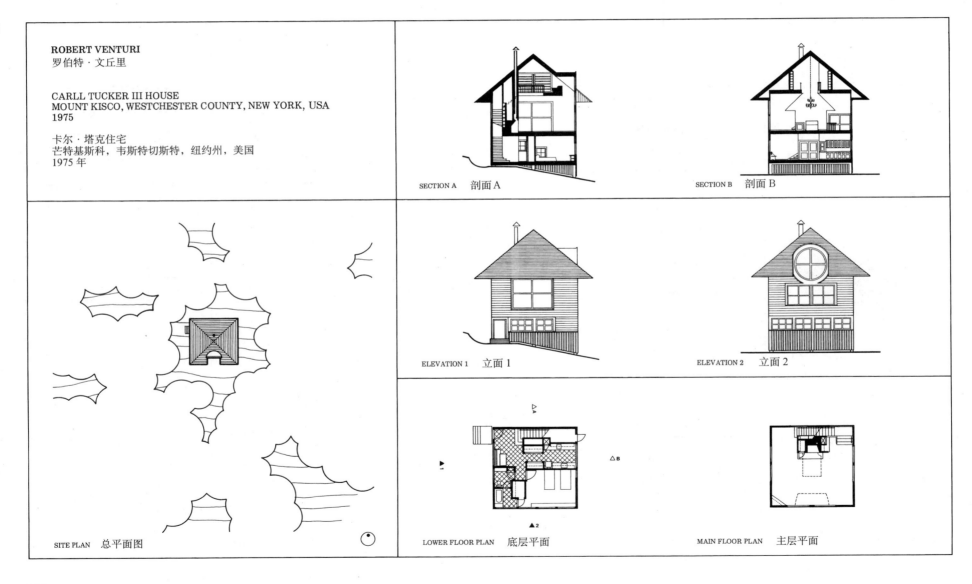

ROBERT VENTURI
罗伯特·文丘里

CARLL TUCKER III HOUSE
MOUNT KISCO, WESTCHESTER COUNTY, NEW YORK, USA
1975

卡尔·塔克住宅
芒特基斯科，韦斯特切斯特，纽约州，美国
1975 年

SECTION A　剖面 A

SECTION B　剖面 B

ELEVATION 1　立面 1

ELEVATION 2　立面 2

SITE PLAN　总平面图

LOWER FLOOR PLAN　底层平面

MAIN FLOOR PLAN　主层平面

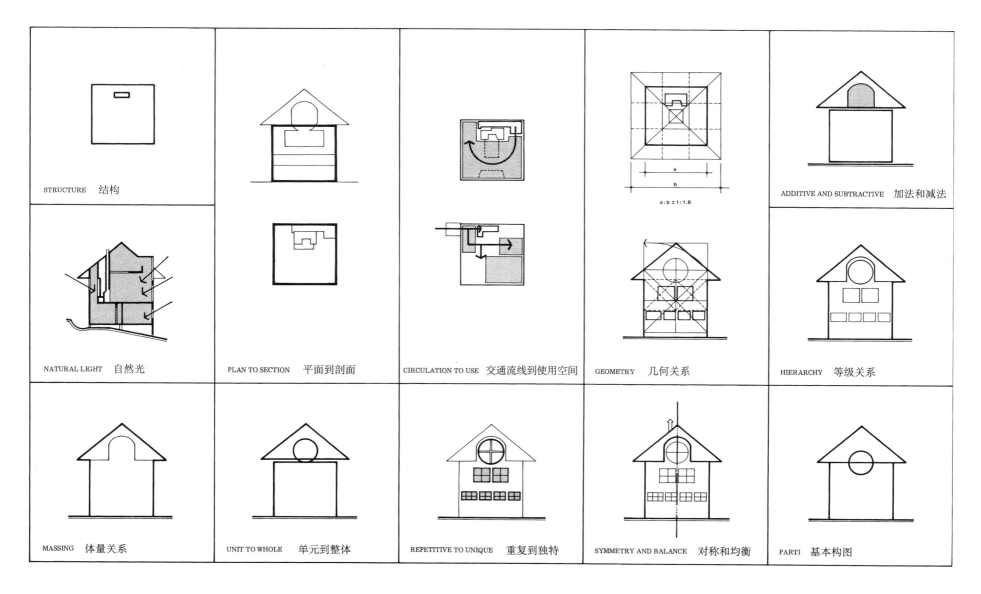

STRUCTURE 结构				ADDITIVE AND SUBTRACTIVE 加法和减法
			a:b = 1:1.6	
NATURAL LIGHT 自然光	PLAN TO SECTION 平面到剖面	CIRCULATION TO USE 交通流线到使用空间	GEOMETRY 几何关系	HIERARCHY 等级关系
MASSING 体量关系	UNIT TO WHOLE 单元到整体	REPETITIVE TO UNIQUE 重复到独特	SYMMETRY AND BALANCE 对称和均衡	PARTI 基本构图

弗兰克·劳埃德·赖特
FRANK LLOYD WRIGHT

FRANK LLOYD WRIGHT　　　　弗兰克·劳埃德·赖特

UNITY TEMPLE　　　　　　　联合教堂
OAK PARK, ILLINOIS, USA　　橡树园，伊利诺伊州，美国
1906　　　　　　　　　　　　1906 年

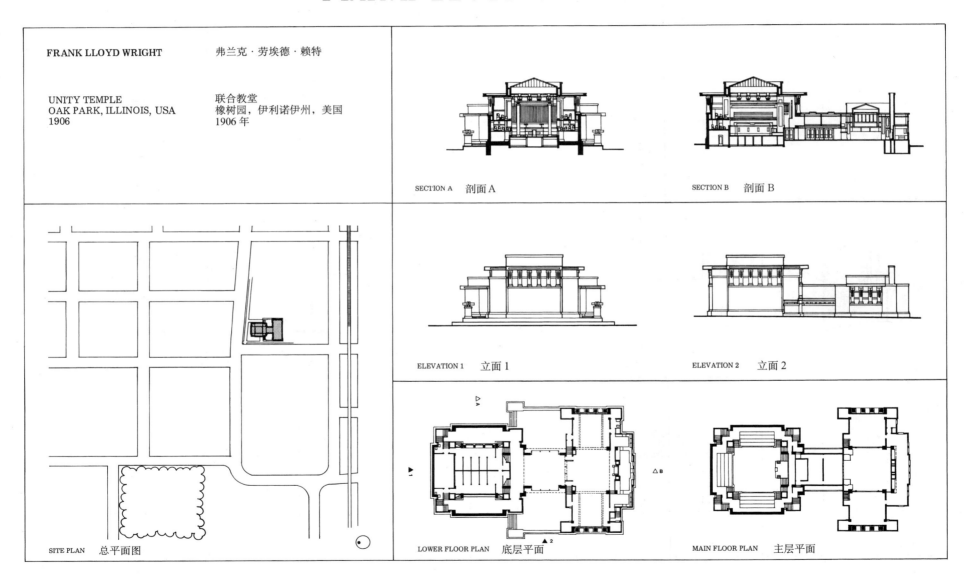

SECTION A　剖面 A

SECTION B　剖面 B

ELEVATION 1　立面 1

ELEVATION 2　立面 2

SITE PLAN　总平面图

LOWER FLOOR PLAN　底层平面

MAIN FLOOR PLAN　主层平面

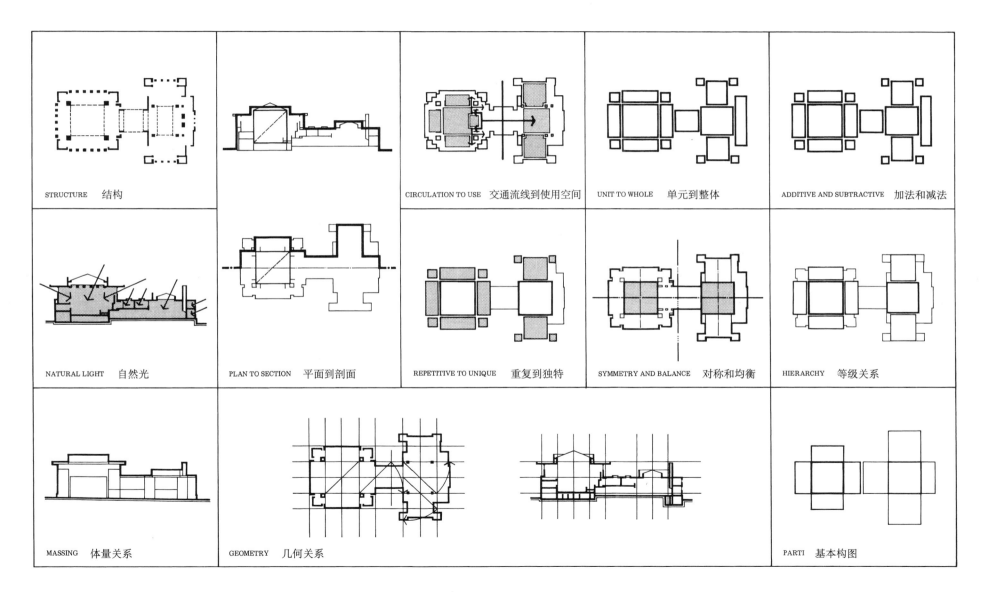

STRUCTURE 结构

CIRCULATION TO USE 交通流线到使用空间

UNIT TO WHOLE 单元到整体

ADDITIVE AND SUBTRACTIVE 加法和减法

NATURAL LIGHT 自然光

PLAN TO SECTION 平面到剖面

REPETITIVE TO UNIQUE 重复到独特

SYMMETRY AND BALANCE 对称和均衡

HIERARCHY 等级关系

MASSING 体量关系

GEOMETRY 几何关系

PARTI 基本构图

FRANK LLOYD WRIGHT 弗兰克·劳埃德·赖特

FREDERICK G. ROBIE HOUSE
CHICAGO, ILLINOIS, USA
1909

弗雷德里克·G·罗比住宅
芝加哥，伊利诺伊州，美国
1909 年

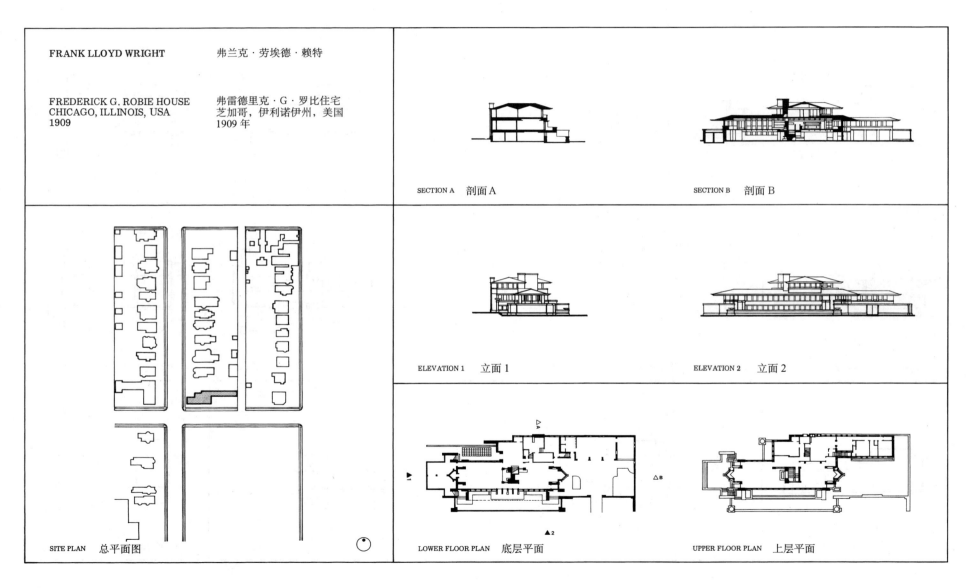

SECTION A 剖面 A

SECTION B 剖面 B

ELEVATION 1 立面 1

ELEVATION 2 立面 2

SITE PLAN 总平面图

LOWER FLOOR PLAN 底层平面

UPPER FLOOR PLAN 上层平面

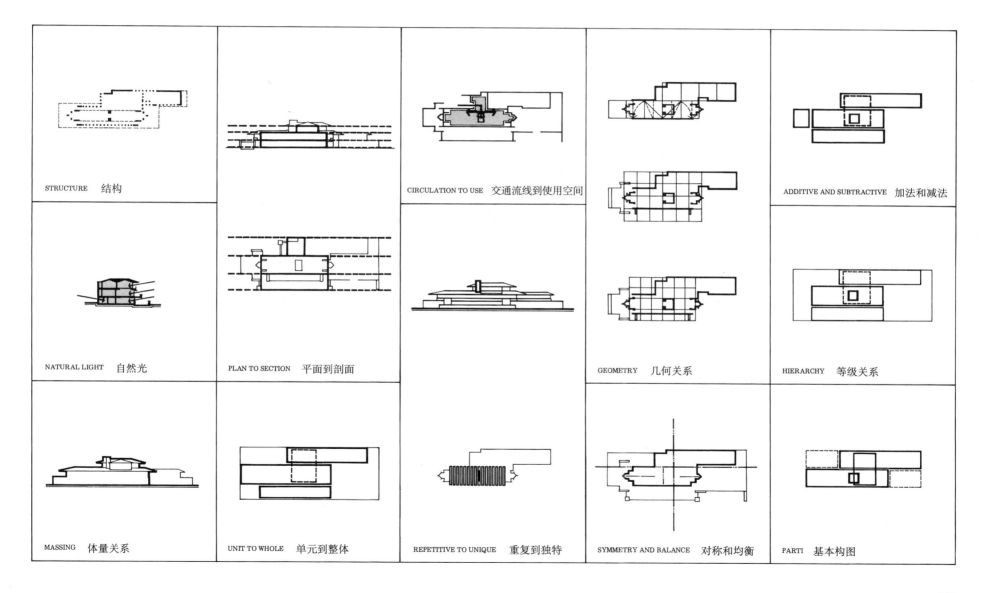

STRUCTURE　结构

CIRCULATION TO USE　交通流线到使用空间

ADDITIVE AND SUBTRACTIVE　加法和减法

NATURAL LIGHT　自然光

PLAN TO SECTION　平面到剖面

GEOMETRY　几何关系

HIERARCHY　等级关系

MASSING　体量关系

UNIT TO WHOLE　单元到整体

REPETITIVE TO UNIQUE　重复到独特

SYMMETRY AND BALANCE　对称和均衡

PARTI　基本构图

FRANK LLOYD WRIGHT
弗兰克·劳埃德·赖特

FALLINGWATER (EDGAR J. KAUFMANN HOUSE)
OHIOPYLE, PENNSYLVANIA, USA
1935

流水别墅（埃德加·J·考夫曼住宅）
俄亥俄派尔，宾夕法尼亚州，美国
1935 年

SECTION A　剖面 A

SECTION B　剖面 B

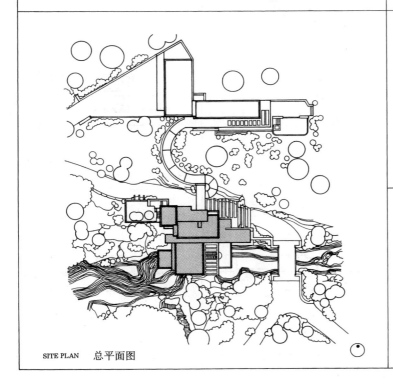

SITE PLAN　总平面图

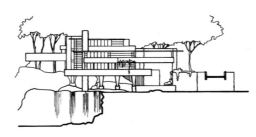

ELEVATION 1　立面 1

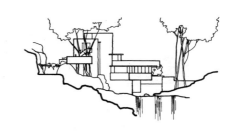

ELEVATION 2　立面 2

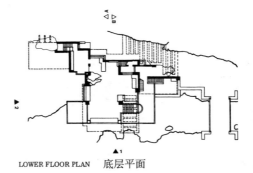

LOWER FLOOR PLAN　底层平面

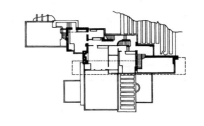

UPPER FLOOR PLAN　上层平面

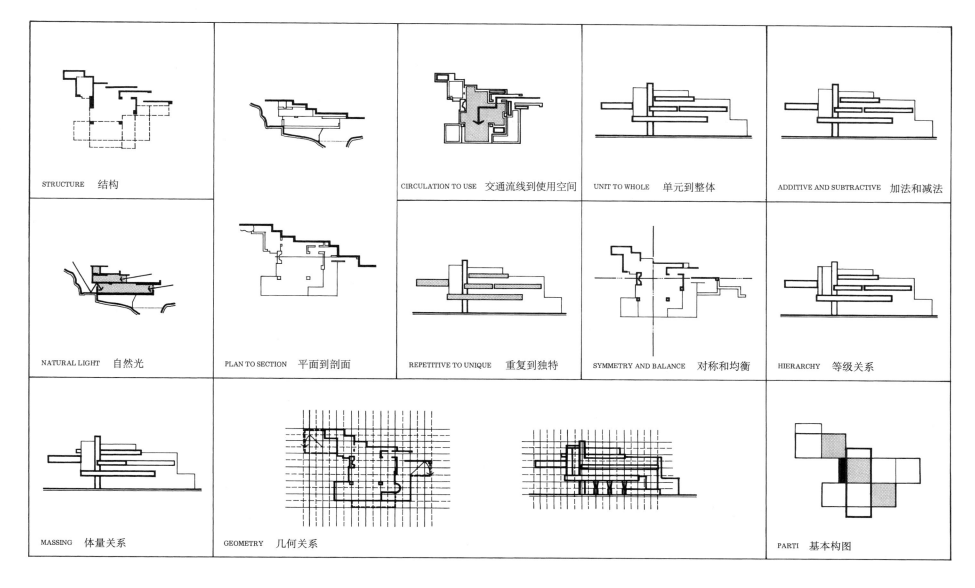

STRUCTURE　结构

CIRCULATION TO USE　交通流线到使用空间

UNIT TO WHOLE　单元到整体

ADDITIVE AND SUBTRACTIVE　加法和减法

NATURAL LIGHT　自然光

PLAN TO SECTION　平面到剖面

REPETITIVE TO UNIQUE　重复到独特

SYMMETRY AND BALANCE　对称和均衡

HIERARCHY　等级关系

MASSING　体量关系

GEOMETRY　几何关系

PARTI　基本构图

FRANK LLOYD WRIGHT　　　　弗兰克·劳埃德·赖特

SOLOMON R. GUGGENHEIM MUSEUM　所罗门·R·古根海姆美术馆
NEW YORK, NEW YORK, USA　　　纽约，纽约州，美国
1956　　　　　　　　　　　　　1956 年

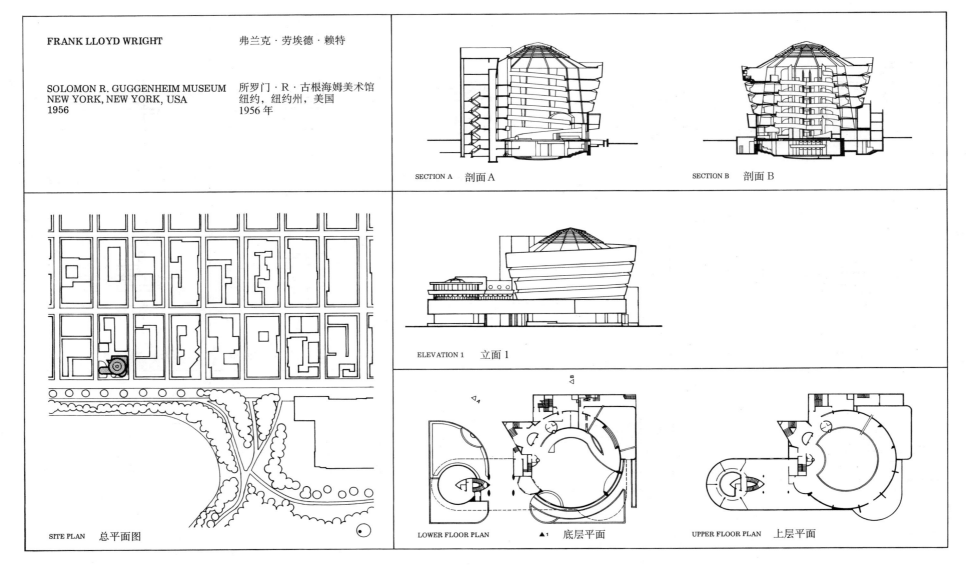

SECTION A　剖面 A

SECTION B　剖面 B

ELEVATION 1　立面 1

SITE PLAN　总平面图

LOWER FLOOR PLAN　　▲1　底层平面

UPPER FLOOR PLAN　上层平面

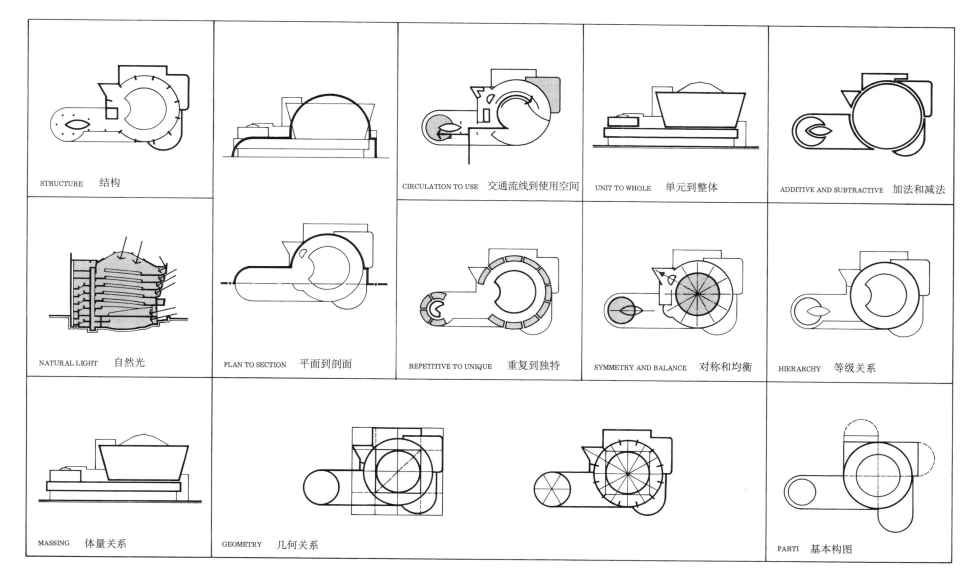

STRUCTURE　结构

CIRCULATION TO USE　交通流线到使用空间

UNIT TO WHOLE　单元到整体

ADDITIVE AND SUBTRACTIVE　加法和减法

NATURAL LIGHT　自然光

PLAN TO SECTION　平面到剖面

REPETITIVE TO UNIQUE　重复到独特

SYMMETRY AND BALANCE　对称和均衡

HIERARCHY　等级关系

MASSING　体量关系

GEOMETRY　几何关系

PARTI　基本构图

245

彼得·卒姆托
PETER ZUMTHOR

PETER ZUMTHOR

彼得·卒姆托

CHAPEL OF ST. BENEDICT
SUMVITG, SWITZERLAND
1987–1988

圣本尼迪克特小教堂
Sumvitg，瑞士
1987–1988 年

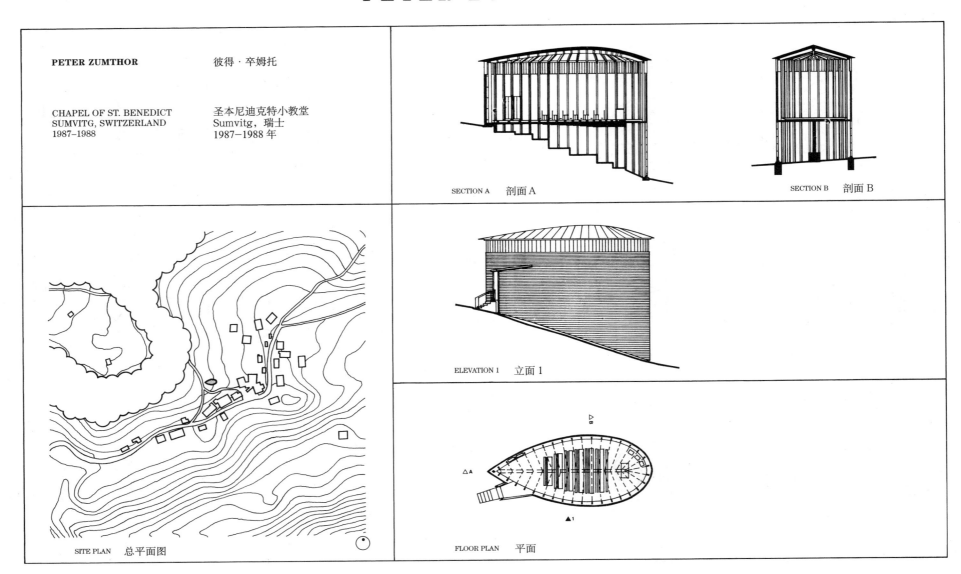

SECTION A　剖面 A

SECTION B　剖面 B

ELEVATION 1　立面 1

SITE PLAN　总平面图

FLOOR PLAN　平面

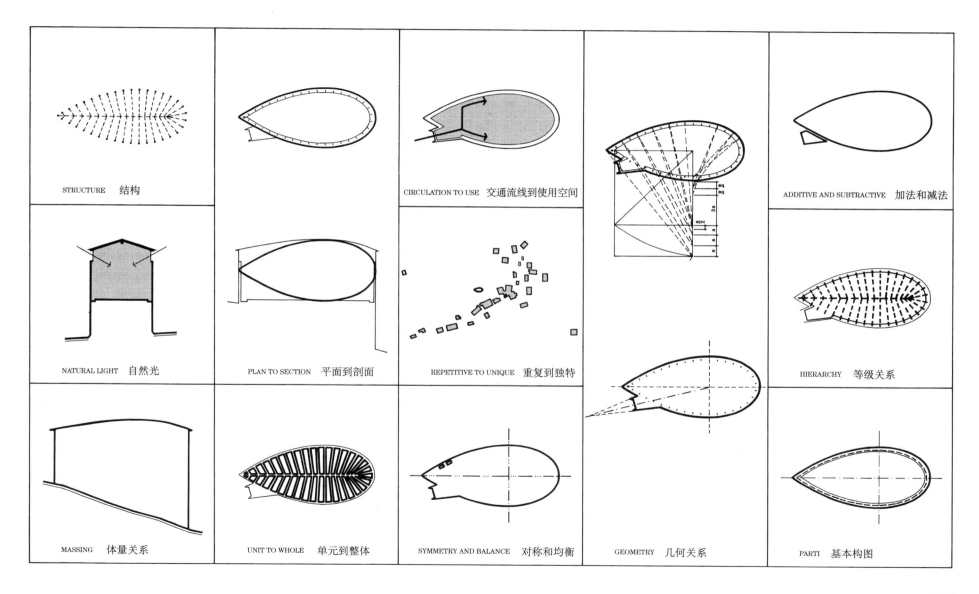

STRUCTURE　结构

CIRCULATION TO USE　交通流线到使用空间

ADDITIVE AND SUBTRACTIVE　加法和减法

NATURAL LIGHT　自然光

PLAN TO SECTION　平面到剖面

REPETITIVE TO UNIQUE　重复到独特

HIERARCHY　等级关系

MASSING　体量关系

UNIT TO WHOLE　单元到整体

SYMMETRY AND BALANCE　对称和均衡

GEOMETRY　几何关系

PARTI　基本构图

PETER ZUMTHOR   彼得·卒姆托

ART MUSEUM (KUNSTHAUS) BREGENZ   布雷根茨艺术馆
BREGENZ, AUSTRIA   布雷根茨，澳大利亚
1990–1997   1990–1997 年

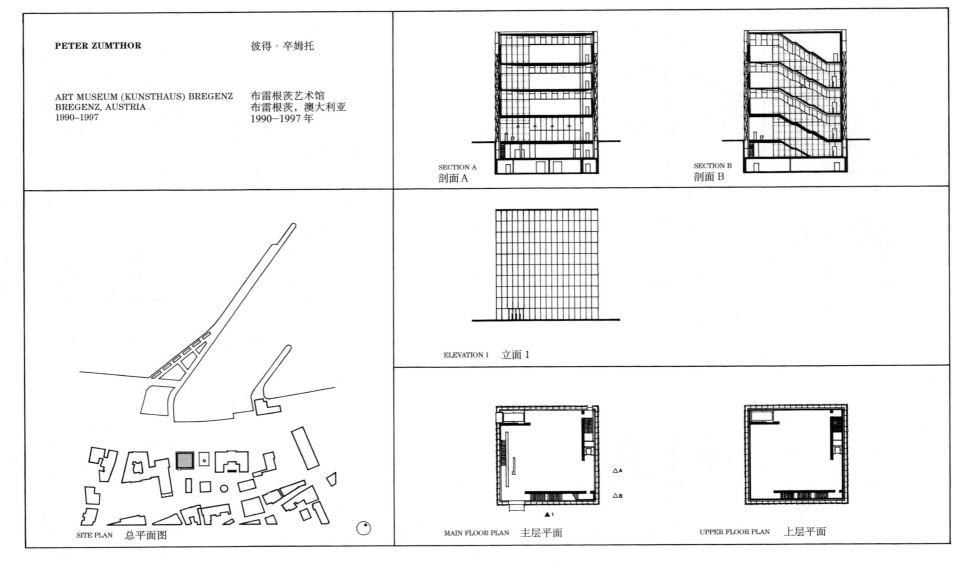

SECTION A   剖面 A

SECTION B   剖面 B

ELEVATION 1   立面 1

SITE PLAN   总平面图

MAIN FLOOR PLAN   主层平面

UPPER FLOOR PLAN   上层平面

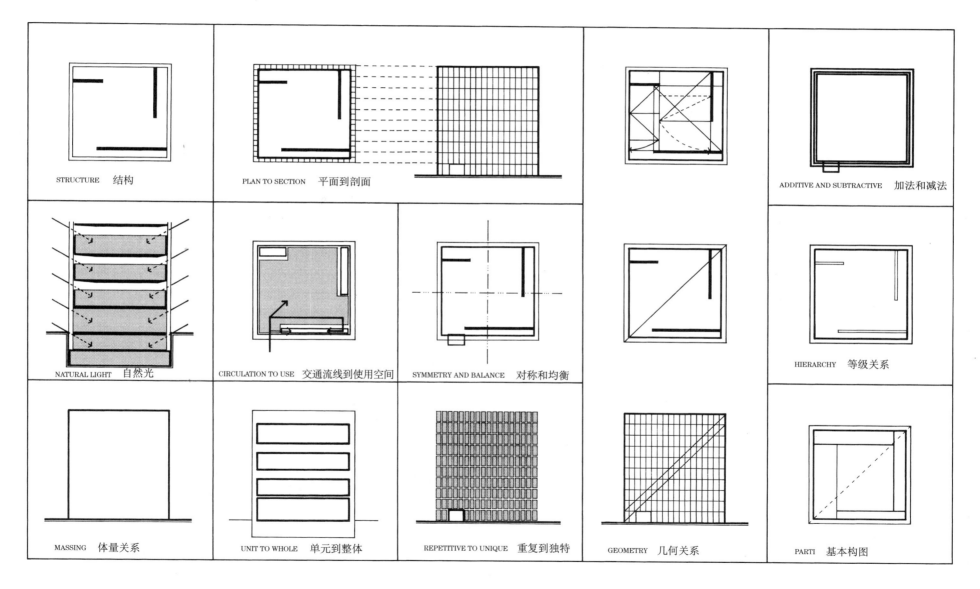

STRUCTURE 结构

PLAN TO SECTION 平面到剖面

ADDITIVE AND SUBTRACTIVE 加法和减法

NATURAL LIGHT 自然光

CIRCULATION TO USE 交通流线到使用空间

SYMMETRY AND BALANCE 对称和均衡

HIERARCHY 等级关系

MASSING 体量关系

UNIT TO WHOLE 单元到整体

REPETITIVE TO UNIQUE 重复到独特

GEOMETRY 几何关系

PARTI 基本构图

形 体 构 思 目 录

形 体 构 思

上一部分 118 个案例的分析里，不同建筑师的设计中，基本模式的应用显而易见。设计中的共通之处会在很多不同建筑师的作品里显现，无拘于时间、风格、位置、功能，以及建筑的类型。基于大的条目和设计方法，这些共通之处可以被分类，并在不同年代的建筑设计之中应用。

建筑构思被理解为一种概念。在设计中，设计师以它赋予或改变设计的形式。不同的构思为决策组织、秩序建构、形态塑造提供了方法。为了使建筑的形态及方式与众不同，设计师会谨慎选择设计构思。不同构思方法的应用，将生成不同的设计结果。

本章中的案例，按照建筑设计形态构思的一系列内在关联编排。每一个概念，基于其构思的共性表现予以分类及研究。阐释性文字后均附有一系列图解，由本书中的一些而非全部案例为佐证。本书不求面面俱到，既非所有构思都会讨论，也非每个案例都将涉及。整体而言，分析部分发展而来的图解，需要其他案例相补充才能说明形态构思。最宜阐释构思的图解均被甄选出，展现了建筑外在表征的多样性；并在最大的时间框架中展示了建筑类型的丰富性。

平面到剖面或立面

作为形体构思方法，平面到立面或剖面，通过构筑建筑水平与垂直方向内在明确的关联性来展开设计。建构这两个范畴的关联，将使得一个领域的决策影响到其他领域的形态。

当平面和剖面相同，即二者的轮廓一致时，它们的关联最为直接。这可以称作"一对一"的关联。例如一个球体，它的平面和剖面都是一个圆。也有可能是图形局部与整体的对应关系。例如，建筑中有可能存在"一对半"的关系。平面上的圆对应剖面中的半圆，生成一个半球形的穹隆。反之亦有可能，平面形式与剖面或立面的一半相同。这两种情形之中，在平面与剖面中各自出现的图形应该尺度相同。在剖面是"半个"平面的那些情形下，剖面图形的两次应用将生成左右对称的平面。当这些图形相同部分彼此重叠则是更为极端的情形，例如圆厅别墅（La Rotonda）的主体空间。

一定比率的比例关系也是平面与剖面或立面的关联纽带。与前文论及的局部与整体关系大相径庭，尽管在尺度上有所变化，比例关系中的平面与剖面，均作为一个整体融入彼此。彼此配对的平面与剖面远非轮廓上的简单关联。符合几何关系的常用比率有 1：2、2：3、1：5。在各个案例中，平面与剖面仅在一个向度上可以有尺度变化。在 1：2 的关系中，平面与剖面形状相同，但在特定向度上一个是另一个的 2 倍。例如，圆形平面可以匹配高度是宽度一半的椭圆剖面。剖面或立面应用中，平面各个部分的缩放比例也不必一致。在尼古拉斯·豪克斯穆尔（Nicholas Hawksmoor）设计的基督教教堂中，一个要素在其他领域出现时可能是缩小，而其他要素可能是放大。

彼此形状大体相近时，平面与剖面（或立面）之间为"相似"关系。这是平面与剖面之间最普遍的关联，且时常只是局部关联而非全部。差异时常表现在形式语言、尺寸与位置、不规则的渐进增长等方面。形式语言变化中，平面或剖面中的直角或许与其他领域的弧形相对应。尺度和位置变化中，一个剖面元素与其相应的平面元素、大小或位置上或有不同。等比缩放中，平面或剖面的相应元素将做比率不同的同类变化。

平面与剖面之间亦存在反向关联，此时一种形态与另一领域中的相反状态对应。例如，构成平面中的组分是大的、简单的、实的、随意的；剖面中的相应部分可能是小的、复杂的、虚的，或规整的，二者之中呈现反向关联。

尽管在上述对等、局部与整体、比例、相似和相反等各种类型的平面与剖面的关系中，双方建立了互相制约、互为影响的形态关联，但仍存在形态关联较弱，属性影响更多的模式。在这种关联类型中，平面或剖面的决策为另一方提供了较宽泛的选择域。

在平面和剖面之间也存在局部与整体的关系。该语境下，某个完整的形态通过缩减，成为了另一形态中的局部。这个完整形态在此关系中非常显眼，其整体形态是另一领域的控制性局部，只是尺度缩减了。矶崎新（Arata Isozaki）所做的矢野住宅（Yano House）是典型案例。在此，平面的完整形态在剖面局部中重现。

当平面与剖面的重要节点相互对应时，平面与剖面之间还可呈现出匹配关系。尽管实际形态变化多样，其关键是校准平面与剖面形态变化所对应的节点。第一部分曾分析过的亨利·霍布森·理查森（H. H. Richardson）做的阿勒格尼县法院（Allegheny Courthouse）就是这样的一处实例。

平面与剖面最后一类关系是有共同的原型或派生物。在此关系中，平面与剖面的构形由同一原型经由不同路径派生而来。例如，在菲利波·伯鲁乃列斯基（Filippo Brunelleschi）所做的圣玛丽亚布道所（San Maria degli Angeli）中，在分析部分中也曾论述，其平面与剖面均由两个旋转45°然后重叠的正方形发展而来。在平面中，两个正方形有共同的中心；在剖面中，正方形的一角与另一正方形一条边的中点重合。平面与剖面均由尺度相同的正方形演化而来，但最后的形态却大相径庭。

单元到整体

单元到整体的关系是一种构形方法，它涉及单元，以及单元如何依靠特定关联生成建筑形式。一个单元是建筑可识别的、主要的组分，或者是构成建筑某个层级的可变体。单元可以在建筑不同的尺度下存在。然而，虽然砖块可以视为墙体的构成单元，但视作建筑构成单元则意义不大，那样所有砖砌的建筑都是同样的单元整体关联了。因此，单元，通常是体量、使用空间、结构元素、体块及其组合体。

当单元等同整体，即二者同为一个实体时，单元和整体的关联最为直接。通常，这发生在建筑物被设计为一个小而完整的几何体。例如，齐奥普斯金字塔（Cheop's pyramid）由大量的石块和面层石片构成。但最终，令人印象深刻的还是建筑清晰可辨的整个形体。即便这印象之中或许也包含了覆面石片单元所形成的表面质感或良好的尺度。类似的，对于某些现代建筑，相对其独立巨石般的形体，紧绷的玻璃表皮终归是第二位的。

单元到整体关系最普遍的形式是单元集合生成整体。单元集合就是并置各个单元以形成可察觉的某种关系。放在彼此接近的位置，使人们能感觉到它们之间存在某种联系。单元之间可界定的关联不一定源自单元间物质上的联系。单元集合生成整体可供选择的方法包括连接、分离与重叠。

连接是最常见的集合方式。在此，单元体是可看见、可感知的实体，而且与其他单元之间存在面对面、面对边、边对边的联系。搭接是面对面连接的一种变体。

单元也可以彼此分离，同时又彼此关联形成整体。分离可以是物理上分隔；也可以在单元之间设置关节，使其形成分离之感。这种关联的精妙之处在于，毗邻单元之间感觉上的分离建构了整体上的关联。

单元还可能以叠合的方式聚集成整体。由于建筑是一个三维现象，单元体可通过交织重叠构成体量。单元体彼此因借形体与空间形成可辨识的整体。叠合的部分，看上去既是一个单元的局部，又为两个单元所共有。

单元也可以包含在建筑的整体之中。这种整体生成的方式与"连接"模式不同，建筑整体是控制性要素，而单元则包含其间不可明辨。这种关系之中，建筑被概念化为一个包裹或容器，单元则通常是空间或结构体量。

建筑整体或许有比单元集合更为丰富的建筑形态，即整体大于部分之和。这里，某些建筑形式作为母体，可以包含、连接，有时或许只是触碰单元。单元则可以是形式或是空间，可见或不可见。在此，图底关系中"实体"（poche，建筑图上涂黑表示的柱子、墙垣的区域——译者注）的概念很关键，它定义了内部空间与外部形态的区别。

249

重复到独特

关联重复与独特建筑要素的设计构形方法，需要建立多次与单独出现要素之间的联系；其基石是理解不同范畴或类属里"独特"的差别。这种差别兼容于重复与独特要素匹配所界定的范畴与类属框架。对于重复而言，独特，是共同要素基础上的差异与识别。例如，体量只能和体量相比较才能辨析出哪个特征使某个单元独特。体量如果与窗户、结构概念进行比较，属性上不具可比性，差异也无从辨识。在一栋建筑中，重复与独特可以在各种尺度、层级中出现。诚如单元到整体的关系那样，什么是主导性方法才是关键。

建筑学领域，重复和独特的要素一般都是三维的，所以，也可以通过平面、剖面表达。大多情况下，重复与独特通常出现在一个向度上，水平或垂直。然而，也有可能在平面上重复而在剖面上独特；反之亦然。伯鲁乃列斯基的圣玛丽亚布道所就是这样一个特性分开的例子。

重复单元通过尺寸、颜色、位置及朝向的改变获得独特性。形状、几何、结合方式的改变也可以塑成单元的独特性。形状还是几何变化，取决于两个图形之间形态差异的程度。如果独

特单元与重复单元有部分是相同的，那么是形态变化。例如，一个正方形的一条边被圆弧代替，这就是形状变化。如果独特单元与重复单元有不同的形式语言，那么就是几何变化。相对一组正方形，圆形就是独特的。工艺改变，是指形状相同做法不同的情形。一组不透明立方体中，透明立方块的工艺就非常独特。

独特单元的变体环绕独特单元布置，其中心感定义出新的独特。独特单元与其变体之间的区别，由独特单元自身界定。在此，独特单元的形态基于重复单元的排列方式。没有重复，独特不存，或者说，至少独特单元的形式会随重复单元排布的方式的变化而改变。

独特要素可以被重复要素包围。在此，独特要素位居中心且形式别致；重复要素环绕其布置。重复单元还可以直接界定中心图案的边缘；否则，重复单元的变化将不会改变被包围单独要素的形态。反之，独特也可以包围重复，独特和重复还可以叠加生成一个新的图案。究竟是重复叠加于独特抑或反之，将由尺度、构形、位置、结合等相对关系决定。总之，后面介入的要素具有主导性。

重复单元叠加可以生成新的独特要素，其叠合部分就是独特的。有时，一个建筑形体中的重复部分先确定下来，剩下的部分成为独特的部分，或许可以说，独特是建筑总体减去所有重复部分后的差。

单元彼此接近将会形成关联，此时，独特可以和重复分离。分离可以是实际的，有时候也可以是感觉上的，就像"单元与整体"里那样。独特单元可以位于一个区域，这个区域里，重复单元的尺度、形态以及之间的关系都是统一的，这使它们形成一个区域或网络，看起来像一个更大的单元。其间，独特单元的异质性破坏了网络或区域的整体感，因而显得格外突出。

位置也可以使单元成为独特体。基本原理可以阐释为，排成一列的单元里突出的那个就是独特。因此，位于图形的中心，路径的终点，行列中的偏移，都将凸显其独特。有时，带型的

两个端点也视作终点，通过中间的重复部分相互关联。

加法和减法

作为构形方法，加法和减法通过形体的增删进行建筑设计。这种构思的基础是，加法设计时，要感知局部的主导性；减法设计时，要感知整体的主导性。加法设计将建筑视作可认识单元的聚合；减法设计时将建筑视作局部被移除的整体。建筑物可以兼容这两种形式，局部增添上去还是局部被移除，判定的依据是哪一种感觉居主导。通常，这些方法主要关注体量，这在建筑形体构思中极为有效。然而，如同一切形体问题，空间的结果也将随之导出。加法和减法，作为形态控制方法，通常用于建筑整体构思，但也适用于其他尺度上（如建筑的局部或者某些房间）的决策。

加法和减法与其他的概念不同，它们是最常用的设计构思方法。建筑构思中两个方法交替使用，可以创造多种可选方案。正如前文所提及的，两种方法配合，设计潜力非常丰富。配合的方法有多种，例如，从一个明确的形体中减去一个局部，生成一个形体；进而再删除一个局部，生成新的形体。每一步形体研究的变化量，由比例感决定；不断地增减变化，形态构思变化无穷。

对称和均衡

对称和均衡是一种形体构思方法，它通过构筑组成建筑各部分之间感觉上的等量关系来设计建筑。建筑学中的对称和均衡是一种观念，认为建筑要素可以以某种可感知的特性定义为"等量"。几何变化中，对称和均衡是这种"等量"可感知的基本属性。对称与均衡都是在潜在的点或线两侧，创造一个相对稳定的状态。通常，均衡以直觉来组织各个元素。当其附加了价值和意义后，均衡则成为概念现象。

对称是一种特殊的均衡，其本质仍是感知。对称轴线两侧的单元完全相同，是对称与均衡的区别。最常见的对称形式是轴对称，即镜像和反射对称，建筑构件如同镜中的显影再现了其他构件。这类对称中，左右单元形态相同，而位置相对。而双轴或双侧对称，则包含两个不同向度的对称性。

对称的第二种形式是单体围绕同一个中心旋转。这种情形中有一个无形的中心，以其形成的对称模式全然不同于以轴线为基准的左右对称。中心点可以位于图形的内部、边缘或外面。如果旋转的中心点在图形之内，会形成一系列的重叠的图形。如果旋转中心在不对称的两个方向上，则形成风车式的对称。除了旋转中心的位置，图形旋转的次数和在旋转过程中的逐级变化也是其重要的变量。

可辨识的形态在特定方向上移动，形成移动式对称。通过多个等量单元的这种对称，可以发展出一种线性集合，这里，任意两个单元之间都具有移动式对称关联。它的形态不局限于直线，可以是一系列的自然曲线。还可以在设计中形成多个系列组合，就如同约翰·伍重所做的庭院式住宅中采用了两组对称关系的单元，它们分别具有各有不同方向。

一条线或一个点的两侧出现相同的单元体称为对称；当两侧单元的某个特性不同时，则为平衡。特性上的不同，将促成形状、朝向、位置、尺寸、搭配和虚实对比。形状上的平衡，是不同形态的图形，比如一个圆与一个方之间的平衡关系。

除了轴对称与中心对称之外，朝向变化的等量单元也可以在潜在的平衡线两侧取得平衡。单元尺寸及其与对称轴的相对位置也可以形成平衡，这与天平上的重量均衡的概念很相似。

单元大小不同，则依据与轴线的某种比率关系来形成均衡。这种均衡关系中，大小的差别可以通过强化较小单元的某个特性来调节，以此保证与轴线距离相等。这是一种特殊情况，当单元被赋予了足够的重要性，仿若一颗宝石时，就能与个头较大但不重要的组件形成平衡。例如，两个尺寸相异的单元，当较小单元采用特殊材料时，则仍可保持对称轴居中的均衡布置。

均衡还有二维和三维均衡的区别。平面或形体的视觉均衡，分别来自对面积或体块的处理。这种区别表现为立面图上是二维的，而建筑上则是三维的。在这种关系中，数量、形状和模式等问题的处理要考虑"开敞－封闭"、"少－多"和"简单－复杂"等一系列的特性。

最后，两个等量体也可以在虚实中产生均衡。由于包含了空间和体块间的等量关系，所以这是最具建筑学本质的应用。这样来看，实的塔与虚的庭院也可以形成均衡。

几何关系和网格

作为构形方法，几何关系利用平面与立体几何学来决策建筑形式。所有建筑中都存在某种特定的几何形式。但作为构形方法，它需要将形式分解为不同的层级，并作出相应决策。

这一方法的基本应用，是组合基本几何形的空间与形态，并决策建筑的整体形式。因此，一栋建筑可以是任何简洁、可描述、易识别的几何图形，如圆形、正方形、三角形、六角形、八角形等等。虽然，几何形不会在建筑的每一个细节里应用，但有必要使其发挥主导作用且易于感知。

建筑可以基于一个几何形生成，也可以由几个几何图形组合生成。一个圆可以与一个正方形叠加来生成建筑。由此类推，两个及以上的基本几何形都可以组合，但各自仍可作为完整几何形认知。这些图形未必真实存在，但可以暗示。组合关系中，两个几何形可以相互包含、毗邻、重叠。当一个置于另一个之内，里面的那个几何形可以是一个物体、房间、院子、界定区域，抑或一个被暗示存在的空间。

建筑中，几何形叠合的特例是正方形及其内切圆。一个或一系列的圆可以与长方形的一条边或一个角重合。这种重合可以生成各种形式，例如长方形长边中轴线两侧的内切圆。在长方形一角的圆圈，可以与两边相交或相切，还可以让圆心正在角点。

不同的几何形可以组合，相同的几何形也可以组合。例如，建筑可以是大小相同或不同的两个圆，三个三角形，或者两个六角形的组合。当同样两个大小的正方形相互组合，将出现一些非常有趣但又极为特殊的现象。

两个相同正方形边缘叠合时，生成2∶1的长方形；重合时，形成小于2∶1的长方形；分开时，则暗示出一个大于此比例的长方形。通常，重合生成的或分离暗示的空间会有诸如入口、建筑主厅之类的特殊目的。两个正方形重合后又以同一个中心旋转，则生成一个八角形。此外，还可以将正方形的一个角与另一个正方形的边相连。

可以成倍叠加或细分，是正方形组合的重要特点。这一特点使得正方形可以组合为一个更大的正方形。当四个正方形集合成一个2乘2的图形时，可以将其视为大的正方形的4等分，也可以是小正方形的4倍。类似的，9个正方形可以是3乘3的正方形组合；以此类推，正方形也可以是16甚至是25个正方形的集合。

一个九宫格中包含3种类型的正方形，各自具有不同特点。位于角部的4个正方形，每个都被两个正方形夹住。位于边上的另4个正方形，各自被3个正方形从3边围合。最后一个位于中心的正方形，被完全包围。正是这个被包围的正方形中心使得九宫格特征独具，令人印象深刻。在四方阵中这个中心是一个点，这种布局总是强调中心广场或中央空间。

九宫格中其他方格不变，去掉部分方格，会导致图形发生显著变化。例如，保留外围8个方格，则形成方环；保留角部和中心方格，则形成一个"X"；利用边上及中心的正方形，可以形成一个加号。去掉两个侧边相对的方形，则得到一个"H"。最后，去掉一个角部及其两边相邻的正方形，则得到一个阶梯。

建筑形式还可以由基本几何形的局部演变而来，简言之，有可能是圆形、方形或三角形的一半或其他的比例的片段。然而，建筑形式也可以是几种基本形组合出来的更为复杂的图形。尽管明显由几何形演变而来，但是建筑形式不能再以简单几何来描述了。另一种几何演变方式，是用一些点位暗示出更大的几何图形。例如，罗伯特·文丘里所做的老年人公寓中，建筑各角的基准线影射出一个大的三角形。

通过特定边长比例的变化，正方形可以推演出三种不同的长方形。这些长方形可由两个正方形组合而成，长宽比小于2：1。第一种是$\sqrt{2}$长方形，其长边是正方形的对角线旋转45°。另一种是1.5：1的长方形，是半个正方形与一个正方形相加。第三种是"黄金分割"比例长方形，其长边等于半个正方形的对角线。上述各种长方形，经常单独或组合使用，生成建筑整体或局部。

可以通过旋转、移位和重合等方式灵活运用各种几何图形来生成另一系列的形状。这些生成方式，均可以描述为潜在的运动，综合应用可以创造更复杂的形体，例如，旋转与重合结合使用。

旋转，是围绕一个中心移动一个或数个部分的过程。各个部分的移动中心点，可以重合也可以分离。旋转自然改变了各个部分的朝向。铰接是旋转得到的一种特殊形式，一个铰接点联系两个不同朝向的单体。在一些案例中，这个铰接点会是一个明确的建筑形象；有时则只是一种暗示。

位移是指各个部分位置的改变，但不同于旋转，其朝向不变。尽管垂直的位移最常见，但沿对角线的位移变化更丰富，因为一次位移将取得两个维度上的改变。移位也可以理解为两个部分的相对滑动。这种情况下，常常会在两个滑动部分之间引入第三个空间，用以调和两者之间的空隙。

重叠具有独特魅力，它是组合两种图形创造新图形的过程。两个简单图形叠合后，将产生一个重合区域，整体上会生成一个相对复杂的形状。基于重合的这一特性，共有面积的新图形可能与原有两个形状大不相同。

放射形、风车形和螺旋形等几何形，其共同特性是它们有一个作为原点的中心。放射形建筑，拥有一组从一个中心发散出的，特征相似的单体。这些射线状的要素可能与同心布置的其他形体互相交叉。螺旋形和风车形比放射形更具动感。螺旋形是从一个中心出发，顺着旋转方向均衡地变化的运动。风车形由几个带形分支组成，它们在一个共同内核处连接，或毗邻一个潜在的中心。构形的各个部分被精心布置，确保其中心线不会交于一点。这些放射状要素之间的间距相等，与其他要素以及中心点的关系相同。螺旋形是具有动感的风车形。

网格是基本几何形的重复。重复是几何形成倍地增殖、叠加、等分、反复的过程。概念上，网格是无限大的区域，其间所有单体的关系相同。网格也可以描述为一组平行线与另一或多组平行线的相交。

平行线的行间距可以重复，也可以变化。最简单的排列是所有行间距一致。如果存在多种行间距，就会形成多种系列的复杂变化。各组相交平行线各自行间距的变化，决定了网络的特征与风格。可以用"a"、"b"、"c"表示一个网格的行间距，如果间距"a"每隔四次就有规律地出现，这种规律就是"a, b, c, a, b, c, a, b, ……"，或者"a, b, b, a, c, c, a, b, ……"，以及"a, b, c, a, c, b, a, b, ……"。

网格另一特性是平行线之间的相互关系。两组平行线之间可能是正交或者斜交。如果是正交，且行间距一致，将形成一个正方网格。如果两组平行线行间距不同，但各组自身的行间距相同，且为正交，就形成长方网格。如果两组正交平行线各自有一个以上的行间距，它们组成一个长方的花网格。两组平行线如果不是正交，则组成平行四边形网格。三角形网格是由三组平行线相交于共同的交点而形成的。相交的平行线的组数，理论上是无限的，但实际数量显然不会太多。

网格之中，系统中两条线的交点是关键构造。但仅仅是交点不足以确切描述一个网络。例如，如果交接方式不同，一个正方格网络的节点也可以描述平行四边形或三角形网络。

线与交点的连接方法，是理解网络最为重要的方法。如前所述，两者在概念上必须存在且被界定，或者被暗示，即两点成一线。范围如果足以显示一个网格的模式，则可删去部分交点或线段，然后填入其他设计内容。交点和线的连接方式，可以凸显重点，或给予网络不同轻重的强调。类似于基本几何形，网格也可以通过旋转、移位和重合的方式灵活运用。

构形模式

作为形体构思方法，构形模式描述了各个部分的相对布局。模式是最基本的主题，它具有创造空间、组织空间或形体群组的潜力。这些基本模式是：中心式、线性式、聚集式、同心式、筑巢式、双中心式和双核式。

中心式可以分为两类：中心主导型与中心组织型，后者是用中心组织其他空间。中心如何处理，是各个案例间的主要区别。前者，人们走向或者环绕这个中心；后者，人们穿越中心。还有本书未收录的第三种可能，中心可能是一个实体，比如一个壁炉。

中心主导型里，中心点放置最重要的使用空间，形成焦点。如果中心是有顶的，最常见的模式是中间高四周低，可以是半球或者穹隆，或者果核形、金字塔形。这样，屋顶或者吊顶强化了中心的概念。中心主导型的首要特征是，这个建筑的体量与形态均由中心生发而来。其空间的主导性可以是功能抑或象征性的。有时，它是神圣的；有时，尽管不神圣，但绝非不重要。这种模式生成的形体可以是单独的，也可以是中心生发的。这些伸展出去的体量，由中心生发，可以生成很复杂的形体。每个伸展出去的中心都在强化中心，弱化自己。过分的伸展会弱化中心。这种构形方法的另一个挑战是，如何塑造入口却不损害中心的主导性。入口设置在中间是难以实现的理想模式，也可以在其外围等距离地连续设置多个开口。

中心主导型里，交通流线可以正对或者围绕中央空间。因此，中央也可能是一个室外空间，通常人们环绕着它，并不穿越。回廊是以虚空作为中心主导空间的范例，它所围合的室外空间可以是一个神坛，或是人们可绕行的多重天井。这一构思中，中央空间无需视觉上的冲击力。

中心型的另一模式，是以中心作为空间的组织者。该模式中，中央的空间可以是一处交通辅助空间，或是解决交通问题的集散枢纽。古典式圆穹建筑就是这类空间。外观上它很显要，形体上可以统一整个建筑；但功能上，只是一个使用空间，并不重要。如同中心主导型那样，其构形或许并不显山露水。它甚至可能是块空地，一处用于交通的庭院或天井。

上述各种构形方法都是由中心的概念衍生出来，而线性模式注重线与运动。它们包含最关键的内容是路径与方向。与中心模式相似，线性模式也可以归纳为两类。二者主要的区别在于使用空间和交通流线之间的关系。第一种模式我们称之为脊骨式，交通流线与使用空间分离；第二种是串联式，交通流线穿越使用空间，就像项链串联起珍珠。

所谓的"脊骨"是一个辅助空间，通往一组独立的空间或房间。通常，公共流线将关联松散的各个部分组织起来。建筑形体上，这个"脊骨"也可以起到主导作用，也可以隐藏其间。后者之中，"脊骨"仅仅是一个单面或双面的走廊。从属性上来讲，脊骨由它所服务的对象确定它的界限，没有等级之分，也没有既定长度。其他建筑要素，例如入口，也将影响"脊骨"的实际构形与空间体验。"脊骨"一般都是直的，但也可以弯曲，形成围合空间，集中视线，缩减视觉长度，抑或回应某些外部条件。一栋建筑之中，可能不止一条"脊骨"。这样，"脊骨"交叉的节点及其属性，将意味着特定的等级和特殊的区域。

线性模式的第二类，单个使用空间被纵向穿越，或者一系列空间被逐一穿越。这样，路径可以穿越或者串联空间。在空间串联模式中，空间之间的开口位置与方式将决定路径的形态与可识别性。倘若空间局部的放大没有喧宾夺主，且进一步强化了通道的带形特征，那将会使通道空间变化丰富。

这类线性空间的构型，使我们有机会探索序列演进的潜力。后续将讨论空间的演进模式，认知空间线性串联模式中的循序渐进至关重要。因此，有可能把重点放在这一序列中的任何一个空间。重点可以在通道的起点、中间、或终点上。

聚集式是大量形体与空间以无特征的方式聚集。无论是形体抑或空间单元，必须彼此靠近，但没有特定的关联规则。聚集式没有预设的规律，其关联的随机性可以容忍单元的不规则。空间可以在一个完整形态内聚集，其聚集方式影响或决定了三维形态。聚集的各种形体可以进行空间再分割，但这种分割并不重要。

同心式的构型就像石子投入水面，是具有同一个中心的一系列不同形状的单元。这种形式也可以看作一种单元相互嵌套的层叠。同心式的特性首先是看它有几个环；进而应当指出，虽然有相同的中心，但其形式语言未必相同。

筑巢模式与同心式有些类同。这两种模式都有单元的相互嵌套，但是筑巢模式中各个单元的中心不同。筑巢单元一般还有其他共有部分，比如共有的一条或数条边线，或者中心线。筑巢模式或同心模式都可以是形体或空间的，且包含潜在的层叠效应。

一个模式具有两个同等重要的中心，即可以称作双中心式。明确的边界或区域对理解双中心式至关重要。这个区域虚实均可。如果是虚的，这个区域可以是一间房、一个大的室内体量；抑或一个室外空间，比如一个庭院或一块地皮。

如果把建筑看作一个体块，那么这个区域就是实的。无论区域的虚实，双中心都在场地内相对存在。这样，如果区域是虚的，双中心由界定空间中的实体生成。如果区域是实的，双中心就是体量里的中空部分，剩余部分可看成是填充体。

双核式的基本特性是有两个等量的主体，这两部分形体组合成建筑的整体。两个形体之间有一条对称线或平衡线。这两个核心部分可以是相同的，但也可以因为形状、朝向、搭配或者状态等方面的变化而有所不同。两个核心形体之间可以用第三个空间连接，但这并不是本质问题。通常这个连接体是个次要或中性空间，它不包含在两个重要部分之内。虽然在偶然情况下它也可能是个重要的使用空间，或者是实的墙体。入口经常是在两个主要部分的中间，但也可以通过其中之一进入另一个空间。

演进

演进是类型学的主题之一，着眼于两个不同形态模式的逐级变化。逐级演进的概念是多层次而非二元的。因此，为了界定这种模式，通常需要两级以上变化。等级、转换、变形及中介等都是本书要讨论的一般演进。这些一般演进与广义演进的重要区别是，前者是限定性的演进。由于演进可以是无限的，但这四种演进是有限的，具有明确的起始边界。这些有限演进的系列中，增量的特性可以通过与相邻增量的变化来描述，而不是孤立的变化。也就是说，这些增量可以由它与边界的关联来描述。一个案例中的大要素，在另一个系列中可能是小要素。

等级体系是指各个部分某种属性的秩序评级。由于重要性评级的不同，不同部分的评级有差异。从神圣到世俗、从大到小、从图形到填充、从中心到边缘，从指挥到服从、从高到矮、从少到多，以及从外到内，无论其中之一或者多种并存，这都是在建筑中的常见等级体系。某些情形中，了解重要性之前需要确定属性。例如，大的未必比小的重要。从大到小和从小到大的排列，在建筑中都比比皆是。

一栋建筑之中，等级体系中的主导部分，常常通过不止一个类型的演进反复强调。例如，位于伊德富（Edfu）的何露斯神庙（Temple of Horus），供奉主神的空间就用了好几种等级体系来强调。这些建筑等级体系支撑着社会中的宗教与社会等级体系。神庙的等级基于神圣与世俗的等级变化，建筑中可通过大小、简繁、明暗、室内外、封闭与开敞进行表现。建筑中不同局部的开放性可以由面向世俗场地的入口来改变，越神圣的房间门越小，越封闭。通过台阶与坡道改变地面的高度，调整自然光的引入方式，甚至走向神圣的行为引导都充满变化。大多数神圣的场地，会以数道墙垣围合，使其与尘世隔绝，而且是神庙中最小、最暗、最封闭、最室内化的场地。这种内庭是专为少数信徒供奉主神用的，与之相对的是神庙其他区域中可见的次神。直接位于主入口后面的是大院子或者"众生之厅"，这些场地大，向天空开敞的空间，也是神庙中最不室内化的地方。

另一些建筑中，建筑师们常常以最丰富的装饰、最浓烈的斑斓色彩、最贵重的材料，以及最繁复的细部和纹理作为等级提升的手段。位置，在中心或者在轴线的末端，似乎都可以强化空间与形式的特殊性。通常，但凡可以用来凸显局部特殊或贵重的特性，都可以视为一种设计手段，强化某个建筑局部的重要性。

转换是有限度的演进，是特性转变而形式不变的变化。典型的转换包括开敞到封闭，室内到室外，简单到复杂，动态到静止，个体到集体，以及尺寸上的变化。与等级体系类似，转换是有界定的；但与之相反，转换的终端状态上没有意义，即简单未必比复杂重要，反之亦然。当终端状态相同，那其间的各个状态也将平权。阿尔多·凡·艾克（Aldo van Eyck）论述的"其内－其间"（in-between）和"孪生现象"（twin phenomena）对理解"转换"及其潜力颇有价值。转化过程包含了一系列的中间状态。转换极端状态之间的中间态，对两级的状态均有提示作用，它们将成为两边不同状态的连接物。

变形是演进中的一种，其对象的边界将发生变化。与转换相似，但其形状变化更为具体，并对二维和三维形态都有影响。一个多形象的参照系颇有必要，如此方能察觉形体间的变化。变形并非两个形体的比较，而是一系列的形体变化，而且各个形体的等级没有区别。

中介式与其他的一般性演进不同，其终端状态在中介建筑之外。建筑被视为不同文脉之间的桥梁或纽带。因此，建筑不能视为独立体，必须置于特定文脉之间考察。作为设计方法的中介式，应对建筑所处的文脉进行阐释或作出判定。通常，达成这个需要抽象思维。理查德·迈耶（Richard Meier）设计的，位于新哈莫尼（New Harmony）的"雅典娜"游客中心就以波浪形墙体抽象了一侧的河道；以正交几何体抽象了

另一侧的城市网络。更进一步，这样的方法至少可以对自然或建成环境中的两种状态分别予以抽象。这样，新建筑可以衔接两个不同的建成环境，或者两套自然系统，或者自然与建成环境。

在此方法中，建筑可被视作大片区中的小局部。通过中介方法，建筑物得以调和场所中既存的不同文脉。建筑中，可以通过一系列的手段造型，使其回应外部环境。当然，也可以在建筑局部造型上重现某种外部环境，或是将建筑作为两个外部环境的中点或过渡。

缩小

"缩小"是一种形体构思方法，是将建筑中的某个形态尺度缩小后再度重现。"缩小"有"整体包含局部"、"大小并置"两种方式。第一种方式中，整体，或曰大比例的整体，尺度缩减后用作建筑的局部。通常，这样的案例中，缩小的部分居于整体之中。另一种方式中，大的单元与至少一个微缩版结合成为建筑或建筑的局部。缩小的单元可以进一步重现或缩小。在此类型中，缩减的局部与大的局部通常毗邻设置，而非置于其间。两种情形之中，尤其整体包含局部的情形之中，缩减的过程可以伴随由实到虚的过程。一种尺度之中，形态是实体的体量，另一种尺度中，可能是虚空的空间。

"整体包含局部"的这类缩减中，其重要特性是见微知著。由于具备传递信息的能力，这类构形方法可以将感知转换为概念。这样，观察一个房间，一个院子或者建筑的一翼，就可以推导出建筑的全貌。这种概念的转换也可能存在于平面和剖面之间。平面或剖面可以在对方之中微缩重现。例如，一个房间或空间的剖面可能应对整个建筑的平面，就像矶崎新做的矢野住宅那样。

另一方面，"大小并置"的缩减模式中，从局部仅可以推知另一个局部，而非全貌。因此，这种类型是纯知觉性的。大量案例中，"大小并置"的"缩小"模式中，大小局部并置，

建筑的次要方面由缩小的局部表达。无论概念上或其他意义上的，这种现象的典型案例是那些辅助空间都较小的建筑。然而，典型的反例更为有趣，"小"也许才有力度，"小"才是一种强调。阿尔瓦·阿尔托（Alvar Aalto）的芬兰赛于奈察洛*（Saynatsalo）市政中心中，尺度较小的局部才是市议会，它才是"大小并置"模式中的主角。

---

*此为芬兰的一个地名，按《世界地名录》翻译。建筑界过去多音译为"珊纳特赛罗"。——编者注

FORMATIVE IDEAS

From the analysis of the 118 buildings in the first section, patterns in the design consideration of various architects were identified. Similarities in design approaches appeared among many of the architects' works, independent of time, style, location, function, or type of building. The similarities can be grouped into dominant themes or formative ideas which were conceivably used in the generation of the building designs.

A formative idea is understood to be a concept which a designer can use to influence or give form to a design. The ideas offer ways to organize decisions, to provide order, and to consciously generate form. By engaging one formative idea instead of another, a designer begins to determine the formal result and the manner in which it will differ from other configurations. The use of different ordering ideas may generate different results.

Presented in this section of the book is a series of connections among architects' designs organized by formative idea. Each concept is defined and explored through the presentation of generic manifestations of the idea. The written description is followed by a set of diagrams which exemplify some, but not all, of the generic alternatives. The inventory is not exhaustive: every idea is not explored, nor is every example included. Generally, diagrams developed in the analysis section are supplemented with other examples to illuminate a formative idea. Diagrams were selected which best illustrate the idea, show a variety of manifestations, and represent the widest range of building types from the broadest time frame.

247 **PLAN TO SECTION OR ELEVATION**

As a formative idea, the relationship of a plan to a section or an elevation entails design by using an identifiable correlation between the horizontal and vertical configurations of the building. Embodied in this is the linking of the two realms so that decisions in one arena determine or influence the form of the other.

The most direct connection between the plan and section occurs when they are the same—when the delineation of the two is equal. This can be described as a one to one relationship. A sphere, for example, is a figure in which the plan and section are represented by one circle. It is also possible to relate part of one configuration to the whole of the other. For example, a one to one-half relationship exists in a building that has a section or elevation equal in figure and dimension to one-half of the plan. In this case, a circle in plan becomes a half-circle in section, creating a hemisphere. The reciprocal condition is also possible, where the whole plan form of a building is the same as one-half of the section or elevation. In either case, the figures that appear in both plan and section are equal in dimension. In those circumstances in which the section is one-half the plan, a laterally symmetrical plan configuration can be achieved by utilizing the section form twice to create the whole plan. A special condition occurs when the same part of each figure overlaps, such as in the definition of the main space at La Rotonda.

A relationship of proportion by ratio can be used to link the plan with section or elevation. Distinct from the part to whole connection just described, the relationship of proportion establishes the plan and the section as the totality of the other, though different in scale. This relationship is predicated on more information in plan and section being paired than just the outline of each. Examples of ratios which are often used because of their compatibility with primary geometry are 1 : 2, 2 : 3, and 1 : 5. In each case, the plan and section have configurations that differ by dimension in one direction only.

In the case of a 1 : 2 relationship, the plan and the section have the same shape, but one is twice the other in one dimension. For example, a circle in plan would be an oval in section, with the height one-half the width. It is not necessary, though, for each of the parts in the plan to be reduced or increased at the same rate when they are utilized in the section or elevation. In Christ Church, by Nicholas Hawksmoor, for instance, while one element is reduced when it appears in the other realm, the other element is increased.

Plan and section or elevation can have a relationship identified specifically as analogous when the information from one is seen to resemble generally the shape of the other. This type of relationship between the plan and section is the most common, and often involves part of the plan and section rather than the entire plan or section form. Differences between the two may be due to a form language change, size or location shift, or irregular increments of change. In a form language change an orthogonal element in plan or section may be paired with a comparable curve form in the other realm. When size and location shifts occur, an element in the horizontal arena is larger or smaller, or in a slightly different location than in the vertical dimension. In increment change the plan or section information changes at one rate while the correlative information in the other changes in a similar way, but at a different rate.

An inverse relationship exists between the plan and section when the configuration of one is paralleled with some opposite condition in the other. For example, when the plan form has components which are large, or simple, or positive, or random, and that correspond to section elements that are small, or complex, or negative, or ordered, respectively, then an inverse relationship exists between the two.

Whereas the relationships of equal, part to whole, proportional, analogous, and inverse establish a link between plan and section in which decisions about one determine the configuration of the other, it is also possible to have a connection that is less deterministic and more influential in nature. In this type of relationship, decisions about the plan or section establish a range of possible configurations for the other.

A part to whole relationship can be created between the plan and the section. In this context, one configuration serves as the whole shape, which, by reduction, becomes a part in the other configuration. The whole is evident in this relationship in its entirety as a part in the other domain, but in reduced dimensions. An example of this form of relationship exists in the Yano House by Arata Isozaki. In this house, the whole configuration of the plan is repeated as part of the section.

Plan and section can also have a coincident relationship when significant points and limits in the plan form coincide with important points in the section. Essential is the alignment of the locations where major changes occur in both plan and section even though the actual configurations are quite varied. The Allegheny Courthouse by H. H. Richardson, which is analyzed in the first part of this book, exemplifies this relationship.

A final alternative to the plan and section relationship is that of common derivation or common origin. In this case, the plan and the section configurations are determined by separate derivation from a common origin. For example, in San Maria degli Angeli by Filippo Brunelleschi, which is also in the analysis section, the plan and section forms are both developed from two overlapped squares that are rotated 45 degrees to each other. In the plan, the two squares have a common center, while in the section, the corner of one square intersects the middle of a side of the other. Both plan and section derive from the same size squares, but the resulting configurations are quite different.

UNIT TO WHOLE

The unit to whole relationship is a formative idea which involves the concept of unit and the understanding that units

248

can be related to other units in specific ways to create built form. A unit is a major recognizable component of a building that generally has a scale that approaches, or is one level removed from, the scale of a whole building. Units can exist within a building at several scales. However, while a brick can be seen as a unit at the scale of a wall, it is less productive to view the brick as a unit at the scale of a building; otherwise, all brick buildings will have the same unit to whole relationship. Units, then, are normally spatial volumes, use-spaces, structural elements, massing blocks, or composites of these elements.

The most direct relationship between a unit and the whole occurs when the two are the same entity—when the unit is equal to the whole. This usually occurs in buildings which are designed as minimal monolithic forms. For example, Cheop's pyramid comprised enormous quantities of stone blocks and cladding pieces. Yet, the dominant perception of this building is that of an identifiable entity. At a greatly reduced level of importance this perception may be qualified to include the surface texture or pattern developed by the fine scale cladding units. Similarly, the glass, tight-skinned cladding on some modern buildings is secondary to the overall monolithic form.

The most prevalent form of unit to whole relationship is the aggregation of units to create the whole. To aggregate units is to put the units in proximity with each other such that some relationship is perceived to exist. The units may or may not be in physical contact with each other for a relationship to be identified. The alternative forms of creating a whole through the aggregation of units are characterized as adjoining, separate, and overlapped.

Adjoining is the most common form of aggregation. In this relationship the units are visible, perceived as entities, and relate to other units through face to face, face to edge, or edge to edge contact. Interlocking is one variation of face to face adjoining.

Units may be separate and at the same time related to other units to form a whole. Separation can occur through physical isolation or through the articulation of the connec-

tion between the units such that the units are perceived to be separate. Essential to this type of relationship is the perceived segregation of units and the proximity of the units so as to establish a compositional relationship.

Units may also aggregate to form a whole through overlap. Since architecture is a three-dimensional phenomenon, the overlap of units in the volumetric realm is by interpenetration. For this to happen the units are identified as entities that partially share form or space with other units. The portion of the overlap is seen as part of each unit and at the same time common to both.

Units can also be contained within a built whole. To distinguish this relationship from units adjoined to form a whole, the building as a whole is the dominant expression with the units contained and not expressed. Embodied in this relationship is the concept of a building as a wrapper or container for units which are usually spatial or structural volumes.

It is possible for a building whole to have more built form than that generated by the assemblage of the identified units. This relationship can be described as one in which the whole is greater than the sum of the parts. In this case, some of the built form serves as a matrix which holds, connects, or at times, just has contact with the units. The units may be formal or spatial, and visible or not. Important to this relationship is the concept of poche, which is the defined difference between interior volume and exterior configuration.

REPETITIVE TO UNIQUE

The formative idea of relating repetitive and unique elements entails the design of built form through the establishment of relationships between components which have multiple and singular manifestations. Fundamental to this idea is the understanding of unique to be a difference within a class or a kind. This distinction allows for the common reference frame of class or kind to couple the domain of the repetitive with the unique. The definition of the unique, in terms of the repetitive, permits the identification of the differences in attributes of common elements. For example, massing units are compared with massing units to determine the differentiating features which make one unit unique. If massing units were compared to windows or structure, the nature of the difference might never be discerned because of the disparity of characteristics to be compared. Repetitive and unique elements can occur at a number of varied scales and levels within a building. As with the unit to whole relationship, the concern is with the dominant manifestation of the idea.

In the realm of architecture, the repetitive and unique elements are usually three-dimensional, and, as such, can be communicated through the conventions of plan and section. In most cases, the repetitive and unique will appear in the same vertical or horizontal arena. However, it is possible for the repetitive elements to occur in plan and the unique element to occur in section, or conversely for the unique to appear in plan and the repetitive in section. San Maria degli Angeli by Brunelleschi is an example of this separation.

A unique element can be developed through the transformation of repetitive units through changes in size, color, location, and orientation. Shape, geometry, and articulation changes can also render an element unique. The distinction between a change in shape and one in geometry is determined by the degree of difference between the two figures. If the unique element is in part the same configuration as the repetitive, then a transformation by shape exists. For example, a square can be transformed into a figure that has three straight, equal length lines at right angles to each other and is closed by an arc of a circle. If the unique component is different in form language from the repetitive, then a transformation by geometry occurs. In this situation, a circle is unique to repeated squares. A change in articulation happens when the same form or configuration is made manifest in two ways. For example, a transparent cube is unique by articulation to a series of opaque cubes.

The unique component can be surrounded by the repetitive. In this case, the unique is central and has its own configuration. The repetitive elements are located around it. It is possible, but not necessary, for the repetitive elements to be coincident with the boundary of the unique. However, a change in the arrangement of the repetitive elements will not change the unique that is surrounded. The counterpart relationship where the unique surrounds the repetitive is also possible.

An alternative to the unique surrounded configuration occurs when the unique is defined by the arrangement of the repetitive. The distinction between this alternative and the unique surrounded model is determined by the manner in which the unique is established. In this case, the unique is dependent upon the configuration of the repetitive elements for its shape or form. The unique does not exist without the repetitive, or, at least, its form will change if the repetitive elements or their arrangement changes.

Unique and repetitive elements can be added together to create built form. The determination of whether repetitive is added to unique or unique is added to repetitive is made perceptually by consideration of relative scales, configuration, location, or some combination. Generally, that which is added to will appear to be dominant.

Unique elements can be formed as a result of overlapping repetitive units where the shared configuration is unique. In some cases, the unique component in a building is the remainder of the built form after the repetitive units have been defined. In this instance, the unique is the difference between the overall building configuration and the sum of the repetitive parts.

If units are in proximity to each other so that a relationship exists, then the unique element can be separate from the repetitive. The nature of the separation can be physical or perceptual, as it is in the unit to whole relationship. Unique elements may also be located within a field in which the repetitive elements have a scale, configuration, and uniformity of relationship that renders them a larger unit that can be identified as a field or network. In this relationship, the difference between the repetitive and the unique is heightened by the disruption of the field by the unique.

Location can establish an element as unique. Singular occurrence in a linear arrangement can be the basis of uniqueness. Therefore, a unit at the center, one which is a terminus to a path, or one that is shifted out of alignment, can be rendered unique. It is also possible in a linear configuration to view the ends as unique units connected by repetitive elements.

ADDITIVE AND SUBTRACTIVE

Additive and subtractive are formative ideas which entail the design of buildings through the aggregation or removal of built form. Basic to these related ideas is the understanding that an additive design has perceptually dominant parts and a subtractive scheme has a perceptually dominant whole. The image a person has of an additive design is that the building is an assemblage of identifiable units. A person engaging a subtractive design understands the building to be a recognizable totality from which parts are removed. Buildings may embody both images, but it is the dominant perception of parts added or parts subtracted from a whole which renders them additive or subtractive, respectively. Generally, these ideas have the greatest bearing on formal considerations of a building, with massing a particular concern. However, as with any formal issue, spatial consequences can result from decisions made in this realm. Although additive and subtractive, as formative ideas, operate at the scale of the building, it is possible to use these concepts to make design decisions at other scales, like parts of buildings and rooms.

Additive and subtractive differ from the other concepts presented in that they are the generic examples of the idea. Alternatives are possible when the ideas are used in conjunction with each other to determine a building design. As noted previously, the potential for design richness is enhanced by the use of the two concepts in consort. This normally occurs when the use of the alternative is sequenced in some manner. For example, the creation of a form by subtracting pieces from a recognized whole, and then after adding parts to form a new whole, subtracting again. The amount of imagery developed by any one step, the dominance of the perception, and the sequence of the processes allow a broad range of alternatives within this formative idea.

SYMMETRY AND BALANCE

Symmetry and balance are formative ideas which entail the design of buildings through the establishment of perceived and conceived equilibrium between components. Intrinsic to an understanding of balance and symmetry in architecture are the notions that elements can be identified as equivalent, and that the nature of the equivalency can be discerned. The generic alternatives for balance and symmetry exist in the nature of these equivalencies. Balance and symmetry both create a stable state relationship between components on either side of an implied line or point. Generally, balance is perceptually based and focuses on the composition of elements. It becomes a conceptual phenomenon when components are given added value and meaning.

Symmetry, as a specialized form of balance, is perceptual in nature. Symmetry differs from balance in that the same unit occurs on both sides of the line of symmetry. The most familiar form of symmetry is referred to as axial, reflected, or mirrored, because the components are oriented such that one unit appears to be reflected in a mirror to create a second unit. In this type of symmetry, the elements are equal in configuration and opposite in handedness. That which occurs on the left side of one element will be on the right side of the other. Biaxial or bilateral symmetry is reflected symmetry that occurs in two directions.

A second form of symmetry is developed through the rotation of components about a common center. Implied in this situation is the central point, which by definition establishes patterns that are different from those developed by symmetry about a line. The central point can be located within, at the edge of, or outside the figure. If the point of rotation is within the figure, a series of overlapping forms will be created. This type of symmetry might also result in pinwheel configurations if the center of rotation is asymmetrically located in both directions. Besides the location of the center of rotation, other important variables are the number of times the figure is rotated and the increments between the rotations.

Symmetry by translation occurs when elements with identical shape and orientation are shifted. This symmetry allows for the development of linear organizations through the aggregation of multiple, equal units, where the symmetrical relationship exists between any two components. Configurations are not limited to straight lines, and can be serial in nature. It is also possible to incorporate more than one sequence of translation into a design. For example, the atrium housing by Jørn Utzon utilizes two sets of symmetrically related units, each with a different orientation.

While symmetry is predicated on equal units occurring on each side of a line or point, balance exists when the units on each side are different in some identifiable way. Differences in attributes which can create a balanced situation between elements include geometry, orientation, location, size, configuration, and a positive-negative reversal. Balance by geometry results from the relationship of equivalent units that vary in form language. For example, one element could be circular and the other rectilinear.

Equal units that have an orientation difference other than those stipulated in reflected and rotational symmetry can be balanced about an implied line. Unit size and relative distance from the line of equilibrium determine balance by location, which closely parallels the concept of balance by weights on a scale.

Units that vary in size can be equidistant from the line of balance when balanced by ratio. In this relationship, the difference in size is balanced by an intensification or concentration of other attributes within the smaller unit, such that the line of balance is created midway between the two. This occurs when a special condition, given importance, like a jewel, balances a much larger, less significant component. For example, two dissimilar size units can be related to a balance line midway between them through the utilization of special materials on the smaller unit.

Balance can also be developed through configuration differences in two and three dimensions. Visual equilibrium on a surface or in a form is achieved by the manipulation of area or mass, respectively. This distinction applies to a building elevation which can be understood in two dimensions, and to architecture which is a three-dimensional phenomenon. In this relationship, the issues of number, shape, and pattern are engaged through consideration of ranges of attributes like open-closed, few-many, and simple-complex.

Finally, balance can occur when two equivalent components exist in positive and negative form. It is this type of balance that can utilize the very essence of architecture, for it embodies equilibrium between mass and space. In this context, the positive tower form balances the void of the courtyard.

GEOMETRY AND GRID

252

As a formative idea geometry entails the use of the tenets of both plane and solid geometry to determine built form. Geometry in one form or another exists in all buildings, but as a formative idea it must be knowingly central to decisions regarding form at several levels.

The most fundamental use of this idea incorporates the basic figures of geometry as form or space to determine the overall configuration of a building. Thus, a building might be a circle, a square, a triangle, a hexagon, an octagon, or any other singular describable and recognizable geometric form. While the geometric figure may not totally incorporate every piece of the building, it is necessary that the basic figure be dominant and perceptible.

Although architecture might be developed from one geometric figure, these forms can also be combined to generate a building; that is, a circle and square can be added together to create a building. Similarly, any two or more other basic forms might be combined, providing each is perceptible as a whole figure. The forms do not have to physically exist, but each must at least be implied. Within the realm of combinations, it is possible to locate one geometry that is within, contiguous to, or overlaps the other. When one geometry is located inside the other, the inner geometry might be an object, a room, a courtyard, a defined precinct, or an implied space.

A specialized form of geometric overlap prevalent in architecture is the combination of a rectangle and a smaller circle. A circle or a series of circular forms can overlap the rectangle at a side or corner. The overlap can result in a number of specific configurations, including the circle engaged on the centerline of the major side of the rectangle. A circle at the corner of the rectangle can overlap both sides, can have its center at the corner, or can be tangent to one of the sides.

As differing geometries are assembled, so too can similar geometries be combined. For example, buildings may consist of two circles, three triangles, or two hexagons of the same or different size. When square figures of the same size are combined in specific ways, some interesting and very particular phenomena occur.

Two identical squares combined with one congruent face create a rectangle with a 2:1 proportion. However, these same squares can be overlapped to make other rectangles smaller than 2:1, or separated to imply rectangles larger than this proportion. Normally, the space formed by the overlap or the space implied by the separation is used for special purposes, like entrances, or the main hall of building. Two squares can also be overlapped and rotated about a central point such that an eight-cornered figure is developed. It is also possible to unite two squares by attaching the corner of one to the face of the other.

Particular combinations of squares have the characteristic of being either multiples or equal subdivisions of a square. The distinguishing characteristic of these combinations is that they actually form another larger square. When four squares are assembled into a two-square by two-square configuration, the result is a figure that can be viewed as a four-part subdivision of the larger square or as a multiple of the smaller square. Similarly, nine squares can be assembled into a three-square by three-square configuration. By extension, squares can be assembled into 16-square and 25-square constructs.

In a nine-square configuration there are three types of squares, each with its own characteristics. Four of the squares are located on the corners and are bounded by two other squares. Four others are located on the sides and are bounded on three sides by other squares. The final square is located in the center and is completely surrounded. This bounded center square makes the nine-square format an identifiable and unique configuration. Whereas this arrangement emphasizes a central square or space, the four-square format articulates a central point.

Identifiable variations within the nine-square configuration are possible by removing certain squares, while maintaining others in their normal location. Thus, by using only the eight squares on the perimeter, a square ring is created. An "X" form is possible by using only the corner and center squares. By utilizing the middle square on each side and the square in the center, a "plus" configuration is made. Leaving out two side squares opposite each other results in an "H" shape. Finally, a stepped configuration is possible by removing one corner and the two contiguous side squares.

Forms can also be derived by using parts of the basic geometric shapes. In the simplest terms, this might be one-half or some other fraction of a circle, square, or triangle. However, more complex configurations are possible through combinations of forms derived from several geometric shapes. Though clearly derived from geometry, these configurations are not describable in simple geometric terms. Another geometric derivation is the implying of a larger geometric shape by points located within the architectural configuration. For example, at the Guild House by Robert Venturi, the corners of the building align to project a large triangle.

Certain derivations from a square result in three different rectangles with sides of particular proportions. The proportions are all less than the 2:1 proportion that results from combining two squares. The first, the square-root-of-two rectangle, is derived from the 45 degree rotation of the diagonal of a square, to form the long side. A 1.5:1 rectangle can be formed by adding one-half of a square to a square. The third, the golden-section rectangle, is derived from the rotation of the diagonal of one-half the square to form the major side of the figure. The center of rotation in this case is the midpoint of one of the sides of the square. Each of these rectangles, used either alone or in combinations, is frequently utilized to form buildings or parts of buildings.

Another series of configurations can be developed through the manipulation of geometries by rotation, shift, and overlap. These manipulations, all described by a process of implied movement, can be used in combinations to create more complex forms: for instance, rotation used in conjunction with overlapping.

Rotation is the conceptual process of moving a part or parts about a center. This center of rotation may be, but is not necessarily, the same for all the parts. Rotational movement naturally changes the orientation of the part involved. A particular configuration that results from rotation is the hinge in which two linear and connected elements are normally oriented in different directions. In some examples, the pin of the hinge or connector actually appears as a figure in the building; in other cases, it is implied.

When the manipulation by shifting occurs, the parts move, but unlike rotation, the orientation of the parts remains the same. While the shifting is often orthogonal in nature, a diagonal shift can create added richness by resulting in change in two directions through movement in one. Shifting might also be understood as sliding of two parts against one another. When this occurs, a third space or form is usually introduced between the shifting parts to neutralize the fissuring.

Overlap has the unique characteristic of creating a third figure from the combining of two other figures. The overlap of relatively simple shapes can result in a common space, as well as a total configuration, that is quite complex. Depending upon the nature of the overlap, the figure of the common area might be quite different from either of the overlapping figures.

The geometric configurations of radial, pinwheel, and spiral share the common attribute of originating from a center. Buildings that can be considered radial have dominant multiple elements that extend from a center. These raylike elements may be intersected with other elements that are in a concentric arrangement. Both spiral and pinwheel configurations are more dynamic than radial. Spirals move away from a center at a constant rate of change and in a rotational direction. Pinwheels consist of offset linear elements that are connected to a common core or abut to form an implied core. The parts of this configuration are positioned so that the centerlines of the elements do not intersect at a common center. These elements do, however, occur radially at regular intervals, and have similar relationships to the core and to each other. Spinning is the implied dynamic of a pinwheel configuration.

Grids are developed from the repetition of the basic geometries. Multiplication, combination, subdivision, and manipulation are the processes used to create the repetitions. Conceptually, grids are infinite fields in which all units relate equally to all other units. A grid can be described as a series of parallel lines that intersects at least one other series of parallel lines.

The intervals between lines can repeat or vary. In the series' simplest form, all intervals would be equal. The complexity of the series can be altered by increasing the number of intervals that occur within it. The frequency with which a particular interval occurs, and its relationship to another interval and its frequency, will determine whether a discernible patterns exists, and the nature of that pattern. Thus, if "a," "b," and "c" represent intervals on a grid, and if "a" is to occur at the frequency of every fourth interval, then the pattern might be "a, b, c, a, b, c, a, b, . . ."; but it might also be "a, b, b, a, c, c, a, b, . . ." or "a, b, c, a, c, b, a, b, . . ."

Another aspect of grid is the relationship between one series and another. Two series might or might not be orthogonal to each other. If the relationship is orthogonal, with all intervals in both series equal, a square grid results. A regular rectangular grid occurs when two series, each with a different interval, are orthogonal, and the intervals within each series are equal. Two orthogonal series, each with more than one equal interval, create a rectangular, plaid grid. Two nonorthogonal series of lines constitute a parallelogram grid. A triangular grid is formed by three intersecting series of lines which have common points of intersection. The number of series of lines which might exist coincidentally is conceptually infinite, but practically, the number is significantly lower.

Within the grid, a critical construct is the intersection created by any two lines in the series. However, intersections alone do not provide enough information to describe a grid accurately. For instance, a series of intersections arranged in what is apparently a square grid configuration also can describe a parallelogram or triangular grid if the intersections are connected differently.

Important to the total understanding of a grid is the method of articulating both the line and the intersection. As discussed, both must exist conceptually and be defined, but either may be implied by the existence of the other; that is, at least two points or intersections must exist in order to imply a line. If enough of the field exists so that an expected pattern can be perceived, then it is also possible for an intersection or part of a line in the grid to be removed. Expectations, then, complete or fill in the implied piece. Articulation of the lines and intersections can establish importance or give major and minor emphasis to the grid. Like the basic geometric figures, grids can be combined or manipulated through the processes of rotation, shift, and overlap.

CONFIGURATION PATTERNS

As a formative idea, patterns of configuration describe the relative disposition of parts. The patterns are essentially themes that have the potential for making space and organizing groups of spaces and forms. The terms that describe these basic patterns are: central, linear, cluster, concentric, nested, double-centered, and binuclear.

Central configuration patterns can be classified as those that are central-dominant and those in which the central space is used to organize other spaces. How the center is engaged is the primary difference in each of these cases. In the first, one goes to or around the center while in the second, one goes through the center. A third model of central configuration, but one that is not included in this study, is that of a central solid, such as a fireplace.

In the central-dominant model, the center is the focus with the most important use-space located in that position. If this space is covered, it is very often done so by forms that are higher in the center than at the edges—a hemisphere or dome, a cone, or a pyramid. Thus, the idea of center is reinforced by the roof or ceiling configuration. A primary characteristic of central-dominant space is that the center appears to generate the entirety of volume and form. This space can be functionally or symbolically dominant. In some cases it is considered sacred; in others it is less sacred, but no less important. The configuration of this pattern may suggest a singular volume or a spatial composition that extends from the center. These volumetric extensions, which might create complex patterns, emanate from the center. Each successive volume reinforces the center, but lessens its own importance. Excessive extensions will at some point diminish the importance of the center itself. A fundamental difficulty in this configuration is maintaining the center focus or dominance while introducing entrance. Ideally, though it is not usually feasible, the entrance should be introduced at the center or through a continuous series of openings equally spaced around the perimeter.

Circulation within the central-dominant configuration is either to or around the center space. Therefore, the central space can be an outdoor space that one walks around, but generally not through. A cloister, in which the outdoor space is a sanctuary, or a multistoried atrium that one walks around, might be examples of voids that are central-dominant. Within this idea, the central space does not necessarily have to have external visual impact.

The other model of central configuration employs the center as an organizer of spaces. In this case, the center space can be considered a servant space that is used for circulation and as a clearinghouse that resolves circulation problems. The classic rotunda is an example of such a space. It may have great significance externally, and formally may unify the building, but functionally it is not important as a use-space. This configuration, like the central-dominant organization, does not necessarily have to be expressed externally. It can be a void, such as a courtyard or atrium, that is used for circulation.

Whereas the previous configurations developed from the concept of center, linear configuration patterns focus on line and movement. They entail the critical issues of path and direction. As with central configuration, linear patterns are classified into two types. The primary distinction is identified by the relationship of use-space and how one engages it through circulation. In the first model, the circulation is separate from the use-space, and can be referred to as a spine. In the second type, circulation is through the use-space and the spaces are linked, much as the chain of a necklace links beads by passing through them.

The spine is a servant space that provides access to a series of independent parts or rooms. Often, the common circulation route allows parts that have no direct relationship to each other to be grouped. The spine may be dominant in the form of the building, or it may be hidden within. In the latter case, the spine is reduced to a single- or double-loaded corridor. Symmetrical or asymmetrical arrangements of parts is possible along the spine. By nature, a spine is not hierarchical, nor is it of a given length, but what it serves may begin to determine its limits. Other architectural issues, like entrance, also influence the actual spine configuration and the way it is experienced. Normally, spines are assumed to be straight, but they can be bent to create enclosed space, to focus view, to reduce its apparent length, or to respond to some exterior situation. Within a building there also may be more than one spine. In these instances, spines that cross and the nature in which they cross might suggest hierarchy or special areas.

A use-space that is traversed longitudinally, or a series of spaces that are linked to suggest movement from one to another, describes that second type of linear configuration. Thus, a path is either through the space or from space to space. In the space to space circumstance, the pattern of the location of openings between spaces will determine the configuration and the legibility of that path. Volumetric extensions may enrich the path if the extensions are rendered secondary to the primary space and are located in a manner that reinforces the linear quality of the space.

In this type of linear configuration exists the opportunity to exploit the potentials of serial progressions. While progressions themselves are discussed later, it is important to realize that space to space linear configurations are normally engaged sequentially. Therefore, it is possible to place importance on any space in the sequence. Accent can be at the beginning of, along, at the center of, or at the end of the path.

Cluster organizations refer to groupings of spaces or forms in which there is no discernible pattern. The units, whether forms or spaces, need to be in proximity to one another, yet the relationship between these units is irregular. While not a prerequisite for clustering, the random character of the relationships may permit the units to be irregular. Spaces can cluster within an overall form and in a way that influences or determines three-dimensional forms. Forms that cluster may have spatial subdivisions that are not important or dominant within them.

The concentric configuration pattern is analogous to the pattern created by dropping a pebble into water. The pattern is concentric when a series of units of differing sizes have the same center. This configuration can also be viewed as layering in which one element is viewed in the context of another. A characteristic of concentric organizations is that several rings are necessary to begin to see the pattern. However, it is important to note that the rings, though they share a common center, may not be of the same form language.

Nested configuration patterns share certain characteristics with concentric patterns. While both patterns have units inside one another, in nested patterns the center of each unit is different. Nested units can have other parts, such as one or more sides or a centerline, in common. Both nested and concentric patterns can be created at the formal or spatial level, and both imply layering.

A configuration pattern with two equally important foci is called double-centered. Prominent to the understanding of double-center is the idea of a precinct or field that has definite boundaries. The precinct can be either solid or void. If a void, the field can be a room, a large interior volume, or an outside space, like a court or a discernible area.

If the building is considered a mass, then the precinct is a solid. In either case of precinct as void or as solid, the double-centers are rendered opposites within the field. Thus, if the precinct is void, the double-centers refer to objects within a defined space. If the precinct is considered solid, the double-centers are spaces that are hollowed from the mass, and the remainder is considered poche.

Binuclear configuration patterns have the primary attribute of two equally dominant parts, which, as forms, comprise the general building configuration. The two forms establish a line of symmetry or balance. While the nuclear parts may be the same, they also may be different through changes of geometry, orientation, configuration, or state. A third form may create a link between the nuclear forms, but it is not essential. Normally, this connector is a secondary or neutral space which is exclusive of both dominant parts. On occasion, though, it can be a major use-space or a solid in the form of a wall. The dominant parts are often engaged by entering between them, or by entering into one and then proceeding to the other in a linear fashion.

PROGRESSIONS

The archetypal themes that comprise the formative idea of progressions focus on patterns of incremental change that occur between one condition and another. Progressions embrace ideas of multiplicity, rather than duality. Therefore, to discern a pattern, more than two increments of change are normally necessary. Hierarchy, transition, transformation, and mediation are the generic progression types discussed in this study. An important distinction between these generics and the overall progression category is that the generics are bounded subsets of progressions. Whereas progressions can be infinite, the four generic examples are finite, with definite beginnings and ends. In these bounded sets, the characteristics of the increment are describable in relation to the next increment, rather than as an isolate. Similarly, the increment can be understood in relation to the boundaries. Something large in one instance, for example, is actually small in another.

Hierarchy refers to the rank ordering of parts relative to a common attribute. This ranking differentiates among the parts by assigning importance. Sacred to profane, large to small, figure to poche, center to edge, servant to served, tall to short, few to many, and inclusive to exclusive are some of the hierarchies often found, either alone or in any number of combinations, in architecture. In some instances, it is necessary to determine more about the attribute before knowing the importance. Large, for instance, is not necessarily more important than small. Rank orderings from large to small and from small to large are both evident in buildings.

The dominance of hierarchy within a building is often reinforced through the layering of more than one progression type. The Temple of Horus at Edfu, for example, employs several architectural hierarchies to reinforce the importance of the room for the main god. These architectural hierarchies support the religious and social hierarchical beliefs of the society. The Temple's hierarchies are based upon the importance of the sacred to the profane, and are architecturally rendered as small to large, one to many, dark to light, rooms to areas, and closed to open. The openings between the various precincts of the building change with gates in the profane areas and with the openings of increasingly smaller size that are closed by doors at the more sacred rooms. Changes in floor height through steps and sloping floors, even though slight in nature, also signal the movement to the sacred. The most holy space, which is protected and separated from the outside world by a series of walls, then, is the smallest, darkest, most enclosed, and roomlike precinct in the Temple. This sanctuary is for a few worshippers and the main god, as opposed to the many lesser gods found in other areas of the Temple. Immediately behind the large entrance gate is the great court or "hall for the masses." This precinct is large, open to the sky, and the least roomlike area in the Temple.

In other buildings, evidence indicates that the most important increment in a hierarchy is often rendered by architects with the most ornament, the most intense polychromy, the most precious materials, or the highest level of detail and texture. Location, as in the center, or at the end of an axis,

might also reinforce the specialness of a space or form. In general, those qualities which make something special or precious in relation to others suggest the devices which are available to create importance in a piece of architecture.

Transitions are bounded progressions in which change takes place in an attribute without a change in form. A change from open to closed, inside to outside, simple to complex, movement to rest, individual to collective, and one size to another are typical transitions. As with hierarchy, transitions have definite limits, but as opposed to hierarchy, there is no value placed on the end condition of the limits; that is, simple is not seen as being more important than complex, or vice-versa. While the end states are seen as equal, the individual conditions between those ends must also be equal. Aldo van Eyck's discussions of the "inbetween" and "twin phenomena" are of value in understanding transition and its potentials. Within a transition there is necessarily a series of intermediate steps. Each of the increments between the extreme conditions of the transition will suggest what is on either side, and thus will form a link for the conditions on either side.

Transformation is a progression in which changes in form take place within the boundary of the object itself. It is similar to transition, but more specific in that the attribute being changed is the configuration. This configuration change may have impact on either the two or three dimensional form. A reference frame of multiple images is necessary so the change from one form to another is perceptible. Transformation is not, then, a comparison between two forms, but a series of form changes, with each form in the series hierarchically undifferentiated.

Mediation is distinct from the other generic progressions in that the end states are conditions which exist outside the building itself. The building is viewed as a bridge, or a piece of connective tissue, between conditions that exist in the context. Thus, the building cannot be considered autonomous, but must be seen in relation to its context. In order to utilize mediation as a formative idea, a position is taken or a statement is made about the context in which a building is to exist. Generally, this is achieved by a certain amount of abstraction. For example, Richard Meier in the Atheneum at New Harmony abstracts the river on one side as a wavy wall and the grid of the town on the other side as orthogonal geometry. Preferably, such a position entails at least two conditions which might be in either the natural or the built context. Thus, the new building might mediate between two built situations, between two circumstances in the natural environment, or between a built condition and a natural one.

Within this idea, the building is seen as a fragment of a larger piece. Through mediation the building reconciles differences that exist in the context. In the building, a series of gestures might be made which modulate the form to reflect the external conditions. Alternately, one condition can be repeated in some form in part of the building and then altered to be more like the other external condition. Another possibility is that the building is a midpoint or series of intermediates between the two external circumstances.

259 **REDUCTION**

Reduction is a formative idea in which a configuration is repeated at a lesser size within the building. This miniaturization can occur in two ways: part of the whole, and large to small. In the first type, the whole, or a large portion of the whole, is reduced in size, and utilized as a part. Normally, in this case, the reduced piece is located within the whole. Alternately, a large unit and at least one reduction of that unit are combined to form a building or part of a building. The reduced unit may be repeated or reduced further. In this type, the reduced piece is usually located next to, rather than within, the larger unit. In either case, but particularly in the part of the whole type, the reduction may involve a positive to negative state change. At one size, for instance, the configuration might be a solid or mass while at the other size the configuration might be a void or space.

A unique quality of the part of the whole type of reduction is that an observer can learn about the whole by encountering a part. With this capacity to inform the observer, this type transcends the perceptual to the conceptual. Thus, by observing the configuration of a room, a court, or a wing of a building, it is possible to infer the configuration of the entire building. The conceptual transference of information may also take place between the plan and the section. In this case, the whole of the plan or section may be repeated in miniature form in the other position. For example, the section of a space or room may correspond to the configuration of the plan of the entire building, as in the Yano House by Isozaki.

On the other hand, in large to small type reduction, comprehending one part may inform about only another part, and not the whole. Therefore, this type remains purely perceptual. In many cases, large to small reductions are incorporated into buildings with major and minor parts so that less important aspects of the building occur in the reduced piece. Typical examples of this are the several buildings in which servant spaces, literal and otherwise, are the small parts. An interesting reversal to this more typical interpretation, though, is that small might mean intense, and thus more important. Alvar Aalto's Town Hall at Saynatsalo is an example where the small piece, which is the town meeting space, is the more important in the large to small reduction.

260 **平面到剖面或立面**

　　平面、剖面和立面，都是表现所有建筑的水平方向和垂直方向所惯用的。确定其中的一个方面会决定或影响另一方面的形体。以下举例说明它们之间相同的、一对二分之一的、比例关系的、相反的或相似的各种关系。

相　　同

　　平面和剖面或立面一样时，是它们最直接的一种关系。这种关系的最简单的形式是仅仅只包括整体建筑形状。在阿斯普隆德所做的斯内尔曼住宅（1）中，主楼的长方形变成立面的图形，屋顶除外。与此类似，在老圣器堂（5）中整个平面的长方形在立面的主要体块上重复出现。理查德·迈耶在史密斯住宅（2）中对平面和剖面都采用了 1.4 长方形。它的一个小外屋是个立方体，在平面和剖面中都以同样的方式与主楼相连。在万神庙（3）中，平面中主要空间的圆形决定了同一空间的室内形状。这一空间的穹顶是一个半球形，它的冠顶的高度与平面圆形的直径相等。这个空间可能是实际能达到的最接近于球形的形体。由罗伯特·文丘里所做的塔克住宅（4）把屋顶除外就是一个立方体。勒·柯布西耶所做的斯坦别墅（6）的平面和立面外形是相同的，不仅在整个形体上相同，而且在它们的花格式网格分割方面也是相同的。

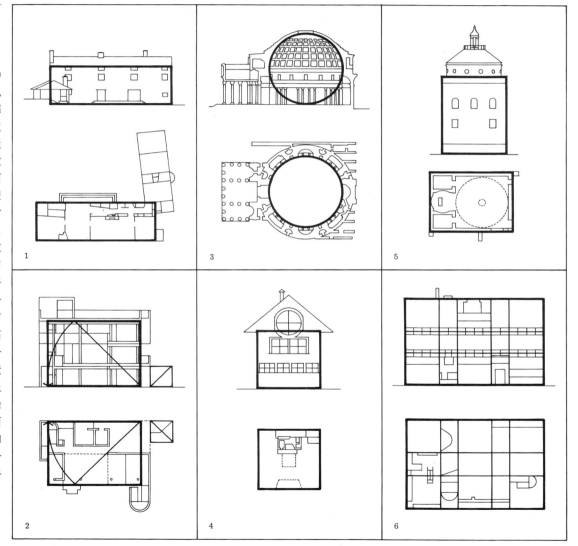

1. **STOCKHOLM EXHIBITION HALL**
   LE CORBUSIER
   1962
2. **NAKAYAMA HOUSE**
   ARATA ISOZAKI
   1964

3. **CHAPEL AT RONCHAMP**
   LE CORBUSIER
   1950–1955
4. **YALE HOCKEY RINK**
   EERO SAARINEN
   1956–1958

5. **RUSAKOV CLUB**
   KONSTANTIN MELNIKOV
   1927
6. **ST. JOHN'S ABBEY**
   MARCEL BREUER
   1953–1961

7. **SAN GIORGIO MAGGIORE**
   ANDREA PALLADIO
   1560–1580
8. **LA ROTONDA**
   ANDREA PALLADIO
   1566–1571

一对二分之一

整个平面或剖面的形状可能与另一方的一部分相同，如由柯布西耶所做的斯德哥尔摩展览馆（1），剖面的墙与平面的一半相同，在矶崎新所做的中山住宅（2）中，大的、主要部位的正方形和较小的正方形天窗组成立面的重要部分，它又重现在平面的一部分中。高地圣母教堂（即朗香教堂）（3）平面的一半基本上又变成立面，平面中的厚墙与屋顶相当。沙里宁在耶鲁大学冰球馆（4）中采用与屋顶中间拱肋一模一样的弧形做出平面两个边的外形。在梅尔尼科夫所做的鲁萨科夫俱乐部（5）和布劳耶的圣约翰修道院（6）中，平面的二分之一与剖面的总体外形近似。在帕拉第奥的圣乔治亚·马焦雷教堂（7）中，主要顶棚的形状等于这一空间的平面形状的一半。在圆厅别墅（8）中，平面的一半与主要立面外形相似。

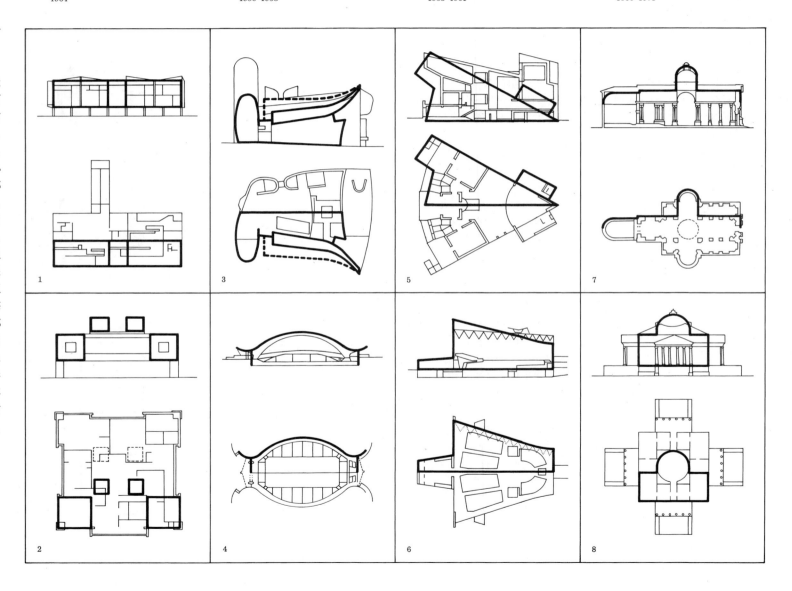

1. **FLOREY BUILDING**
   JAMES STIRLING
   1966
2. **ADULT LEARNING LABORATORY**
   ROMALDO GIURGOLA
   1972
3. **CAMBRIDGE HISTORY FACULTY**
   JAMES STIRLING
   1964
4. **THE PALACE OF ASSEMBLY**
   LE CORBUSIER
   1953–1963
5. **TEMPLE OF THE SCOTTISH RITE**
   JOHN RUSSELL POPE
   1910
6. **POPLAR FOREST**
   THOMAS JEFFERSON
   c. 1806
7. **THE FORD FOUNDATION BUILDING**
   ROCHE-DINKELOO
   1963–1968
8. **EXETER LIBRARY**
   LOUIS I. KAHN
   1967–1972

262 相　似

当平面与剖面中一方的形状大致和另一方的相似时，它们之间是相似关系。这种关系中形体语汇、尺寸、位置的差别，或不规则的逐级变化等都说明这只是相似关系而不是相同关系。弗洛雷大楼（1）和成人学习研究实验室（2）都有 U 形的平面和剖面。在苏格兰教堂（5）、杨树林住宅（6）、萨吕泰兴府邸（9）和沙利文做的国家农民银行（16）中，平面和立面尺寸大小不同。在海因斯住宅（13）中，两个方向都出现尺寸差别，在福特基金会大楼（7）、流水别墅（14）、沃尔夫斯堡文化中心（15），恩索－古特蔡特公司总部（17）和法国贝桑松剧院（18）中，平面和剖面的区别是由于出现了逐级变化的形式。在埃克塞特图书馆（8）、塞弗楼（10）和里丹托教堂（11）中，平面和剖面的区别在于形体语汇。圣克莱门·达内斯教堂（12）中位置的移动使平面与剖面略有差别。在议会大厦（4）中平面和剖面在形体语汇和尺寸上都有变化。形体语言和逐级变化使剑桥大学历史系图书馆（3）的平面和立面只是相似而不是相同。

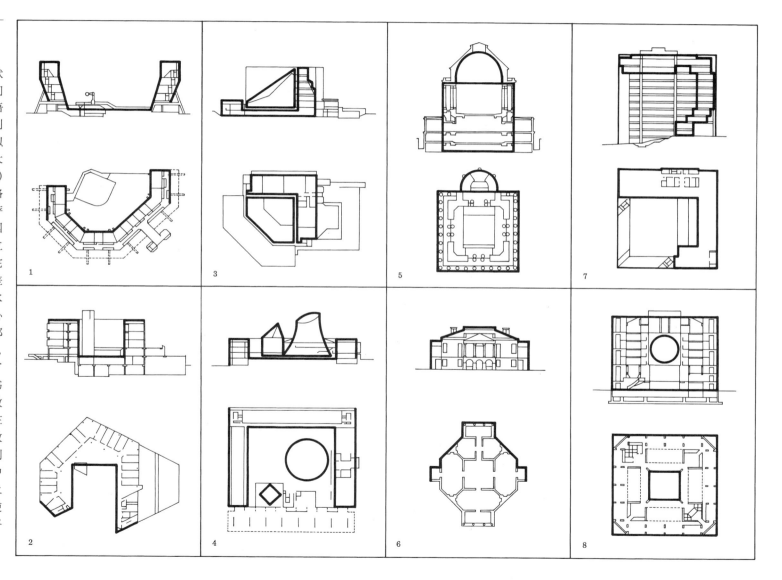

9. **THE SALUTATION**
EDWIN LUTYENS
1911

10. **SEVER HALL**
HENRY HOBSON RICHARDSON
1878–1880

11. **REDENTORE CHURCH**
ANDREA PALLADIO
1576–1591

12. **ST. CLEMENT DANES**
CHRISTOPHER WREN
1680

13. **HINES HOUSE**
CHARLES MOORE
1967

14. **FALLINGWATER**
FRANK LLOYD WRIGHT
1935

15. **WOLFSBURG CULTURAL CENTER**
ALVAR AALTO
1958–1962

16. **NATIONAL FARMERS' BANK**
LOUIS SULLIVAN
1907–1908

17. **ENSO-GUTZEIT HEADQUARTERS**
ALVAR AALTO
1959–1962

18. **THEATER IN BESANÇON FRANCE**
CLAUDE NICHOLAS LEDOUX
1775

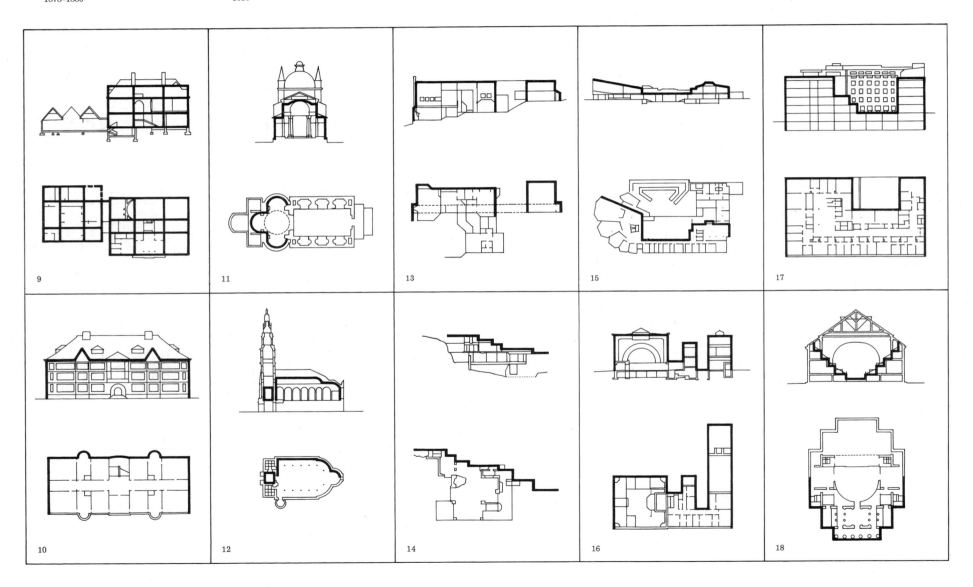

在平面和剖面的比例关系中，平面和剖面或立面中的双方都是以完整的整体相对应，但是在一个方向上的尺度改变了。这两方面的联系不仅仅涉及平面和剖面的外轮廓线形状。大多数的例子中，剖面形状一律都小于平面，但是公寓大楼（5）和卡代纳佐的住宅（10）却是例外。在卡森·皮里与斯科特百货大楼（11）的平面中各部分间的逐级变化在剖面上缩小了，但逐级变化的次数却增加了。在基督教堂（7）中，平面和剖面的比例变化是相反的现象。它的内层剖面形状在平面上扩大

了，而外层剖面形状在平面上却缩小了。胡纳尔别墅（13）的不同部分的平面和立面间的变化比率不同。布兰特住宅（14）和利斯特县法院（15）在平面或剖面上形体语汇都有些修改。范斯沃斯住宅（1）、蒙莫朗西旅馆（2）、萨伏伊别墅（3）、辛克尔做的柏林的住宅（4）和查罗夫住宅（6）的平、剖面比例关系中都是剖面小于平面，但只是内部形状的某些部分如此，而不是所有的都如此。类似这种例子还有圣玛丽·伍尔诺思教堂（8）、兰氏音乐中心（9）和萨尔克生物研究所（12）。

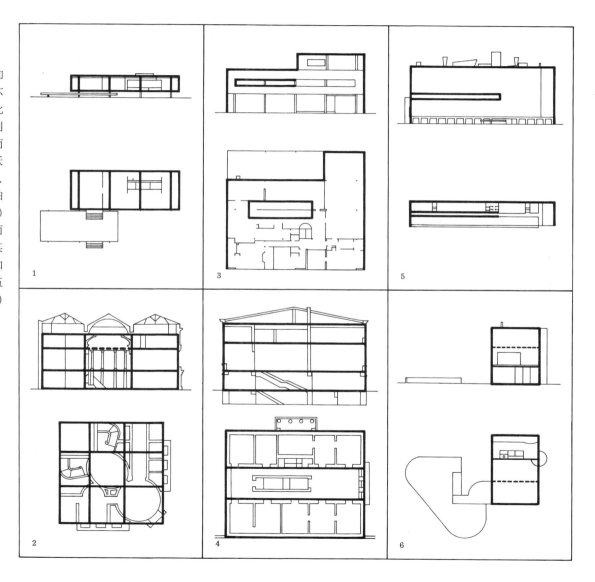

1. **FARNSWORTH HOUSE**
   LUDWIG MIES VAN DER ROHE
   1945–1950
2. **HOTEL DE MONTMORENCY**
   CLAUDE NICHOLAS LEDOUX
   1769
3. **VILLA SAVOYE**
   LE CORBUSIER
   1928–1931
4. **RESIDENCE IN BERLIN**
   KARL FRIEDRICH SCHINKEL
   1823
5. **UNITE D'HABITATION**
   LE CORBUSIER
   1946–1952
6. **CHAROF RESIDENCE**
   GWATHMEY-SIEGEL
   1974–1976

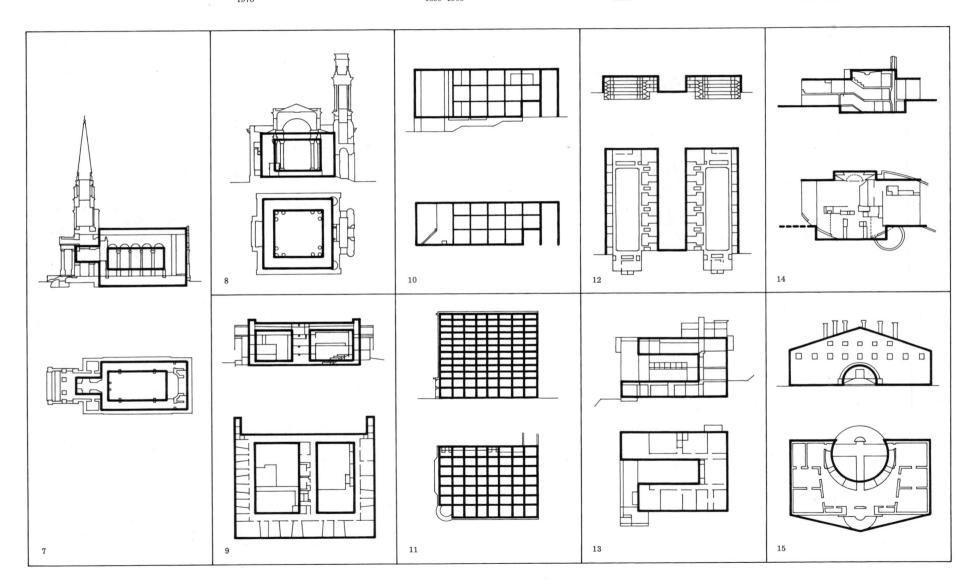

1. **FIRE STATION NUMBER 4**
   ROBERT VENTURI
   1966
2. **ST. MARY LE BOW**
   CHRISTOPHER WREN
   1670–1683

3. **LEICESTER ENGINEERING**
   BUILDING
   JAMES STIRLING
   1959
4. **STOCKHOLM PUBLIC LIBRARY**
   ERIK GUNNAR ASPLUND
   1920–1928

5. **VOUKSENNISKA CHURCH, IMATRA**
   ALVAR AALTO
   1956–1958
6. **WEEKEND HOUSE**
   EDWARD LARABEE BARNES
   1963

7. **KIMBELL ART MUSEUM**
   LOUIS I. KAHN
   1966–1972
8. **ANNEX TO OITA MEDICAL HALL**
   ARATA ISOZAKI
   1970–1972

266 相　反

当平面或剖面中一方的形状
与对方的某些相反的形状相联系
时，平面和剖面是相反的关系。
在消防站4号（1）和圣玛丽·勒·博
教堂（2）中，一个较小的平面
形状在剖面或立面上是很主要的
部位。这种重要部位的相反关系，
在莱斯特大学工程馆（3）中两次
出现的大的平面形状在立面上并
不显要，而立面上最主要的构成
部分在平面上却是很小的。在斯
德哥尔摩公共图书馆（4）中这种
相反关系的表现方式是虚实对比；
它的立面中央的突起的圆鼓形在
平面上是个圆形凹室。在伊马特
拉的伏克塞涅斯卡教堂（5）有一
连串三个逐个变大的弧形平面形
状，与之相应的是剖面中逐个缩
小的形状。周末别墅（6）平面上
的长边在剖面上是矮的，而短边
倒是高的。在金贝尔艺术博物馆
（7）中，简单的平面形状与相反
的复杂剖面形状联系在一起。在
日本大分医学楼（8）中，立面中
的两个形状，一个简单的弧线，
另一个逐级衔接的直角形，在平
面上它们的特征完全相反。

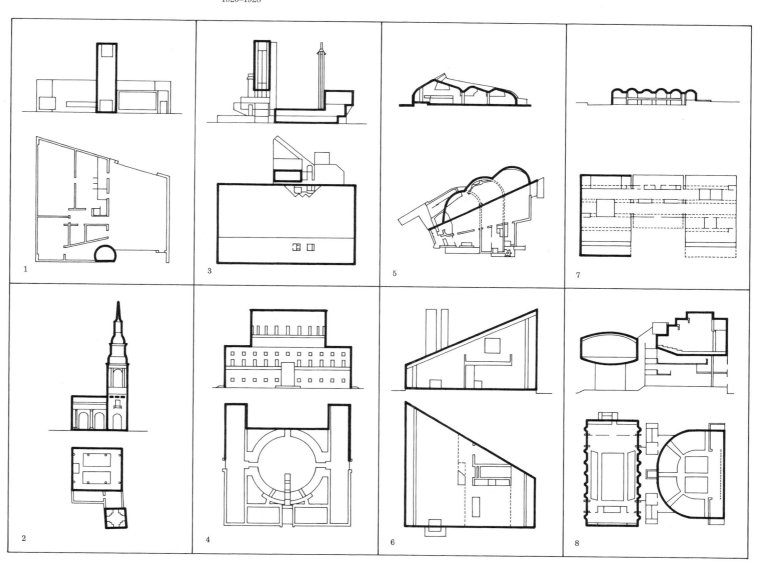

1. **PYRAMID OF CHEOPS**
   ARCHITECT UNKNOWN
   c. 3733 B.C.
2. **RUFER HOUSE**
   ADOLF LOOS
   1922
3. **FROG HOLLOW**
   STANLEY TIGERMAN
   1973–1974

4. **HOUSE AT WEISSENHOF**
   LE CORBUSIER
   1927
5. **UNITED NATIONS PLAZA**
   ROCHE-DINKELOO
   1969–1975
6. **KRESGE AUDITORIUM**
   EERO SAARINEN
   1955

7. **RESIDENCE IN RIVA SAN VITALE**
   MARIO BOTTA
   1972–1973
8. **ELPHINSTONE TOWER**
   ARCHITECT UNKNOWN
   16th CENTURY
9. **SMALL OLYMPIC ARENA**
   KENZO TANGE
   1961–1964

267　**单元到整体**

单元到整体的关系是一种形体构思，它以一定的方式把一些单元和另一些单元或整体联系起来，从而创造出建筑形体。以下举例说明单元等于整体、单元包含在整体之内、单元小于整体、以及单元集合构成整体等关系。

单元等于整体

单元和整体之间最直接的关系是单元就等于整体。在齐奥普斯金字塔（1）和鲁费尔住宅（2）中，表面材料、色彩和形体把单元处理得如同整体一样。在"蛙穴楼"（3），运用黑色把屋顶、墙和窗统一为一个实体。统一的网格像一件外套一样使联合国广场（5）既是一个单元同时又是一个整体。像球体的一片切块似的克雷斯吉礼堂（6）也是既是单元同时又是整体。勒·柯布西耶所做的魏森霍夫住宅（4）和马里奥·博塔所做的在瑞士的圣维达尔河村的独户住宅（7）是删繁就简的整体形状。埃尔芬斯通塔楼（8）的简简块状形体外沿的厚墙，用了统一的材料和颜色，使它的单元与整体相等同。一个独特的雕塑般的形体使东京代代木小体育馆（9）的单元等同于整体。

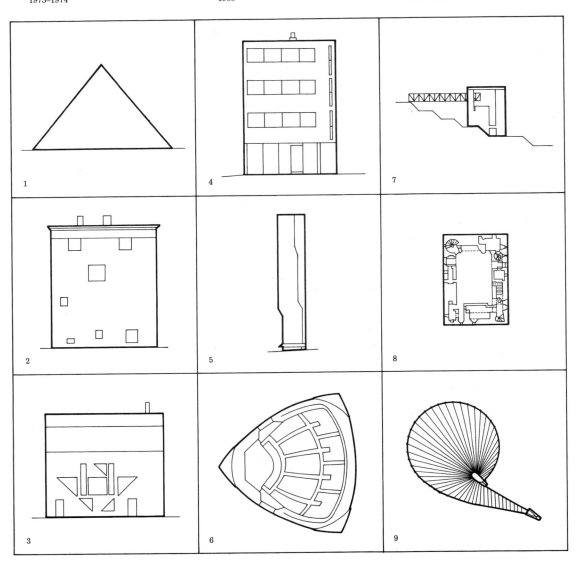

268　单元包含在整体之内

　　在单元包含在整体之内的关系中，单元是指结构构件、使用空间或使用空间的组块。整体是处于主导地位的形象，而单元则在外观上并不显示出来。基督教堂（6）和圣乔治亚·马焦雷教堂（1），圣灵教堂（3）和里丹托教堂（4）中有各种形状的空间单元隐含在内，这些单元都强调出了在形体方面的大的分区。学生俱乐部（2）和卡森·皮里与斯科特百货大楼（5）是以结构模数单元来设计的。在会堂大厦（7）中，单元是使用空间的组合块，这些组块一般分成不同类型的用途。在圣玛丽·伍尔诺思教堂（8），把较小的空间单元组织在中央主体空间的周围。在老圣器堂（9）中，构成主体单元的是空间的立体体积，而圆穹状的吊顶则形成一些次要的单元。在场长住宅（10），单元一般与使用空间和交通空间相吻合。在兰氏音乐中心（11）和圭马尔旅馆（12）中的单元是一些大的房间和小房间组团。

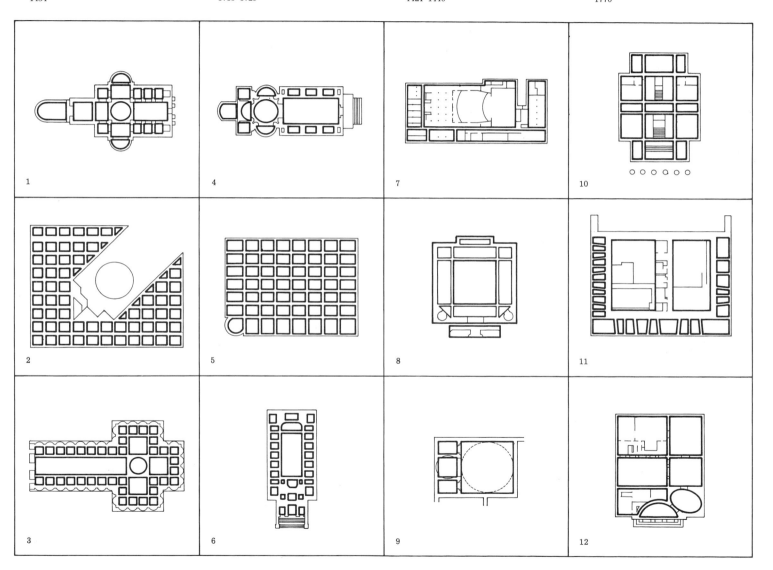

269　整体大于单元之和

在这种关系中，整体所组合的建筑形体比单元的组合要更多。埃克塞特图书馆（1）的中央空间不是使用空间，所以不是一个单元。特里迪弗林公共图书馆（2）的整体要大于以结构框架构成的主要使用空间。在蒙莫朗西旅馆（3）、天德林府邸（4）、摩根图书馆（5）、爱尔兰"安娜格丽"住宅（6）和芬兰地亚会议厅（12），构成单元的是各种形状的主要使用空间，而次要的、辅助性的空间只是填充体。伊德富的何露斯神庙（7）中的单元，是几个大的建筑组块，这些组块都放在以围墙形成的一个整体之内。围墙减去这些单元之后剩下的是一块室外空间。在议会大厦（8）中的单元是两个位于中间的独特的形体，各组使用空间放在周边，以剩下的内部庭院作为过渡。在流水别墅（9）中，表现在立面上的单元是阳台和烟囱的形体，这些形体在与建筑的其余部分的对比中显现出来。以墙来确定的莫斯古姆村舍（10）的整体大于各个单元的组合。在西兰奇一期共管住宅（11）中，单元是居住空间，而整体还包括了中央空间以及附加到各个住宅上的次要空间。

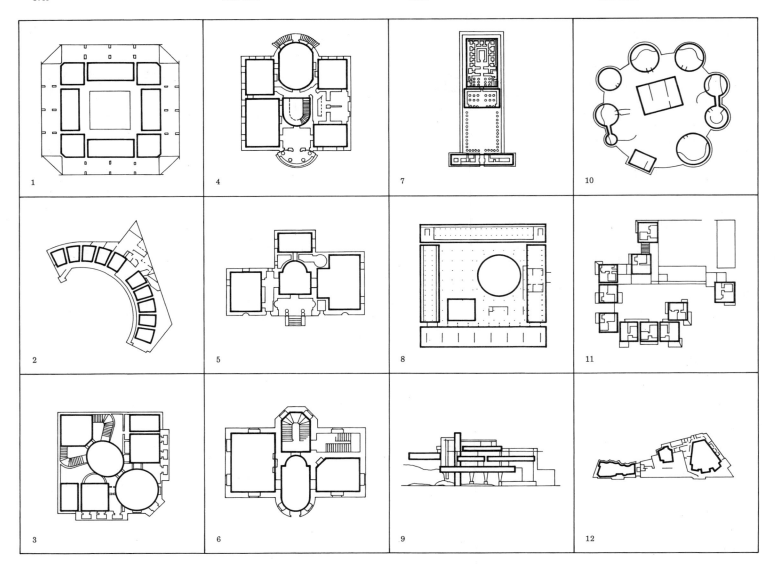

单元集合构成整体

单元集合起来形成整体，各个单元之间的布局应当靠近，以使人觉察到它们之间的联系。这种关系是以连接、隔开和重合等方式形成的。

单元连接

单元连接构成整体时，其中的单元是可以看得见，能够感觉到它们的存在，并且相互以面相接。连接的方式表现在詹姆斯·斯特林的近期作品中，范例包括莱斯特大学工程馆（1）、剑桥大学历史系图书馆（2）和弗洛雷大楼（3）。在伊斯顿·内斯顿府邸（4）和纳希顿府邸（7），聚集起来的单元使古典式的中央入口更为突出。在圣乔治教堂（5）中是建筑形体和空间单元的结合。贝桑松剧院（6）和圣三一教堂（8）是

单元围在一个主要的中央形体周围。在理查森所做的阿勒格尼县法院（9）和阿尔托所做的赛于奈察洛市政中心（13）中，单元是使用空间组群，这些组群连接围着一个中央庭院；而阿尔托所做的伏克塞涅斯卡教堂（12）、沃尔夫斯堡文化中心（14）和帕米欧疗养院（15）中，这些组群连接起来就是建筑本身。在联合教堂（10）中，把两套附加的单元组合起来。一大一小两个单元用第三个单元连接起来，构成了古根海姆美术馆（11）。在圆厅别墅（16），单元对称地围在中央空间的四周，而在卡尔斯基尔希教堂（19）中各个使用空间对称地连接着。几个大体积和一些组成体组成了消防站4号（17）和布兰特住宅（18）的整体，而在斯德哥尔摩公共图书馆（20）和圣玛丽亚布道所（21）中，单元围绕在中央的形体或空间的周围。在萨伏伊别墅（22）和金贝尔艺术博物馆（23）中，连接起来的是结构单元。在路易斯·康所做的多米尼加修女会修道院（24）中有一系列的形体，这些形体中分别包含着的单元是一些空间组合体。

1. **LEICESTER ENGINEERING BUILDING**
   JAMES STIRLING
   1959
2. **CAMBRIDGE HISTORY FACULTY**
   JAMES STIRLING
   1964
3. **FLOREY BUILDING**
   JAMES STIRLING
   1966
4. **EASTON NESTON**
   NICHOLAS HAWKSMOOR
   c. 1695–1710
5. **ST. GEORGE-IN-THE-EAST**
   NICHOLAS HAWKSMOOR
   1714–1729
6. **THEATER IN BESANÇON, FRANCE**
   CLAUDE NICHOLAS LEDOUX
   1775
7. **NASHDOM**
   EDWIN LUTYENS
   1905–1909
8. **TRINITY CHURCH**
   HENRY HOBSON RICHARDSON
   1872–1877
9. **ALLEGHENY COUNTY COURTHOUSE**
   HENRY HOBSON RICHARDSON
   1883–1888

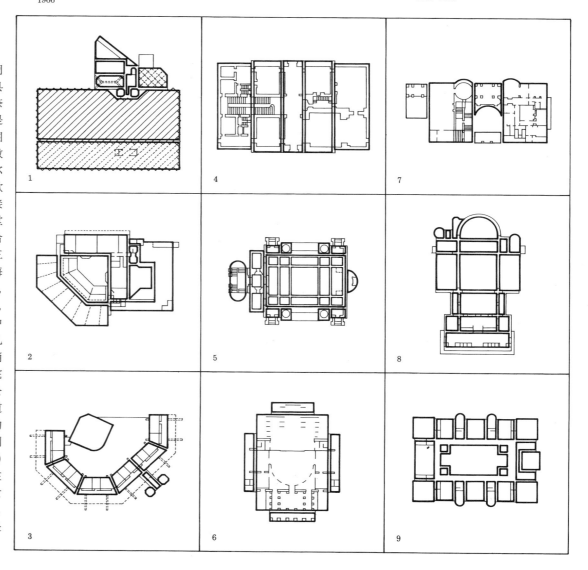

10. **UNITY TEMPLE**
FRANK LLOYD WRIGHT
1906

11. **GUGGENHEIM MUSEUM**
FRANK LLOYD WRIGHT
1956

12. **VOUKSENNISKA CHURCH, IMATRA**
ALVAR AALTO
1956–1958

13. **SAYNATSALO TOWN HALL**
ALVAR AALTO
1958–1962

14. **WOLFSBURG CULTURAL CENTER**
ALVAR AALTO
1956–1962

15. **PAIMIO SANITORIUM**
ALVAR AALTO
1929–1933

16. **LA ROTONDA**
ANDREA PALLADIO
1566–1571

17. **FIRE STATION NUMBER 4**
ROBERT VENTURI
1966

18. **PETER BRANT HOUSE**
ROBERT VENTURI
1973

19. **KARLSKIRCHE**
JOHANN FISCHER VON ERLACH
1715–1737

20. **STOCKHOLM PUBLIC LIBRARY**
ERIK GUNNAR ASPLUND
1920–1928

21. **SANTA MARIA DEGLI ANGELI**
FILIPPO BRUNELLESCHI
1434

22. **VILLA SAVOYE**
LE CORBUSIER
1928–1931

23. **KIMBALL ART MUSUEM**
LOUIS I. KAHN
1966–1972

24. **CONVENT FOR DOMINICAN SISTERS**
LOUIS I. KAHN
1965–1968

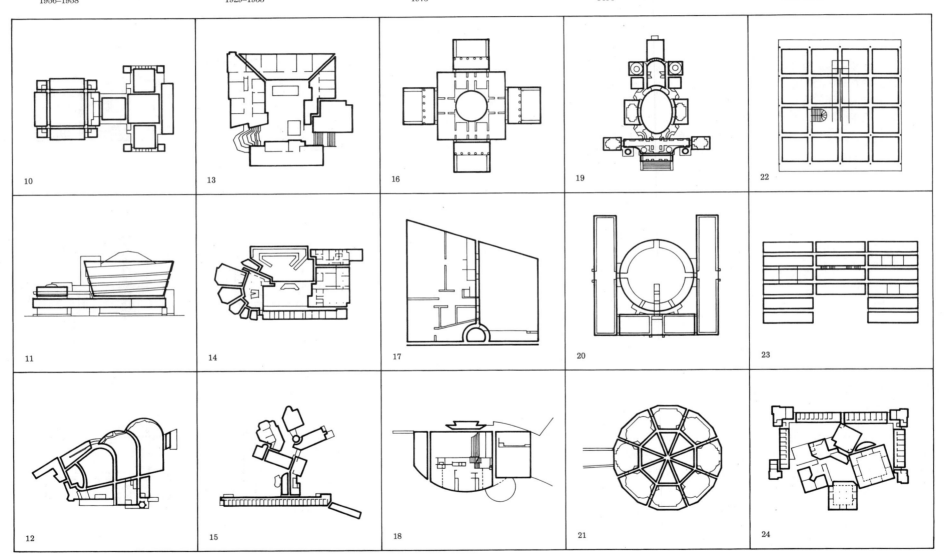

1. **SEVER HALL**
   HENRY HOBSON RICHARDSON
   1878–1880
2. **FREDERICK G. ROBIE HOUSE**
   FRANK LLOYD WRIGHT
   1909
3. **YALE ART AND ARCHITECTURE**
   PAUL RUDOLPH
   1958

4. **LISTER COUNTY COURTHOUSE**
   ERIK GUNNAR ASPLUND
   1917–1921
5. **CARLL TUCKER III HOUSE**
   ROBERT VENTURI
   1975
6. **ERDMAN HALL DORMITORIES**
   LOUIS I. KAHN
   1960–1965

7. **OCCUPATIONAL HEALTH CENTER**
   HARDY-HOLZMAN-PFIEFFER
   1973
8. **PRATT RESIDENCE**
   HARDY-HOLZMAN-PFIEFFER
   1974
9. **SALISBURY SCHOOL**
   HARDY-HOLZMAN-PFIEFFER
   1972–1977

10. **RESIDENCE IN BRIDGEHAMPTON**
    GWATHMEY-SIEGEL
    1969–1971
11. **COOPER RESIDENCE**
    GWATHMEY-SIEGEL
    1968–1969
12. **BARCELONA PAVILLION**
    LUDWIG MIES VAN DER ROHE
    1929

272 单元重合

通过立体的互相穿透，单元即重合而成整体。塞弗楼（1）以四个塔楼形成的长条形和主体的长方块重合，在罗比住宅（2）上层的侧厅把下面两个方向与之垂直的体块联系起来。在耶鲁大学艺术与建筑系馆（3）中，一系列重合的槽形确定了内部空间。利斯特县法院（4）的圆形主体有一部分嵌进了中央体块，而在塔克住宅（5）中的圆形把屋顶的三角形和建筑的正方形统一起来。在布林莫尔的厄德曼宿舍楼（6），几个角重合起来使交通路线能够连贯相通。职业保健中心（7）、普拉特住宅（8）和索尔兹伯里学校（9）中，几组方向旋转的形体互相重合。在布里奇汉普顿的住宅（10）中，隐含的两个圆形互相重合，又与一个长方形重合，而在库珀住宅（11）中，互相重合的形体构成了空间的分割并且暗示出一个不完全的风车形状。在巴塞罗那国际博览会德国馆（12）中有一连串复杂的垂直相交的穿插和隐含着的空间体积。

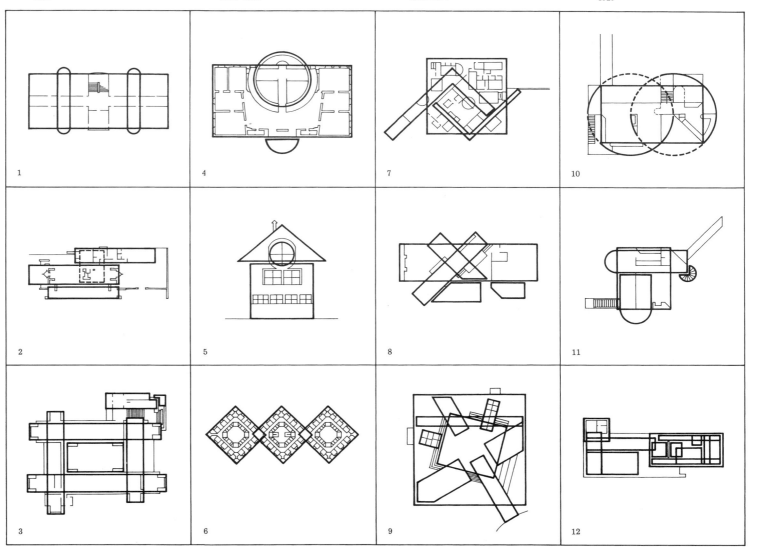

1. **DEERE WEST OFFICE BUILDING**
ROCHE-DINKELOO
1975–1976

2. **OLYMPIC ARENA**
KENZO TANGE
1961–1964

3. **RESIDENCE IN STABIO**
MARIO BOTTA
1981

4. **COLLEGE LIFE INSURANCE COMPANY**
ROCHE-DINKELOO
1967–1971

5. **RESIDENCE ON MT. DESERT ISLAND**
EDWARD LARABEE BARNES
1975

6. **PAUL MELLON ARTS CENTER**
I. M. PEI
1970–1973

7. **EVERSON MUSEUM OF ART**
I. M. PEI
1968

8. **NATIONAL ASSEMBLY**
LOUIS I. KAHN
1962–1974

9. **CHAPEL AT RONCHAMP**
LE CORBUSIER
1950–1955

单元隔开

互相有关的单元也可以隔开，隔开的方式可以通过完全隔离或者以连接体相接形成感觉上的分隔。在迪尔·韦斯特办公楼（1）中，单元之间以一条明确的玻璃通道和一个内天井隔开。在东京代代木小体育馆（2）和瑞士的斯塔比奥的住宅（3）中，玻璃是用来造成感觉上的分割。学院人寿保险公司（4）是各个隔离的形体，只是在某一层上用一个桥细弱地联系着。一座平台用来统一荒漠岛山住宅（5）的各个隔离的单体，而在保罗·梅隆艺术中心（6）中各个分开的形体共有一个屋顶。玻璃使埃弗森艺术博物馆（7）在感觉上是隔开的，并且它也用来形成了达卡国民议会大厦（8）和朗香教堂（9）中的明确的分隔。

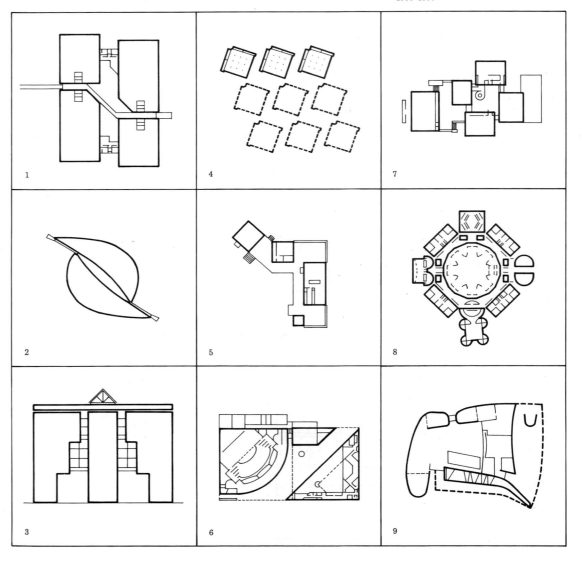

重复到独特

重复与独特相联系的形体构思，是建筑的一种设计方法，这种方法是在建筑中具有单一的和重复的表现形式的各个组成部分之间确定某些关系。以下举例说明重复包围独特、独特由重复区域中的变形来构成、独特加于重复，以及独特由重复来确定等各种关系。

重复包围独特

重复的单体围绕一个独特单元，这时的独特部分是有界线的形体，并且被多个相同的单元所环绕。在辛克尔所做的阿尔泰斯

博物馆（1）、弗洛雷大楼（10）和议会大厦（18）中，独特部分位于由重复单元围成的较大的空间内。在猎庄（2），独特的中心是被围起来的。在成人学习研究实验室（16），它是局部地被包围的。在罗得岛州议会大楼（3）、联合教堂（4）和圣灵教堂（23）中的独特单元都是被包围的，而在古根海姆美术馆（5）、多米尼加修女会修道院（13）、会堂大厦（15）和兰氏音乐中心（17）中，中心都是局部地被环绕。重复的单体在泉北考古资料馆（6）中形成一个风车形，而在胡纳尔别墅（9）和斯德哥尔摩公共图书馆（22）中都是"U"形。圣乔治教堂（7）、剑桥大学历史系图书馆（11）、圣三一教堂（12）和圣灵教堂（23）是两种类型的重复单体的范例。在基督教堂（8）和福斯卡里别墅（20）中，重复单元与独特体的联系不止一种方式。在埃克塞特图书馆（14）、贝桑松剧院（19）和圆厅别墅（21）中，中央的独特体是完全被围起的。圣玛丽亚布道所（24）的独特中心被两组重复单体包围着，一组是空间，另一组是结构。

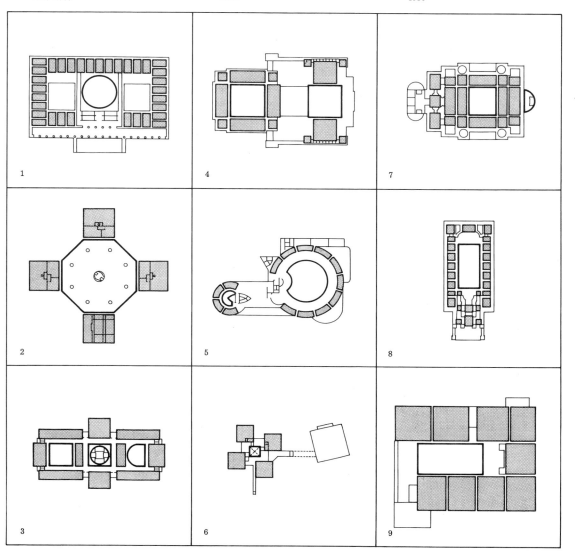

1. **ALTES MUSEUM**
   KARL FRIEDRICH SCHINKEL
   1824–1830
2. **HUNTING LODGE**
   KARL FRIEDRICH SCHINKEL
   1822
3. **RHODE ISLAND STATE CAPITOL**
   McKIM, MEAD, AND WHITE
   1895–1903
4. **UNITY TEMPLE**
   FRANK LLOYD WRIGHT
   1906
5. **GUGGENHEIM MUSEUM**
   FRANK LLOYD WRIGHT
   1956
6. **SHENBOKU ARCHIVES**
   FUMIHIKO MAKI
   1970
7. **ST. GEORGE-IN-THE-EAST**
   NICHOLAS HAWKSMOOR
   1714–1729
8. **CHRIST CHURCH**
   NICHOLAS HAWKSMOOR
   1715–1729
9. **KHUNER VILLA**
   ADOLF LOOS
   1930

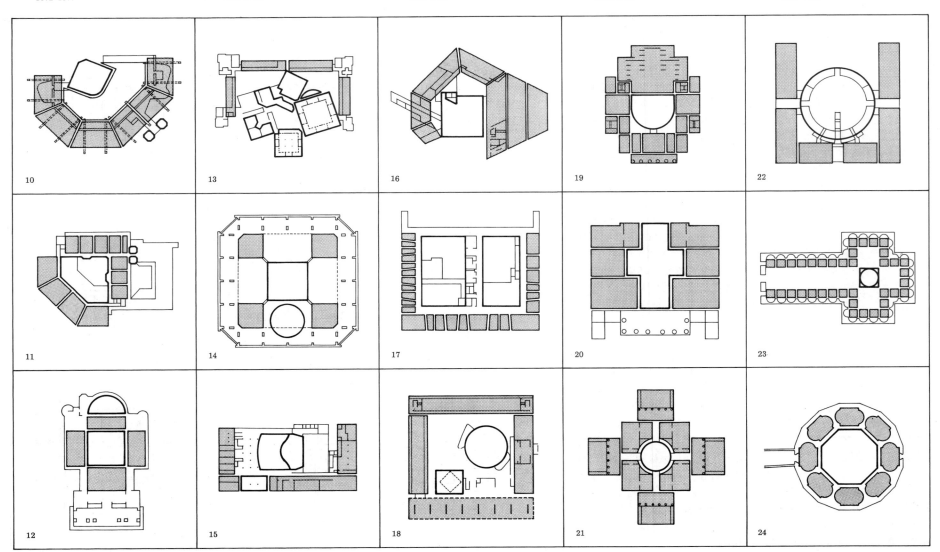

1. **CARSON PIRIE AND SCOTT STORE**
LOUIS SULLIVAN
1899–1903

2. **STEINER HOUSE**
ADOLF LOOS
1910

3. **ALEXANDER HOUSE**
MICHAEL GRAVES
1971–1973

4. **SNELLMAN HOUSE**
ERIK GUNNAR ASPLUND
1917–1918

5. **HOTEL DE MONTMORENCY**
CLAUDE NICHOLAS LEDOUX
1769

6. **TENDERING HALL**
JOHN SOANE
1784–1790

7. **CARLL TUCKER III HOUSE**
ROBERT VENTURI
1975

8. **HOMEWOOD**
EDWIN LUTYENS
1901

9. **EASTON NESTON**
NICHOLAS HAWKSMOOR
c. 1695–1710

10. **MOORE HOUSE**
CHARLES MOORE
1962

11. **FALLINGWATER**
FRANK LLOYD WRIGHT
1935

12. **GUMMA MUSEUM OF FINE ARTS**
ARATA ISOZAKI
1971–1974

276　独特是重复的变形

　　通过尺寸、形状、轮廓、朝向、几何形、色调和连接方式等方面的变化，重复的单体可以变形而产生独特体。形状和几何形的变化是类似并且有联系的，但通常形状所涉及的形体变化不如几何形变化所涉及的多。卡森·皮里与斯科特百货大楼（1）、斯内尔曼住宅（4）和蒙莫朗西旅馆（5）中的独特体都是由几何图形的变化产生的。在天德林府邸（6）、塔克住宅（7）和霍姆伍德住宅（8）中也是同样情况。关于形状的变化可以在斯坦纳住宅（2）和亚历山大住宅（3）中见到。在伊斯顿·内斯顿府邸（9）中，尺寸大小的变化创造出了独特的双层空间。位于奥林达的穆尔住宅（10），重复单元间的不同连接方式构成了独特体。在流水别墅（11）和群马美术馆（12）中的独特部分都是由重复体的方向变化造成的。

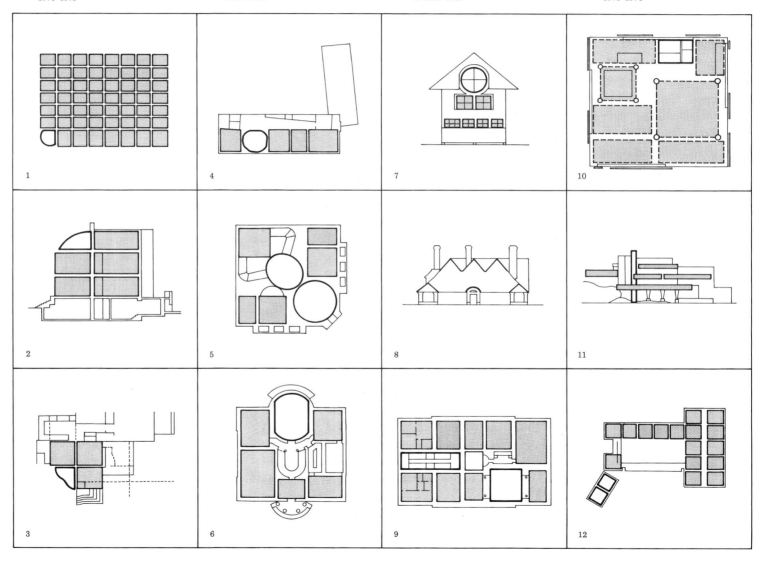

1. **TEMPLE OF ARTEMIS**
   PAEONIUS AND DEMETRIUS
   c. 356 B.C.
2. **KIMBELL ART MUSEUM**
   LOUIS I. KAHN
   1966–1972
3. **INSTITUTE FOR ADVANCED STUDIES**
   GBQC
   1968–1972
4. **STUDENT UNION**
   ROMALDO GIURGOLA
   1974
5. **BROOKLYN CHILDREN'S MUSEUM**
   HARDY-HOLZMAN-PFIEFFER
   1977
6. **OCCUPATIONAL HEALTH CENTER**
   HARDY-HOLZMAN-PFIEFFER
   1973
7. **FREDERICK G. ROBIE HOUSE**
   FRANK LLOYD WRIGHT
   1909
8. **ST. STEPHENS WALBROOK**
   CHRISTOPHER WREN
   1672–1687
9. **VILLA SAVOYE**
   LE CORBUSIER
   1928–1931

277　重复区域内的独特

　　一个由相同的单元以同样的联系组成一个区域或网络，它又被一个独特体插入。在阿丹密斯神庙（1）中,在柱网区内插入了几道墙。在金贝尔艺术博物馆（2）和学生俱乐部（4）中，开敞的庭院插入结构体系中形成了独特的单元。在高级研修学院（3），独特的几何形体位于垂直相交的结构网格中。一个通道在结构网络中旋转方向形成了布鲁克林儿童博物馆（5）中的独特体，而在职业保健中心（6）中，在方格网中转动方向的一个天窗是独特的。结构区域在罗比住宅（7）中被一个炉体断开，在圣斯蒂芬斯·沃尔布鲁克教堂（8）中被一个穹顶插入，以及在萨伏伊别墅（9）中被两个不同的互相垂直的通道切断。

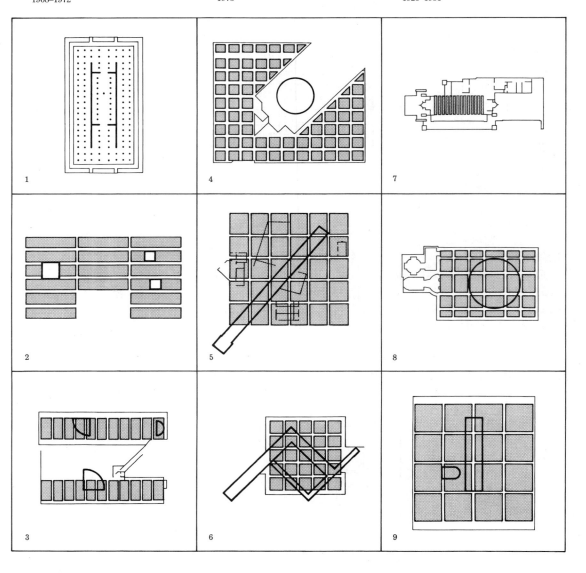

1. **SEINAJOKI TOWN HALL**
   ALVAR AALTO
   1962–1965
2. **KAMIOKA TOWN HALL**
   ARATA ISOZAKI
   1976–1978
3. **BOYER HALL OF SCIENCE**
   GBQC
   1970–1972

4. **CONVENT OF LA TOURETTE**
   LE CORBUSIER
   1957–1960
5. **UNITE D'HABITATION**
   LE CORBUSIER
   1946–1952
6. **WAINWRIGHT BUILDING**
   LOUIS SULLIVAN
   1890–1891

7. **ST. NICHOLAS COLE ABBEY**
   CHRISTOPHER WREN
   1671–1681
8. **OLIVETTI TRAINING SCHOOL**
   JAMES STIRLING
   1969
9. **LEICESTER ENGINEERING BUILDING**
   JAMES STIRLING
   1959

278　独特加于重复

　　有时重复部分的尺度和体量是最首要的，这时独特体看上去是加到重复部分上去的。在塞伊奈约基市政厅（1），独特体加在重复部分的末尾，成为一个终端。日本神冈市政厅（2）的独特体加在一串重复体的中点上。三个独特单元被放在博耶科学馆（3）的一个隐含的活动剧场中，而在拉图雷特修道院（4）中独特和重复结合构成一个回廊。在马赛的公寓大楼（5）的立面上，在主体的顶部和底部加上了特殊的形体。在温赖特大厦（6）加了一个独特的顶部，形成了不同的收头。在圣尼古拉斯·科尔修道院（7）中的独特形体加在前面。在奥利韦蒂专科学校（8）中的两个相连的独特体放在重复的单元中间，而在莱斯特大学工程馆（9）中，两个加上去的独特体紧靠着建筑的主体。

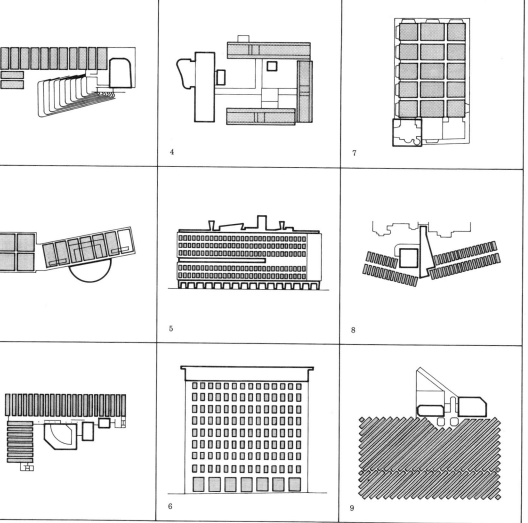

1. **COLOSSEUM**
ARCHITECT UNKNOWN
70–82

2. **ST. LEOPOLD AM STEINHOF**
OTTO WAGNER
1905–1907

3. **ST. ANTHOLIN**
CHRISTOPHER WREN
1678–1691

4. **HOUSE OF THE MENANDER**
ARCHITECT UNKNOWN
C. 300 B.C.

5. **BOSTON PUBLIC LIBRARY**
MCKIM, MEAD, AND WHITE
1898

6. **SAYNATSALO TOWN HALL**
ALVAR AALTO
1950–1952

7. **YALE ART AND ARCHITECTURE**
PAUL RUDOLPH
1958

8. **LARKIN BUILDING**
FRANK LLOYD WRIGHT
1903

9. **ALLEGHENY COUNTY COURTHOUSE**
HENRY HOBSON RICHARDSON
1883–1888

279 **独特由重复确定**

在独特体是由重复体组成的形状来构成时，独特就是由重复确定的。这一类的所有实例中，独特体既有在室内的，也有在室外的。在罗马大角斗场（1）、庞培的米南德住宅（4）和波士顿公共图书馆（5），许多重复单元的布局构成了一个室外大空间。阿尔托所做的赛于奈察洛市政中心（6）和阿勒格尼县法院（9）也是同样情况。在奥托·瓦格纳所做的斯坦霍夫的圣莱奥波尔德教堂（2）和仑（Wren）所做的圣安多林教堂（3）中的独特体是显露在外观上的室内大空间。在耶鲁大学艺术与建筑系馆（7）和拉金公司大厦（8）中，多层的独特空间就像周围的重复空间的聚合中心。

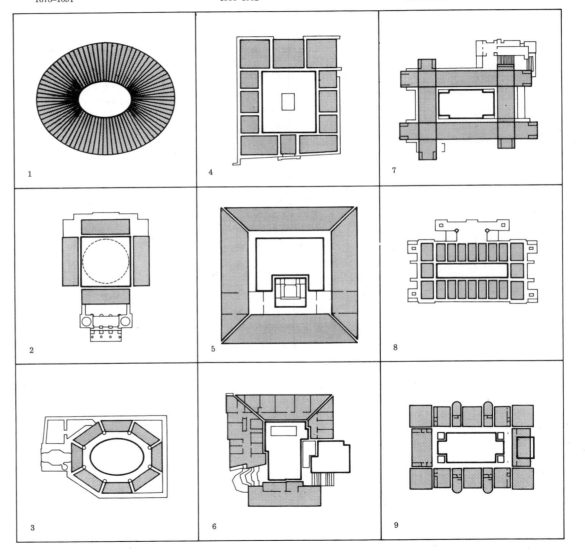

加法和减法

加法和减法，是指把各个部分加起来或减掉一部分，以此构成建筑形体的形体构思方法。在加法的构思法中，各个局部是首要的，而在减法中，整体是首要的。

减　　法

下面所有的实例都是以在简单的直角形中扣掉一部分的办法来做建筑设计。在霍姆伍德住宅（1）中，台地和入口是用减法形成的，而在温赖特大厦（2）中做了一个采光井。惠特尼美术博物馆（3），表示一个剖面被删掉一部分，以使光线能进入较低的各层，并使建筑与街道形成很别致的联系。在萨伏伊别墅（4）中，被删减部分出现在限定的框框以内。在斯德哥尔摩公共图书馆（5）中，在删掉一块后做出的内庭院中加上了一个鼓形。文丘里住宅（6）和恩索－古特蔡特公司总部（7）中都同样以减法确定入口。在恩索－古特蔡特公司总部中还以此将光线引入室内。埃克塞特图书馆（8）的室内中央大空间也是减法的结果，而学生俱乐部（9）的室外大空间，还有入口和较小的一些室外空间，都是由减法的方法做出的。

1. **HOMEWOOD**
   EDWIN LUTYENS
   1901
2. **WAINWRIGHT BUILDING**
   LOUIS SULLIVAN
   1890–1891
3. **WHITNEY MUSEUM OF AMERICAN ART**
   MARCEL BREUER
   1966
4. **VILLA SAVOYE**
   LE CORBUSIER
   1928–1931
5. **STOCKHOLM PUBLIC LIBRARY**
   ERIK GUNNAR ASPLUND
   1920–1928
6. **VANNA VENTURI HOUSE**
   ROBERT VENTURI
   1962
7. **ENSO-GUTZEIT HEADQUARTERS**
   ALVAR AALTO
   1959–1962
8. **EXETER LIBRARY**
   LOUIS I. KAHN
   1967–1972
9. **STUDENT UNION**
   ROMALDO GUIRGOLA
   1974

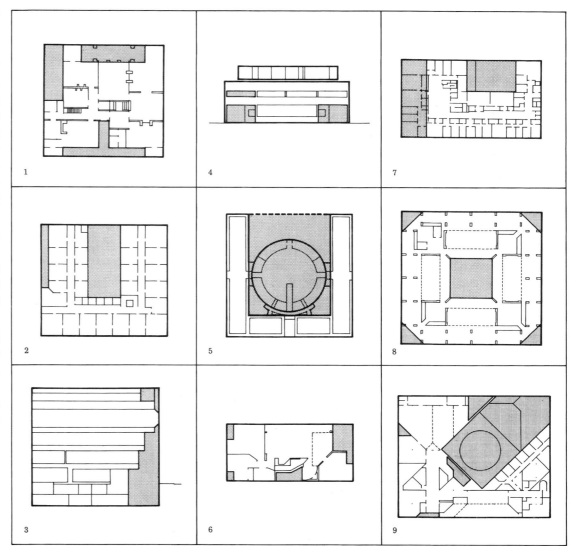

281　加　　法

加法设计显然是把各个分部作为首位来考虑的。在圆厅别墅（1）中，几个分部加到大的中心单元周围。在理查德医学研究中心（2）中出现了一连串加起来的单元，设备塔加在每个研究室一起组成一个组合单元，这些组合单元又和相似的另一些组合单元加在一起，再和中间的设备中心加在一起。在萨吕泰兴府邸（3），服务人员用房是小单元，它和大的主体接在一起。在利斯特县法院（4）中，一个大的使用空间加进到建筑主体的里面。在弗洛雷大楼（5），一连串分块聚集起来构成一个室外空间，在这个空间中又加上一个独特的共用空间。在西兰奇一期共管住宅（6）中，每一个单元都是一个多种形体的集合，并被集合在一个共同的屋顶之下。重复的又互相垂直相交的两组单元连接起来成为联合教堂（7）的两个建筑主体。在阿勒格尼县法院（8），各个分部组合出一个敞开的中央空间。一些小单元集合围绕着圣乔治教堂（9）的中殿，而相当于使用空间的各个组成体聚集起来构成了沃尔夫斯堡文化中心（10）。在圣玛丽亚布道所（11）和圣维塔莱教堂（12）中，一系列小空间环绕着一个大空间。

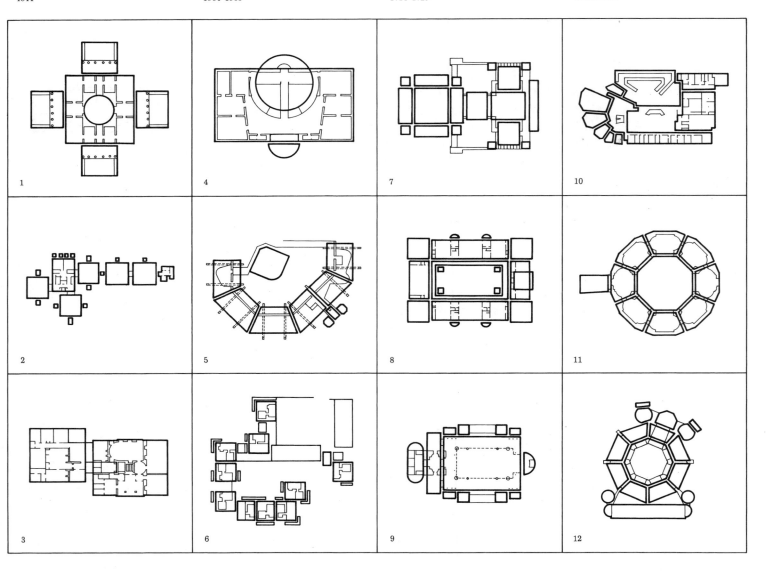

对称和均衡

在对称和均衡的形体构思中，是通过确立各个组成部分之间的直觉性的或概念性的等量关系来创作建筑形体。以下举例说明通过形状、几何形、虚实对比等方法做出的轴线式、双轴式、旋转式和转换式等各种对称和均衡。

对 称

对称是均衡的特殊形式，它是指把相同的单元放在无形的一条线的每一边，或围绕一个点。在萨尔克生物研究所（1），对称

轴线穿过室外的大空间。在场长住宅（2）、联合教堂（3）、基督教堂（4）、里丹托教堂（5）和圣灵教堂（6），中轴线都是穿过主要的使用空间。在圣玛丽亚布道所（7）中，放射式的对称由于放置了两个相对的出入口而转变成了轴线式的对称。在利斯特县法院（8）和斯德哥尔摩公共图书馆（9）中，对称轴线都穿过主要的室内空间。在维纳斯与罗马神庙（10）中，双向对称轴线在主要空间的中间穿过。在埃克塞特图书馆（11），轴线把主要空间对半切开，而在圆厅别墅（12），轴线把交通面积对切。圣马克塔楼（13）旋转式的对称中有四个单元围绕一个点，而蒙特城堡（14）则有八个，圣约翰·内波穆克教堂（15）有五个。在圣伊夫教堂（16）、卡佩尔圣徒朝圣教堂（17）和圣墓教堂（18）中，各自有三个单元旋转对称。在圣安德鲁斯宿舍(19)和博塔所做的莫尔比奥·因费里奥雷中学（20）中，单元是一些房间或房间组合体，它们对称地转变成了长条形状。在伍重所做的庭院式住宅（21）中，两组单元以不同的方向转换着。

1. **SALK INSTITUTE**
   LOUIS I. KAHN
   1959–1965
2. **DIRECTOR'S HOUSE**
   CLAUDE NICHOLAS LEDOUX
   1775–1779
3. **UNITY TEMPLE**
   FRANK LLOYD WRIGHT
   1906
4. **CHRIST CHURCH**
   NICHOLAS HAWKSMOOR
   1715–1729
5. **REDENTORE CHURCH**
   ANDREA PALLADIO
   1576–1591
6. **CHURCH OF SAN SPIRITO**
   FILIPPO BRUNELLESCHI
   1434
7. **SAN MARIA DEGLI ANGELI**
   FILIPPO BRUNELLESCHI
   1434–1436
8. **LISTER COUNTY COURTHOUSE**
   ERIK GUNNAR ASPLUND
   1917–1921
9. **STOCKHOLM PUBLIC LIBRARY**
   ERIK GUNNAR ASPLUND
   1920–1928

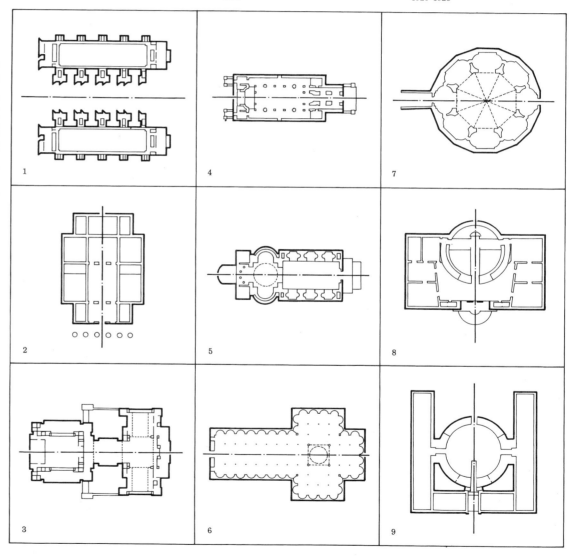

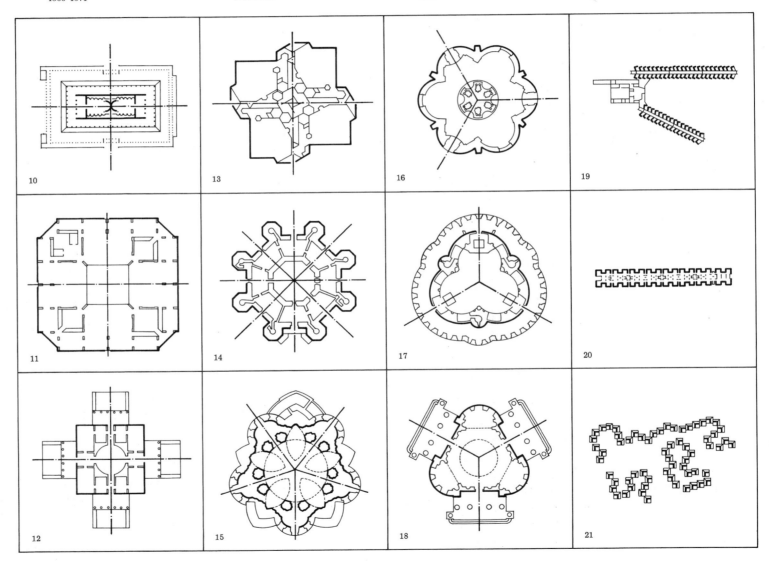

　　形状的均衡是在形状不同的各组成体之间的等量关系。奥利韦蒂专科学校（1）的新建部分与原有的旧建筑均衡，其中长的一翼与短的一翼加上特别形状的空间互相等量。孤儿院（2）是体块均衡的实例，一边有一个空地，另一边有一个附加的单元。在西兰奇一期共管住宅（3）中，对角平衡线的一侧有六个居住单元，而另一侧有四个单元加两个车库。在联合教堂（4）中，两个一样的内核由于加在周围的次要单元的布局不同而变成了不同的形状。在罗比住宅（5）和格莱斯纳住宅（6）中，互相分开的公共部分和私用部分之间产生了一条均衡线。在朱尔戈拉所做的成人学习研究实验室（7）中的均衡是由几何形和体块形成的。圣乔治亚·马焦雷教堂（8）的一个方向是对称的，而另一个方向是简单的形状和复杂的形状之间的均衡，简单和复杂分别反映着神圣和世俗两个不同的区域。布兰特住宅（9）中的形状区别出现在地面和体块的变化上面。在朗香教堂（10）的平面上，和在里奥拉教区中心（14）的剖面上，单个的较大的单元和多次出现的较小的单元互相均衡。流水别墅（11）是在较小的封闭空间和较大的敞开

空间之间取得均衡。利斯特县法院（23）和达利奇艺术馆（13）在一个方向上是对称的；在另一个方向上，利特斯县法院是以公共地面之间的区别构成均衡关系，而达利奇艺术馆则是以尺寸大小差别构成均衡关系。在圭马尔旅馆（15）的外部是对称的，由于三个大的居住空间移动了位置而变成了均衡关系。在弗洛雷大楼（16）中，均衡出现在分量较重的一对塔和另一边的独特的空间之间。在赛于奈察洛市政中心(17)中的一个隔开的特殊空间和其余的部分互相均衡；而在会堂大厦(18)中，塔楼与主体中的空地在两个方向上产生均衡关系。在伊斯顿·内斯顿府邸（19）中，两个独特的双层空间造成形体上的差别。霍姆伍德住宅（20）的均衡关系出现在前后两个布局中的轴线有移动。在斯内尔曼住宅（21）的辅助空间和主要使用空间在两个方向上都有形状变化。在马赛的公寓大楼（22）中，购物街把均衡线定位在减法的底部和加法的顶部之间（原文如此，可能有误——译者注）。在莱斯特大学工程馆（12）中，垂直面和水平面有差别，而在文丘里住宅（24）中对称关系由于窗的格式而变成了均衡关系。

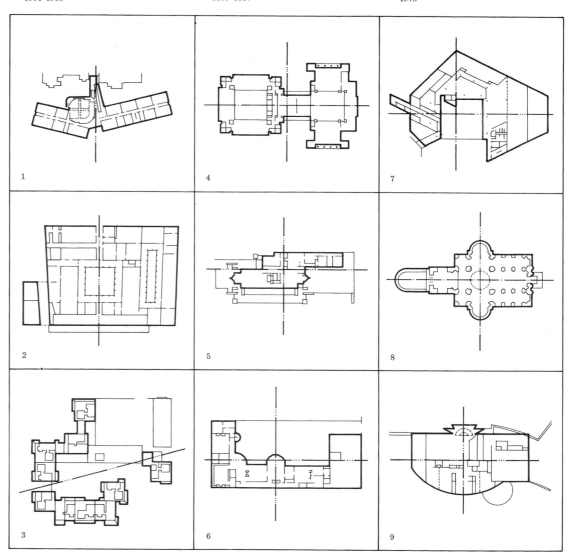

1. **OLIVETTI TRAINING SCHOOL**
   JAMES STIRLING
   1969
2. **OSPEDALE DEGLI INNOCENTI**
   FILIPPO BRUNELLESCHI
   1421–1445
3. **SEA RANCH CONDOMINIUM I**
   CHARLES MOORE
   1964–1965
4. **UNITY TEMPLE**
   FRANK LLOYD WRIGHT
   1906
5. **FREDERICK G. ROBIE HOUSE**
   FRANK LLOYD WRIGHT
   1909
6. **J. J. GLESSNER HOUSE**
   HENRY HOBSON RICHARDSON
   1885–1887
7. **ADULT LEARNING LABORATORY**
   ROMALDO GIURGOLA
   1972
8. **SAN GIORGIO MAGGIORE**
   ANDREA PALLADIO
   1560–1580
9. **PETER BRANT HOUSE**
   ROBERT VENTURI
   1973

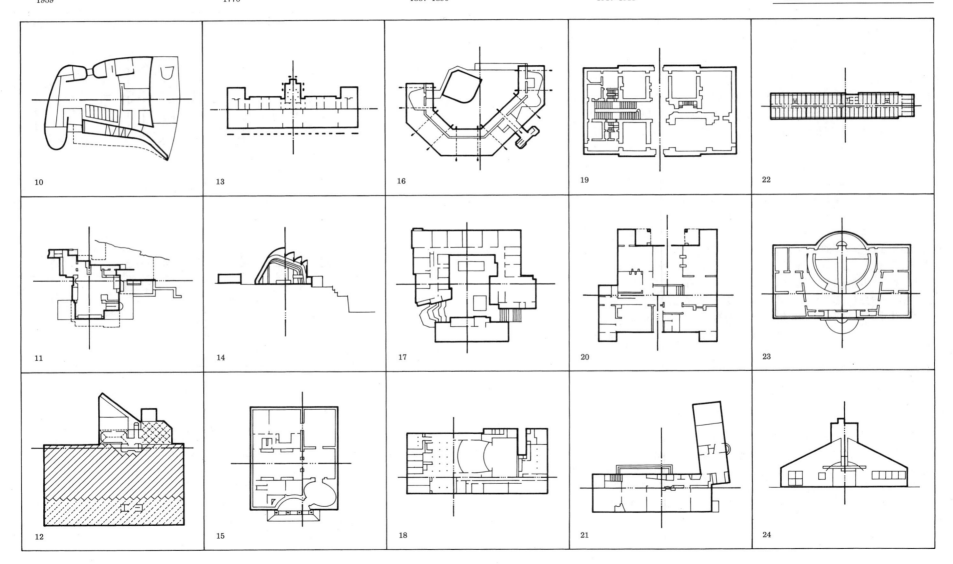

几何关系均衡

在几何关系均衡关系中，均衡两对侧的组成体在形体语汇上有区别。在圣保罗教堂（1）中，一道墙把直角形的辅助空间与半圆形的礼拜堂分隔开。在大分医学楼（2）和梅隆艺术中心（3）中不同的简单几何形互相平衡。在柏林观象台（4）和里丹托教堂（5）中，一个单一的、分割成几块的形体与一连串加在一起的形体互相平衡。圣马尔塔教堂（6）表示了两个不同表现方式的圆形，而在伊马特拉的伏克塞涅斯卡教堂（7）中，两种不同语汇的形体在主要通道上相联系而造成一种延伸的感觉。这种延伸感在阿尔托所做的沃尔夫斯堡文化中心（8）中也是以不同的形体语汇构成的。在特里迪弗林公共图书馆（9）中，弧线几何形和对侧的直线互相均衡。在金屋（10），不同的几何形在两个正交的轴线两侧相互均衡。在圣玛丽亚教堂（11），均衡是以不同的几何形和朝向造成的。伯拉孟特的建筑示意图（12）用两个完整而不同的几何形来举例说明均衡概念的本质。

1. **ST. PAUL'S CHURCH**
   LOUIS SULLIVAN
   1910–1914
2. **ANNEX TO OITA MEDICAL HALL**
   ARATA ISOZAKI
   1970–1972
3. **PAUL MELLON ARTS CENTER**
   I. M. PEI
   1970–1973
4. **OBSERVATORY IN BERLIN**
   KARL FRIEDRICH SCHINKEL
   1835
5. **REDENTORE CHURCH**
   ANDREA PALLADIO
   1576–1591
6. **SANTA MARTA CHURCH**
   COSTANZO MICHELA
   1746
7. **VOUKSENNISKA CHURCH, IMATRA**
   ALVAR AALTO
   1950–1952
8. **WOLFSBURG CULTURAL CENTER**
   ALVAR AALTO
   1958–1962
9. **TREDYFFRIN PUBLIC LIBRARY**
   ROMALDO GIURGOLA
   1976
10. **DOMUS AUREA**
    SEVERUS AND CELER
    c. 64
11. **S. MARIA DELLA PACE**
    DONATO BRAMANTE
    1478–1483
12. **ARCHITECTURAL SETTING**
    DONATO BRAMANTE
    1473

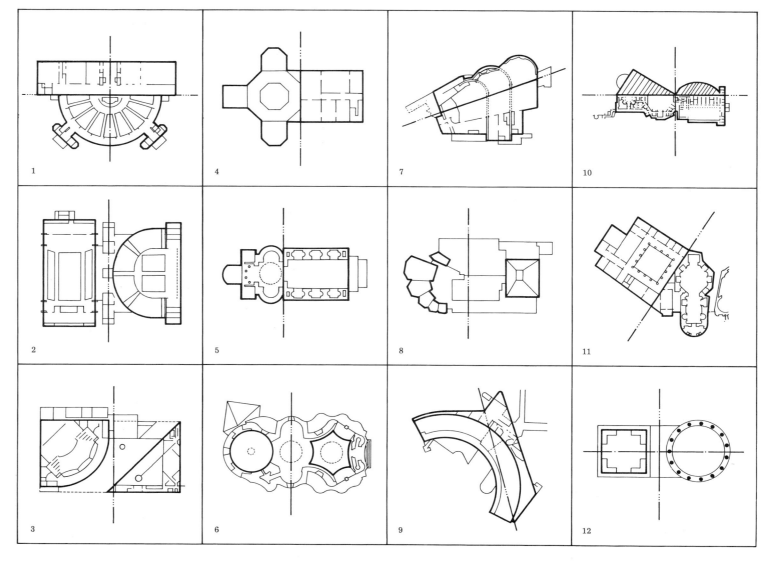

1. **SMITH HOUSE**
RICHARD MEIER
1965–1967

2. **LANG MUSIC BUILDING**
ROMALDO GIURGOLA
1973

3. **WOLFSBURG CULTURAL CENTER**
ALVAR AALTO
1958–1962

4. **HANSELMANN HOUSE**
MICHAEL GRAVES
1967

5. **POWER CENTER**
ROCHE-DINKELOO
1965–1971

6. **WOODLAND CHAPEL**
ERIK GUNNAR ASPLUND
1918–1920

7. **CROOKS HOUSE**
MICHAEL GRAVES
1976

8. **THE FORD FOUNDATION BUILDING**
ROCHE-DINKELOO
1963–1968

9. **VILLA SAVOYE**
LE CORBUSIER
1928–1931

虚与实的均衡

虚实之间的均衡关系中，等量的组成体之间只是在表现方式上不同，比如说实的或是空的。在史密斯住宅（1）中，封闭的私人用面积与开敞的公共面积均衡。兰氏音乐中心（2）中的两个主要使用空间，一个是围筑起来的礼堂，另一个是敞开的大厅。沃尔夫斯堡文化中心（3）的一个方向上是不同形状的均衡；而在另一个方向上，是一个最大的特殊空间和一个庭院互相均衡。在汉泽尔曼住宅（4）、伍德兰礼拜堂（6）和克鲁克斯住宅（7）中，建筑是实的形体，入口庭院是虚的表现方式。在鲍尔中心（5）也有类似情况，它的建筑是实的，而相邻的公园是虚的。在福特基金会大楼（8）中，室内的玻璃暖房是虚的，而办公用房是实的。在萨伏伊别墅（9），别墅的室内和室外居住面积的差别确定了它的虚实均衡线。

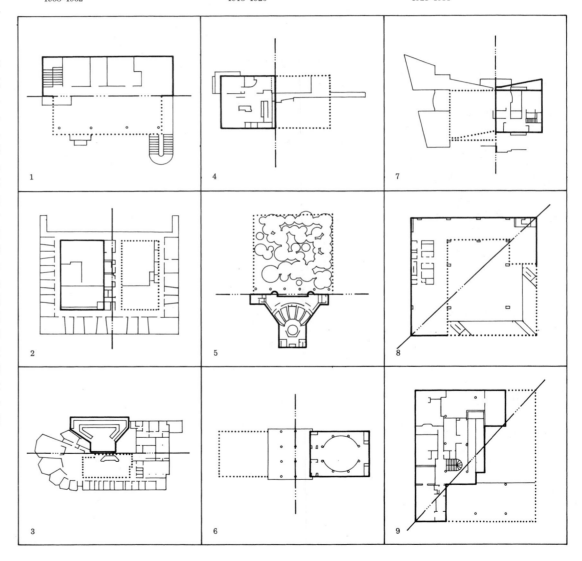

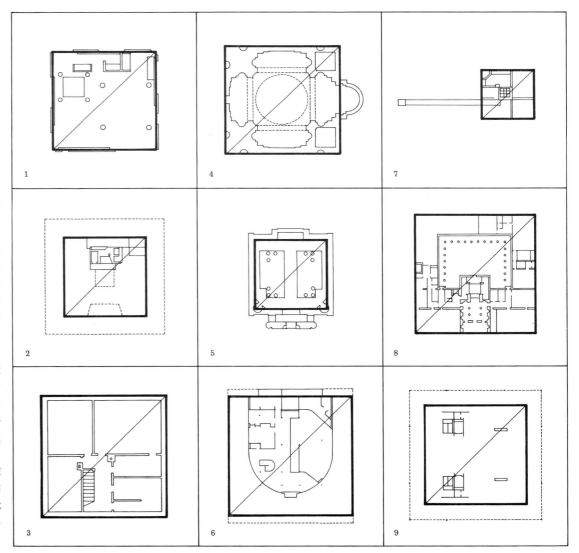

1. **MOORE HOUSE**
CHARLES MOORE
1962

2. **CARLL TUCKER III HOUSE**
ROBERT VENTURI
1962

3. **RUFER HOUSE**
ADOLF LOOS
1922

4. **SANT' ELIGIO DEGLI OREFICI**
RAPHAEL
1509

5. **ST. MARY WOOLNOTH**
NICHOLAS HAWKSMOORE
1716–1724

6. **VILLA SAVOYE**
LE CORBUSIER
1928–1931

7. **RESIDENCE IN RIVA SAN VITALE**
MARIO BOTTA
1972–1973

8. **BOSTON PUBLIC LIBRARY**
McKIM, MEAD, AND WHITE
1898

9. **NEW NATIONAL GALLERY**
LUDWIG MIES VAN DER ROHE
1968

288 **几何关系**

几何关系作为一种形体构思，是运用平面和立体几何学的概念来确定建筑形体。在基本的几何关系之外，我们还将阐述几何关系的组合、倍增、演变和缩小等实例。还包括网格的例子。

基本几何关系

确定建筑形体最基本的图形包括正方形，在穆尔住宅（1）、塔克住宅（2）、鲁费尔住宅（3）、圣埃利吉奥教堂（4）以及圣玛丽·伍尔诺思教堂（5）中都是运用了正方形。正方形也同样用于设计萨伏伊别墅（6）、瑞士的一座私人住宅（即圣维达尔河村的独户住宅）（7）、波士顿公共图书馆（8）和密斯·凡·德·罗所做的新国家美术馆（9）。圆形是圆庙（10）、M.I.T. 教堂（克雷斯吉礼拜堂）（11）、圣康斯坦扎教堂（13）和罗马万神庙（15）的原始形状。托马斯·杰斐逊（Thomas Jefferson）用圆形设计了弗吉尼亚大学圆厅图书馆（14）。康斯坦丁·梅尔尼科夫在他家（12）的设计中采用了两个圆形，而在鲁萨科夫俱乐部（16）中采用的基本形状是三角形。三角形也用来确定了阿里纳楼（17）、芬兰许温凯教堂和教区中心（18）。六边形运用于北基督教会教堂（19）、内盖夫沙漠犹太教堂（20）和菲弗礼拜堂（21）。位于意大利拉韦纳的东正教洗礼堂（22）、杨树林住宅（23）和圣玛丽亚布道所（24）都是从八角形中发展形成的。

303

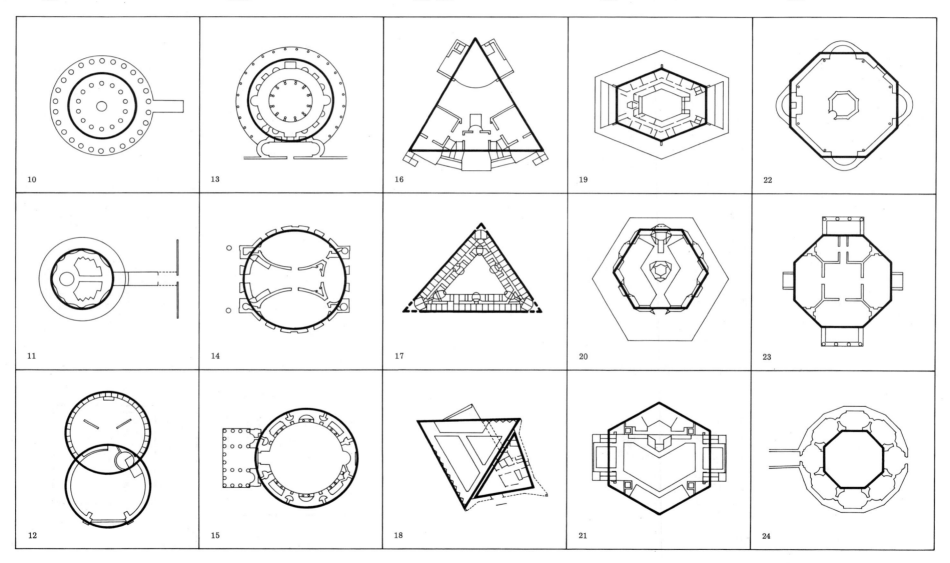

10   13   16   19   22

11   14   17   20   23

12   15   18   21   24

圆形和正方形

圆形和正方形最直接的组合出现在圆厅别墅（1）、老圣器堂（2）、圣彼得罗教堂（3）和霍布金斯大学大厅（4），在这些组合式中的圆形和正方形都是完整的形状，而且很容易认出来，它们还有同一个中心。在伍德兰礼拜堂（6）中包含着完整的圆形，而在斯德哥尔摩公共图书馆（5）包含一个暗示得很清楚的正方形和一个完整的圆形。在查理五世王宫（7）中的圆形是个庭院，在梅泰拉陵墓（8）中是个内核，而在埃克塞特图书馆（9）中是室内立面上的窗孔。在圣彼得大教堂（10）和美国海关大厦（11）中，正方形被包含在更大的形体之中，而在圣玛丽主教堂（12）中的正方形邻接在一个圆上。在杜塞尔多夫艺术博物馆（13）中，斯特林采取了两个圆形和正方形。阿纳姆展览馆（14）和议会大厦（15）中的正方形中包含了一个圆。哥伦布骑士团总部（16）中的正方形四角上加了四个圆，而在蒙莫朗西旅馆（17）的正方形内含有圆和圆的变形。东京代代木国立综合大体育馆（18）和塔尔奎尼亚陵墓（19）是圆中含方的例子。阿尔托所做的设计工作室（20）是在正方形中移动圆的位置形成的，而在斯福尔扎礼拜堂（21）是个精心装修的圆形抱着一个正方形。圣母怀胎告知主教堂（22）、塔克住宅（23）和文丘里住宅（24）都是圆形、正方形和三角形组合的范例。

1. **LA ROTONDA**
   ANDREA PALLADIO
   1566–1571
2. **OLD SACRISTY**
   FILIPPO BRUNELLESCHI
   1421–1440
3. **TEMPIETTO OF SAN PIETRO**
   DONATO BRAMANTE
   1502
4. **JOHNS HOPKINS UNIVERSITY HALL**
   JOHN RUSSELL POPE
   c. 1930
5. **STOCKHOLM PUBLIC LIBRARY**
   ERIK GUNNAR ASPLUND
   1920–1928
6. **WOODLAND CHAPEL**
   ERIK GUNNAR ASPLUND
   1918–1920
7. **PALACE OF CHARLES V**
   PEDRO MACHUCA
   1527
8. **TOMB OF CAECILIA METELLA**
   ARCHITECT UNKNOWN
   c. 25 B.C.
9. **EXETER LIBRARY**
   LOUIS I. KAHN
   1967–1972

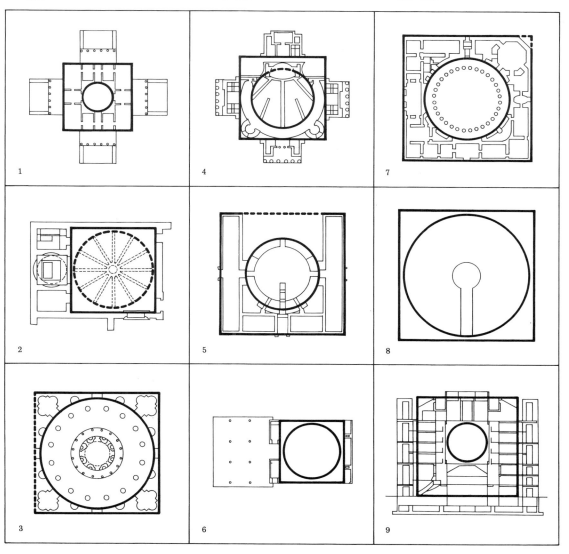

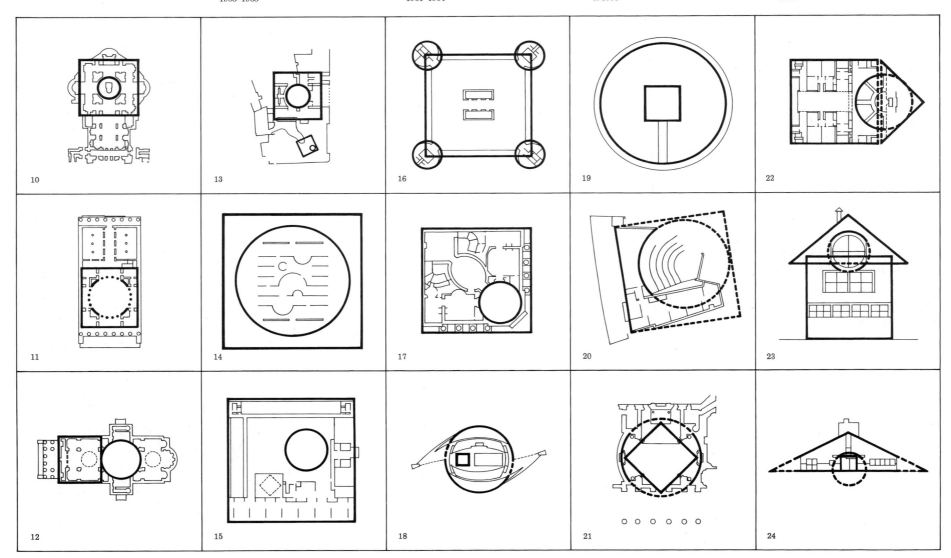

292 长方形与圆形重合

　　长方形和一个较小的圆形重合是几何形组合的一种具体方式。利斯特县法院（1）、护林者住宅（2）、罗马赌场（3）、卡拉卡拉浴场（4）和斯旺住宅（5）都是把一个作为主要使用空间的圆形半嵌入一个正方形的长边的中线上。博塔所做的马萨尼奥的住宅（6）中，同样的圆形是个楼梯，尺寸上缩小了。在卡西尔贾尔府邸（7）中，长方形与椭圆形在中心线上重合，而在天德林府邸（8）中一个圆和一个椭圆与长方形重合。在奥斯汀大厅（9）的两个长方形被两个圆插入，并且还有第三个圆形在入口处重合。在理查森所做的希金森住宅（10）有两个圆在对角上正好暗示一条对角线，而在日本立科天文馆（11）则两个圆出现在同一边上。在天轮教堂（12）中，大小各一对圆形重合在长方形上。在雷特城堡（13）和皮特菲基城堡（14）中，圆形重合在两条边的交角上，而在尚堡府邸（15）中，好几个角都与圆形重合。

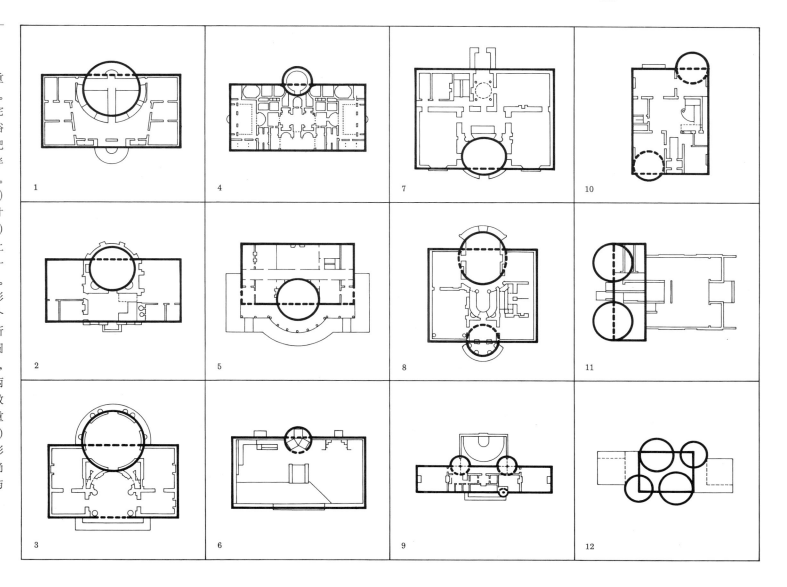

293

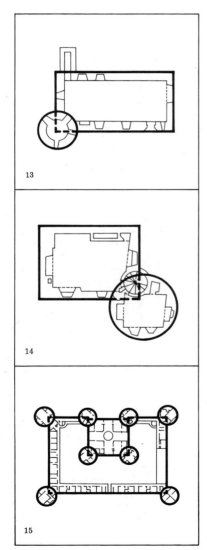

两个正方形

塞弗楼（1）、基督教堂（2）和文丘里住宅（3）的外边界线都是直接由两个正方形邻接而构成的。在布兰特住宅（4）中，两个正方形共有的一条边是平面上大圆的半径，并且这两个正方形确定了整个平面的边界。两个正方形可以重合起来做出特殊的共用面积。在伊斯顿·内斯顿府邸（5）的两个正方形共有的部分是中央大厅，在阿勒格尼县法院（6）的重合处放上了塔楼。帕拉第奥所做的特里西诺别墅（7），和德雷顿大厅（8）都是以两个重合的正方形确定中央的使用空间和入口。在法尔尼斯府邸（9），两个邻接的正方形确定了主立面的边界线。

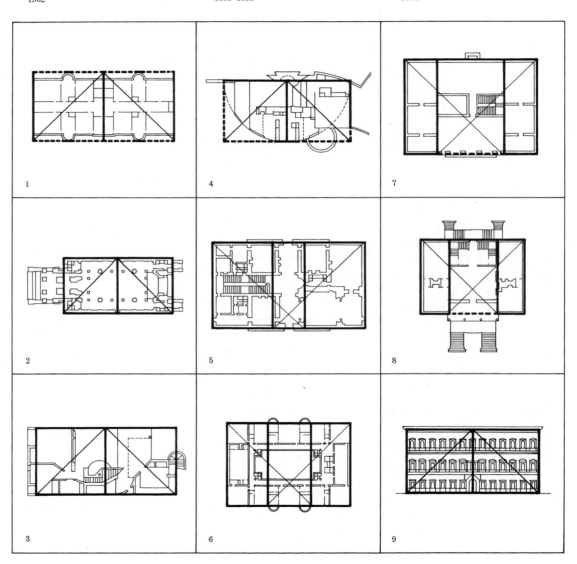

1. **LA ROTONDA**
ANDREA PALLADIO
1566–1571

2. **CHISWICK HOUSE**
LORD BURLINGTON
1729

3. **YORK HOUSE**
WILLIAM CHAMBERS
1759

4. **ST. LOUIS DES INVALIDES**
JULES HARDOUIN MANSART
1676

5. **SANTA MARIA DI CARIGNANO**
GALEAZZO ALESSI
1552

6. **HAGIA SOPHIA**
ANTEMIUS OF TRALLES
532

7. **HOTEL DE MONTMORENCY**
CLAUDE NICHOLAS LEDOUX
1769

8. **SAO FRUTUOSO DE MONTELIOS**
ARCHITECT UNKNOWN
665

9. **THE CAPITOL AT WILLIAMSBURG**
ARCHITECT UNKNOWN
1701

10. **UNITED STATES SUPREME COURT**
CASS GILBERT
1935

11. **WEEKEND HOUSE**
LE CORBUSIER
1935

12. **EXETER LIBRARY**
LOUIS I. KAHN
1967–1972

294　九方阵

　　把每行各有三个邻接正方形的三行连起来，构成一个更大的正方形。这是一种经典式的几何形体。这种三格乘三格的排列通常被称为九方阵形，虽然其中的各个小格不一定是正方形。圆厅别墅（1）、基斯威克府邸（2）、约克住宅（3）、因瓦立德圣路易教堂（4）和卡里尼亚诺圣玛丽亚教堂（5）都是这种经典式形体的范例。圣索菲亚大教堂（6）和蒙莫朗西旅馆（7）表示长方形的九方阵式排列。在九格布局中选出其中一部分组合起来，可以创造出特别的格式。蒙特利奥圣弗鲁托索教堂（8）是把四个角隐退而变成"十"字形。在中央格子的两侧各排上三个格就形成"H"形，如像威廉斯堡议会大厦（9）那样。在美国最高法院（10）中，由几个连接的大院和中心的格子构成的格式是"X"形。勒·柯布西耶在巴黎附近所做的周末住宅（11）展示了阶梯式的形状，而在埃克塞特图书馆（12）可以看到中间空了一格的方环形。

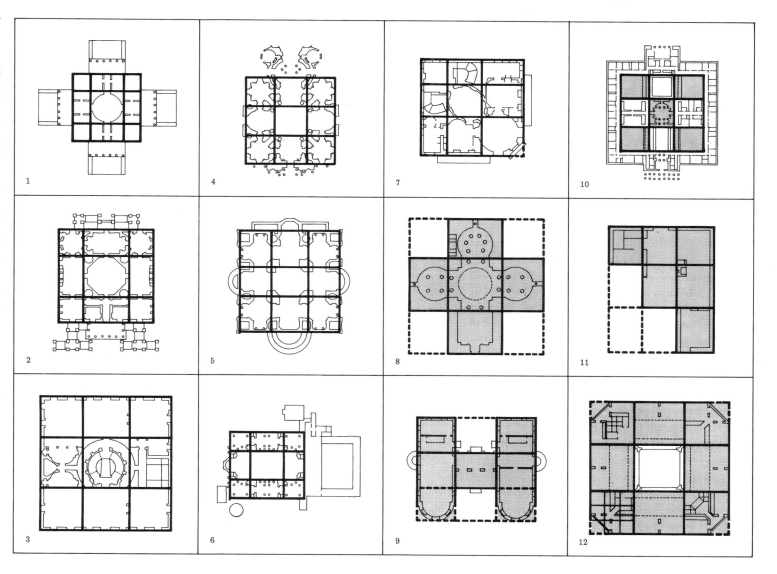

四方阵

1. **THEATER IN BESANÇON FRANCE**
   CLAUDE NICHOLAS LEDOUX
   1775
2. **ST. GEORGE-IN-THE-EAST**
   NICHOLAS HAWKSMOOR
   1714–1729
3. **ADULT LEARNING LABORATORY**
   ROMALDO GIURGOLA
   1972

4. **VILLA SAVOYE**
   LE CORBUSIER
   1928–1931
5. **MUSEUM OF DECORATIVE ARTS**
   RICHARD MEIER
   1981
6. **TRUBEK HOUSE**
   ROBERT VENTURI
   1972

7. **ELIA-BASH HOUSE**
   GWATHMEY-SIEGEL
   1971–1973
8. **VILLA MAIREA**
   ALVAR AALTO
   1937–1939
9. **VIKING FORTRESS**
   ARCHITECT UNKNOWN
   c. 1000

10. **YALE CENTER FOR BRITISH ART**
    LOUIS I. KAHN
    1969–1974
11. **SALK INSTITUTE**
    LOUIS I. KAHN
    1959–1965
12. **HOMEWOOD**
    EDWIN LUTYENS
    1901

四方阵是两格乘两格，并共同相接在一个中心的几何图形。维金要塞（9）是最恰当的例子。勒杜所做的贝桑松剧院（1）和萨伏伊别墅（4）的整体平面是四方阵形，而在圣乔治教堂（2）的一个内部空间是以这种形式组成的。在朱尔戈拉所做的成人学习研究实验室（3）和法兰克福的装饰艺术博物馆（5）中也采用了四方阵组合，其中原有的建筑正好是四方阵中的四分之一，而这个四方阵又转而成为另一个更大的四方阵中的四分之一。四方阵中连接的四个方格不一定是相同的。例如在特鲁贝克住宅（6）中就有两套不一样大的格子。伊利亚-巴什住宅（7）含有四个四分之一的方格，相接在一个明确的中心，梅雷亚别墅（8）中的三个方格是建筑形体，而第四格是个花园。在路易斯·康所做的耶鲁不列颠艺术中心（10）中，整个平面是由九方阵和四方阵组合在一起形成的，平面的整体是由两个重合的九方阵构成的，其中每一格又再分割成四方阵；而在萨尔克生物研究所（11）中的情况是反过来的。在霍姆伍德住宅（12）中，九方阵内藏在四方阵之内，两者共有两个边线。

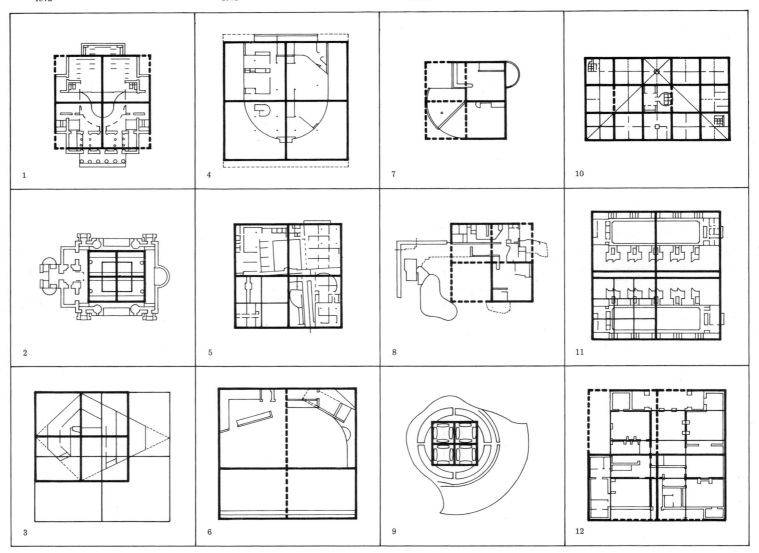

296

1.4 和 1.6 长方形

以正方形的对角线为半径，旋转 45°作为长边，就做出了 1.4 长方形。这个形状确定了谢姆贝格住宅（1）、老圣器堂（2）、兰氏音乐中心（3）和圣詹姆斯教堂（4）的整体平面或内部空间的边界线。一个正方形的两条对角线由于旋转而构成的形状确定了利斯特县法院（5）和纳希顿府邸（6）的平面。由正方形一半的对角线旋转做出的 1.6 长方形，构成了辛克尔所做的阿尔泰斯博物馆（7）、圣米格尔教堂（8）和米利都会议厅（9）的整个平面。斯坦别墅（10）除了附加部分之外，是在一个长方形范围内构成的。勒·柯布西耶也用 1.6 长方形确定了拉图雷特修道院（11）的内庭院。在威尼斯的德尔蒙多剧院（12）平面中，有两个相互为 1：1.4 比率关系的同心正方形，较大的正方形确定了除台阶以外的整个形状；较小的正方形是座位席的边界线。

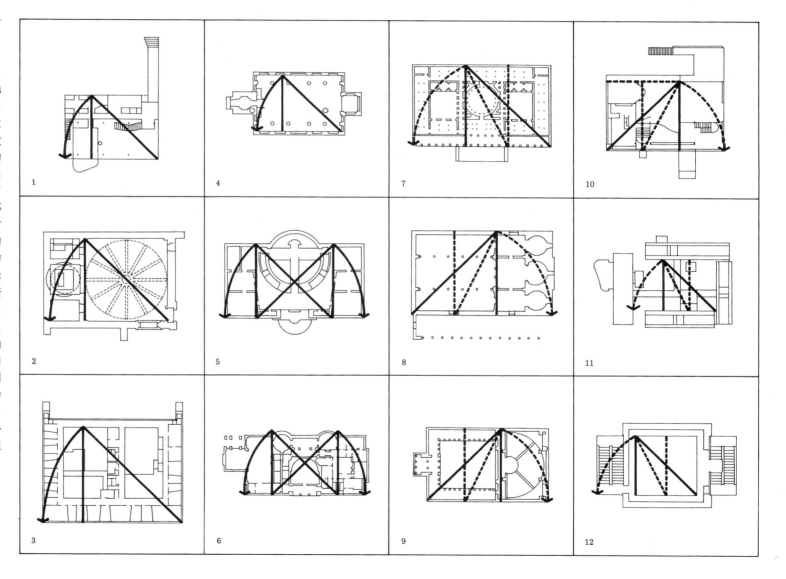

1. **SNELLMAN HOUSE**
   ERIK GUNNAR ASPLUND
   1917–1918
2. **HOTEL GUIMARD**
   CLAUDE NICHOLAS LEDOUX
   1770
3. **ONE-HALF HOUSE**
   JOHN HEJDUK
   1966

4. **NEW LUTHERAN CHURCH**
   ADRIEN DORTSMAN
   1668
5. **HERBERT JACOBS HOUSE**
   FRANK LLOYD WRIGHT
   1948
6. **WIES PILGRIMAGE CHURCH**
   JOHAN & DOMINIKUS ZIMMERMAN
   1754

7. **ORIVESI CHURCH**
   HEKKI SIREN
   1961
8. **SAN CARLO ALLE QUATTRO FONTANE**
   FRANCESCO BORROMINI
   1638–1641
9. **SYDNEY OPERA HOUSE**
   JØRN UTZON
   1957–1968

10. **POST OFFICE SAVINGS BANK**
    OTTO WAGNER
    1904–1906
11. **GUILD HOUSE**
    ROBERT VENTURI
    1961
12. **ROYAL CHANCELLERY**
    ERIK GUNNAR ASPLUND
    1922

297　几何图形的衍生

　　基本的几何图形，通过组合、分割或取其局部，可以衍生出多种多样的形体。三个正方形连接构成斯内尔曼住宅（1）的平面，而两个正方形和四个1.4长方形确定了圭马尔旅馆（2）的边界线。一室半住宅（3）的设计是把半个圆、一个横向对切的正方形和一个斜角对切的正方形组合在一起。新卢特兰教堂（4）和雅各布斯住宅（5）是从两个同心圆的局部演变成的，而维斯教堂（6）是从两个不同中心的圆形产生的。两个圆形重合的共有面积确定了奥里韦西教堂（7）的平面。波罗米尼以四个圆形的局部演变成的形状设计了圣卡洛四喷泉教堂（8）。在悉尼歌剧院（9）中采用了一系列复杂的形体，它们都是一层层球体的切块。邮政储蓄银行（10）、老年人公寓（11）和王室官邸（12）是从三角形演变出来的。后两者中的三角形是以建筑各个角的虚线来暗示的。王室官邸的设计又是两个三角形的组合。

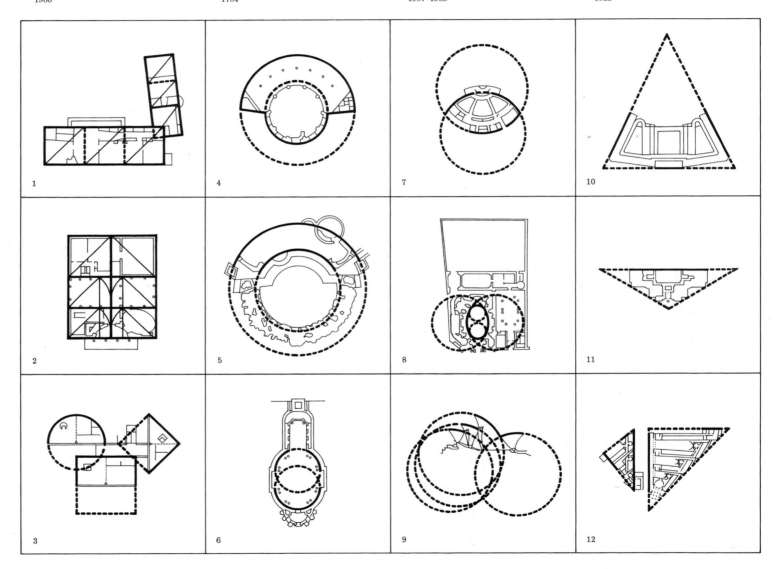

298　旋转、移位和重合

　　旋转、移位和重合是建筑创作中常用的几何形表现方式。在圣玛丽亚布道所（1）中，两个同心正方形旋转45°。在圣灵教堂（2）中采用了顺序的三组，每组都是两个旋转的正方形。在职业保健中心（3）中，两个不同的直角形又旋转又重合，而费希尔住宅（4）的平面是两个相似而又旋转的形体，它们之间只有很少的一点连接。在兰德银行（5）中，一个圆体成为两个另外的形体旋转的中枢。纽约赫拉尔德大楼（6）、圣安德鲁斯宿舍（7）、库诺住宅（8）和斯内尔曼住宅（9）都是铰接式的例子——带形体围绕某个共同的重合空间旋转。交通流线的改变强烈地表现出了迪尔·韦斯特办公楼（10）相对于一个公共空间的移位。在卡彭特中心（11）中，两个相似的形状翻转过来，并相对于一交通坡道移位。通过对角线的移位和重合，斯特林做出了剑桥大学历史系图书馆（12）的主要使用空间。其他关于重合几何形的例子是梅尔尼科夫住宅、德雷顿大厅、伊斯顿·内斯顿府邸以及耶鲁不列颠艺术中心。

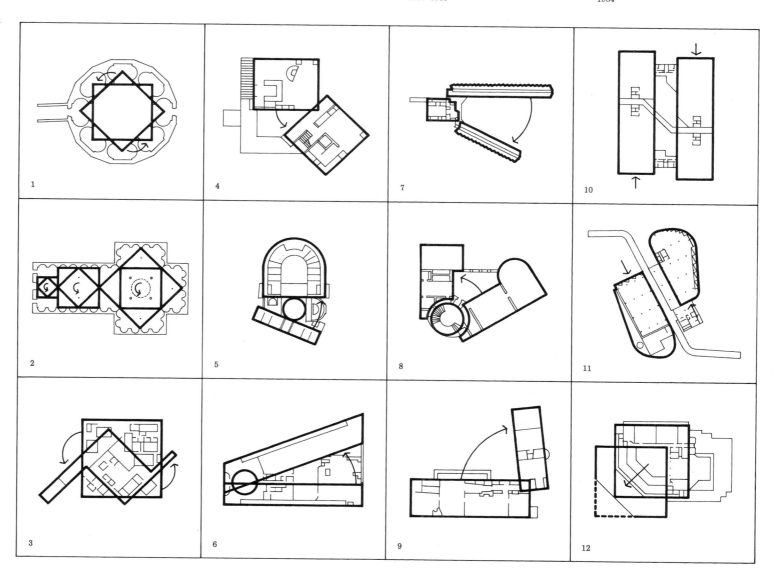

风车形、放射形和螺旋形

　　风车形、放射形和螺旋形这三种形体或空间的形状，都具有一个共同的中心或原点。风车形是带形体围绕一个明确的内核的有规律的布局，例如"展翅"住宅（1）；或者是围绕一个无形的内核，例如古根海姆美术馆（2）。在纽帕克楼（3）中，相邻的空间像风车似地环绕着一个小的交通核心。理查德医学研究中心（4）中，三个较复杂的单元围绕一个设备中心，构成一个风车形。在艾哈迈达巴德博物馆（5）中包括两个风车形，一个在画廊里面，第二个是由三个建筑形体构成并与主体建筑相连。一系列的单体，明确的或是暗示的，从一个中心发出，就表现为放射形。弗洛雷大楼（6）的放射中心有两个，而奥古斯都陵墓（7）是最经典式的放射形。沃尔夫斯堡教区中心教堂（8）的结构是从一个中心发出的，而诺瓦尔公寓（9）的墙是由几个中心放射出的。螺旋形出现在代代木小体育馆（10）和圣安东尼乌斯教堂（11）。新英格兰水族馆（12）是由两个螺旋形构成的，中央的一个是圆的，周边的一个是直线的。

1. **WINGSPREAD**
   FRANK LLOYD WRIGHT
   1937
2. **GUGGENHEIM MUSEUM**
   FRANK LLOYD WRIGHT
   1956
3. **NEW PARK**
   ARCHITECT UNKNOWN
   c. 1775

4. **RICHARDS RESEARCH BUILDING**
   LOUIS I. KAHN
   1957–1961
5. **MUSEUM AT AHMEDABAD, INDIA**
   LE CORBUSIER
   1953–1957
6. **FLOREY BUILDING**
   JAMES STIRLING
   1966

7. **MAUSOLEUM OF AUGUSTUS**
   ARCHITECT UNKNOWN
   c. 25 B.C.
8. **WOLFSBURG PARISH CENTER CHURCH**
   ALVAR AALTO
   1960–1962
9. **NEUR VAHR APARTMENTS**
   ALVAR AALTO
   1958–1962

10. **SMALL OLYMPIC ARENA**
    KENZO TANGE
    1961–1964
11. **ST. ANTONIUS CHURCH**
    JUSTUS DAHINDEN
    1966–1969
12. **NEW ENGLAND AQUARIUM**
    CAMBRIDGE SEVEN ASSOCIATES
    1962

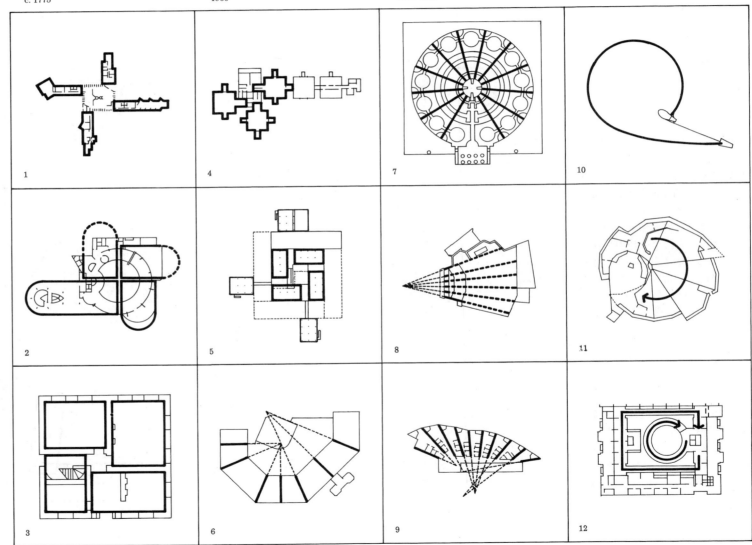

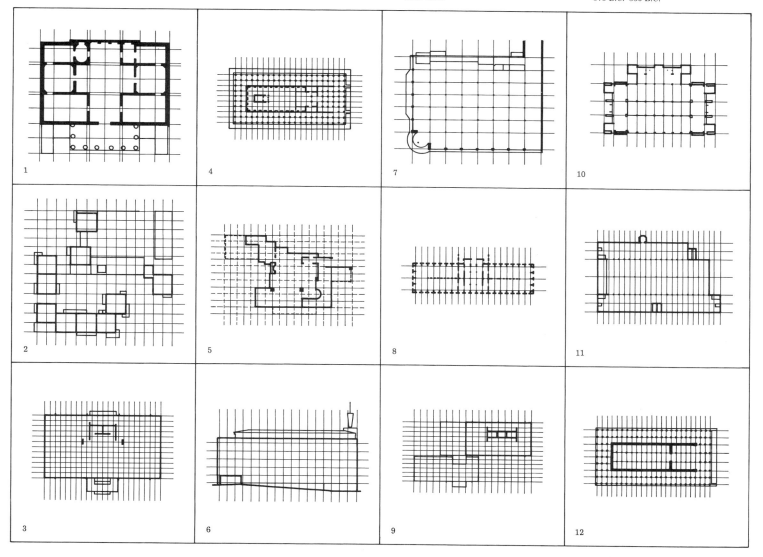

1. VILLA FOSCARI
ANDREA PALLADIO
c. 1549–1563

2. SEA RANCH CONDOMINIUM I
CHARLES MOORE
1964–1965

3. CROWN HALL
LUDWIG MIES VAN DER ROHE
1950–1956

4. TEMPLE OF APOLLO
PAEONIUS AND DAPHNIS
c. 310 B.C.

5. FALLINGWATER
FRANK LLOYD WRIGHT
1935

6. ENSO-GUTZEIT HEADQUARTERS
ALVAR AALTO
1959–1962

7. CARSON PIRIE AND SCOTT STORE
LOUIS SULLIVAN
1899–1903

8. SAINTE GENEVIEVE LIBRARY
HENRI LABROUSTE
1838–1850

9. FARNSWORTH HOUSE
LUDWIG MIES VAN DER ROHE
1945–1950

10. LARKIN BUILDING
FRANK LLOYD WRIGHT
1903

11. A.E.G. HIGH TENSION FACTORY
PETER BEHRENS
1910

12. FOURTH TEMPLE OF HERA
RHOIKOS OF SAMOS
575 B.C.–550 B.C.

300　网　　格

　　网格是由重复的基本几何形构成的。福斯卡里别墅（1）、西兰奇一期共管住宅（2）、克朗楼（3）和古希腊米利都的阿波罗神庙（4）都源出于正方形网格。用在流水别墅（5）的平面图和恩索－古特蔡特公司总部（6）的立面图上的网格中有主次轻重不同的区别。卡森·皮里与斯科特百货大楼（7）、圣日纳维夫图书馆（8）和赫拉第四神庙（12）的直线网格是和结构一致的。直线网格出现在范斯沃斯住宅（9）、拉金公司大厦（10）和德国通用电气公司高拉力车间（11）。金贝尔艺术博物馆（13）、特林顿浴室（14）和圣塞巴斯蒂亚诺教堂（15）是花网格的例子。内布拉斯加州议会大楼（16）产生于一个三元花网格。巴黎圣母院（17）和菲瑟住宅（18）也是如此。布莫尔住宅（19）和唯一神教派教堂（21）是等边三角形网格，而国家美术馆东馆（20）是等腰三角形。莱斯特大学工程馆（22）、会堂大厦（23）和图伦·萨诺马办公室（24）这几个例子都是表示在主体或空间的接合处的网格出现了位移。威尔斯学院图书馆（25）所在的网格区是由网格的旋转和重合构成的。安克楼（26）和群马美术馆（27）都是旋转网格的例子。

13. KIMBELL ART MUSEUM
LOUIS I. KAHN
1966–1972

14. TRENTON BATH HOUSE
LOUIS I. KAHN
1955–1956

15. SAN SEBASTIANO
LEON BATTISTA ALBERTI
1459

16. NEBRASKA STATE CAPITOL
BERTRAM GOODHUE
1924

17. NOTRE DAME CATHEDRAL
ARCHITECT UNKNOWN
1163–c. 1250

18. VISSER HOUSE
ALDO VAN EYCK
1975

19. JORGINE BOOMER RESIDENCE
FRANK LLOYD WRIGHT
1953

20. EAST WING OF NATIONAL GALLERY
I. M. PEI
1975–1978

21. UNITARIAN CHURCH
FRANK LLOYD WRIGHT
1949

22. LEICESTER ENGINEERING BUILDING
JAMES STIRLING
1959

23. AUDITORIUM BUILDING
LOUIS SULLIVAN
1887–1890

24. TURUN SANOMAT OFFICES
ALVAR AALTO
1927–1929

25. WELLS COLLEGE LIBRARY
SKIDMORE-OWINGS-MERRILL
1968

26. THE ANKER BUILDING
OTTO WAGNER
1895

27. GUMMA MUSEUM OF FINE ARTS
ARATA ISOZAKI
1971–1974

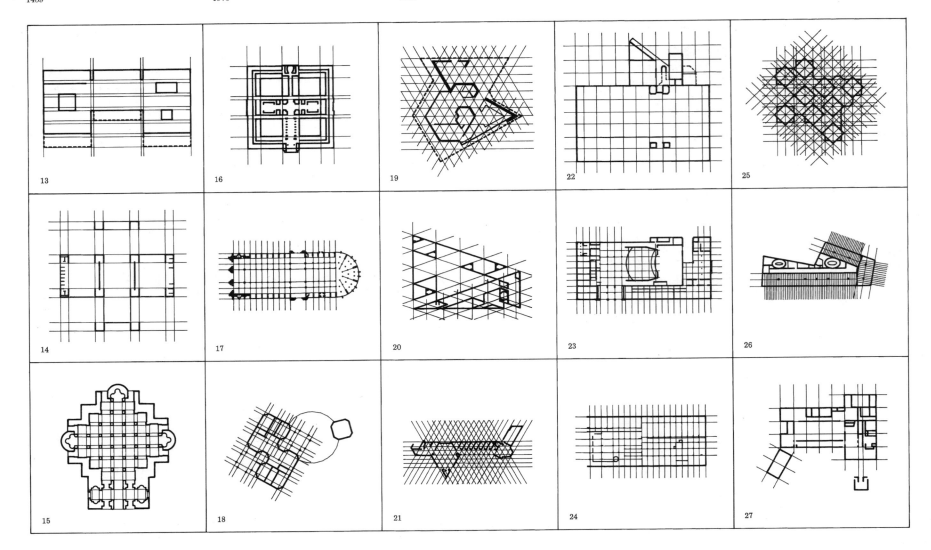

1. **TEMPLE AT TARXIEN, MALTA**
ARCHITECT UNKNOWN
2100 B.C.–1900 B.C.
2. **SOLOMON'S TEMPLE**
ARCHITECT UNKNOWN
1000 B.C.
3. **HOTEL DE MONTMORENCY**
CLAUDE NICHOLAS LEDOUX
1769

4. **TEMPLE OF HORUS**
ARCHITECT UNKNOWN
237 B.C.–57 B.C.
5. **TOMB OF SETNAKHT**
ARCHITECT UNKNOWN
13th CENTURY B.C.
6. **DULWICH GALLERY**
JOHN SOANE
1811–1814

7. **HOUSE IN CENTRAL PENNSYLVANIA**
HUGH NEWELL JACOBSEN
1980
8. **REDENTORE CHURCH**
ANDREA PALLADIO
1576–1591
9. **LAURENTIAN LIBRARY**
MICHELANGELO
1525

302 **构形模式**

构形模式描述的是建筑各部分的相对布局，在空间的设计以及空间和形体组群的组织方面，模式是基本的主题。以下举例说明带状、中心、双中心、聚集、筑巢、同心和双核等各种模式。

带状模式：使用性的

通道穿过使用空间的形状有两类，其中的通道都形成一条带状的形式。第一类，空间和空间是互相连接的，交通流线由空间通向空间。第二类，交通流线纵向地含于一个空间之内。在马耳他的塔尔欣神庙（1）中，空间是连接在横轴线上的，因而把每个纵向空间变成了三个隐含的空间。所罗门神庙(2)和伊德富的何露斯神庙（4）这两个实例都是表示在通过一连串空间的运动中，建筑中的高潮是在通道的起点和终点上。在勒杜所做的蒙莫朗西旅馆（3）中，通道在第二层上原线返回，所以通道的起点和终点是彼此上下相对应的。在塞特纳赫特陵墓（5）中带状的各个空间既有纵向的也有横向的。正是这种变化产生了高潮。在索恩所做的达利奇艺术馆（6）中，入口位于带状相连的空间的中点上。在雅各布森所做的中宾州住宅（7）里，互相连接的空间的中心是实心的，而交通线路是沿着边缘的。里丹托教堂（8）和劳仑齐阿纳图书馆（9）这两个实例都是把单个的空间以一条带的方式组织起来。里丹托教堂与塞特纳赫特陵墓类似，高潮是在通道的沿线上。

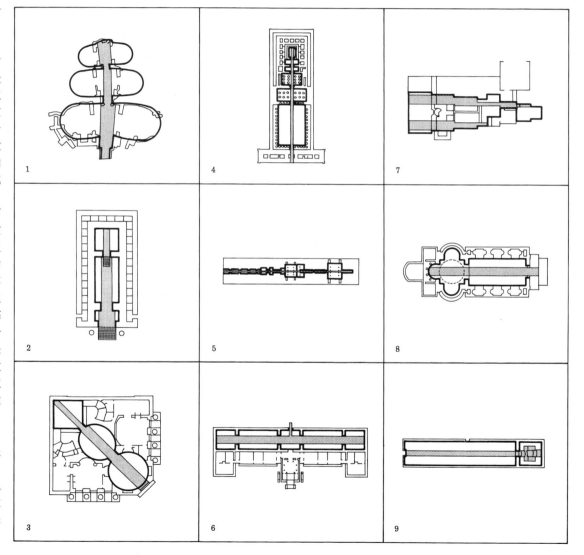

1. **STOA IN SIKYON, GREECE**
   ARCHITECT UNKNOWN
   c. 300

2. **PHYSICAL EDUCATION FACILITY**
   KALLMAN-McKINNELL
   1970

3. **BAGSVAERD CHURCH**
   JØRN UTZON
   1973–1976

4. **FORT SHANNON**
   ARCHITECT UNKNOWN
   1800–1835

5. **SNELLMAN HOUSE**
   ERIK GUNNAR ASPLUND
   1917–1918

6. **BAKER DORMITORY**
   ALVAR AALTO
   1947–1948

7. **UNITE D'HABITATION**
   LE CORBUSIER
   1946–1952

8. **FLOREY BUILDING**
   JAMES STIRLING
   1966

9. **ST. ANDREWS DORMITORY**
   JAMES STIRLING
   1964

10. **STERN HOUSE**
    CHARLES MOORE
    1970

11. **CENTRE BEAUBOURG**
    PIANO AND ROGERS
    1972–1977

12. **PEARSON HOUSE**
    ROBERT VENTURI
    1957

303 带状模式：交通性的

在带状中通道若是从使用空间中分隔开的，这就是脊椎式或走廊式。古希腊建筑的拱廊（西基翁的柱廊建筑）（1）是这种模式的最简单的形体，而埃克塞特体育训练馆（2）所表现的是典型的脊椎式方案。在这类情况中，脊椎是决定形体的首要因素，在伍重所做的鲍斯韦教堂（3）中，包含着重复的形体语汇，并以此排列出各个使用空间的位置。爱尔兰住宅（即香农要塞）（4）和斯内尔曼住宅（5）是单面走廊的例子。阿尔托所做的贝克学生宿舍（6）说明带状交通路线并不一定要直的或两边对称的，而勒·柯布西耶所做的马赛公寓大楼（7）的交通流线在剖面上十分重要。在斯特林所做的弗洛雷大楼（8）和圣安德鲁斯宿舍（9）中的交通流线都可以在室外看到，它还说明通道可不是直的。两个脊椎式的通道可以同时存在，如像在穆尔所做的斯特恩住宅（10）中那样，这里的两条脊椎是互相交叉的。在博布中心（11），两条脊椎是平行的，一条用于竖向交通，另一条用于横向交通。文丘里所做的皮尔森住宅（12），采用了两种类型的带状模式。其中私人用房是以一条隔开的通道联系起来的，而公共面积的交通流线则隐含并穿越其间。

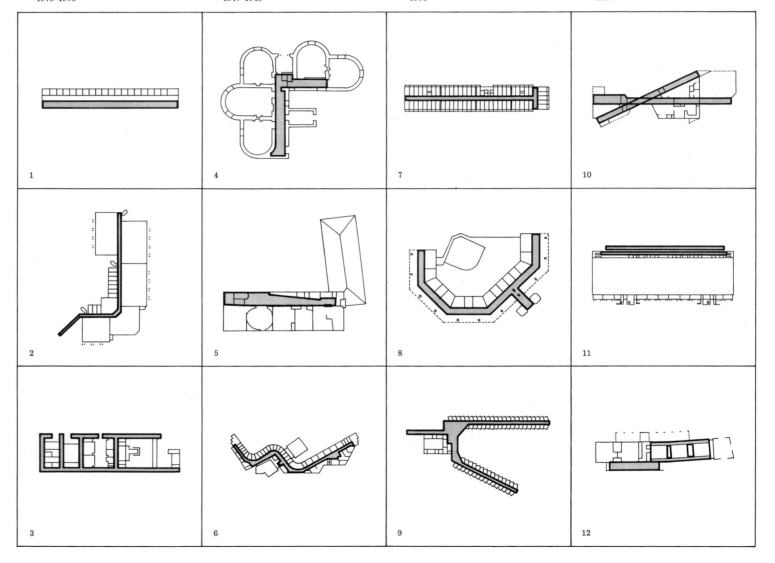

1. **FIRST UNITARIAN CHURCH**
   LOUIS I. KAHN
   1959–1967
2. **WOLLATON HALL**
   ROBERT SMYTHSON
   1580–1588
3. **SHAKER BARN**
   ARCHITECT UNKNOWN
   1865

4. **HUNTING LODGE**
   KARL FRIEDRICH SCHINKEL
   1822
5. **PALACE OF CHARLES V**
   PEDRO MACHUCA
   1527
6. **FARNESE PALACE**
   ANTONIO DA SANGALLO
   1534

7. **ST. COSTANZA**
   ARCHITECT UNKNOWN
   c. 350
8. **TRINITY CHURCH**
   HENRY HOBSON RICHARDSON
   1872–1877
9. **ST. MARY WOOLNOTH**
   NICHOLAS HAWKSMOOR
   1716–1724

10. **SECOND BANK OF THE U.S.**
    WILLIAM STRICKLAND
    1818–1824
11. **STOCKHOLM PUBLIC LIBRARY**
    ERIK GUNNAR ASPLUND
    1920–1928
12. **SAN MARIA DEGLI ANGELI**
    FILIPPO BRUNELLESCHI
    1434–I 436

304 　中心模式：使用性的

　　把最重要的空间放在中心位置上，再用通向或者环绕这个空间的不同模式与之衔接，这就是中心模式。在路易斯·康所做的唯一神教派第一教堂（1）和沃莱顿府邸（2）中，中央大厅从上部照亮并在形体上占有主导地位，它是被一些较小的空间和隔开的通道包围起来的。在谢克谷仓（3）中的交通道围绕着中央的干草堆，这个干草堆具有象征意义的、功能方面的和形式上的重要性。在辛克尔所做的猎庄（4）中，中央八角形大厅的周圈是交通道，而较小的空间只在四个边上。在查理五世王宫（5）和法尔尼斯府邸（6），中央空间是个庭院，以柱廊作为它的交通线路。在圣康斯坦扎教堂（7）的中心是最神圣的地方，而在圣三一教堂（8）和圣玛丽·伍尔塔思教堂（9）中，中心是放在一个更大的空间之内。斯特里兰所做的美国第二银行（10）中最重要的空间位居中央，有隐含的交通线和较小的空间在两侧。斯德哥尔摩公共图书馆（11）的通道是在中央空间的周圈。伯鲁乃列斯基所做的圣玛丽亚布道所（12）有一个主要的中央空间，被围在较小的空间之中，交通线通过这些小空间，既环绕着同时又都通向中央空间。

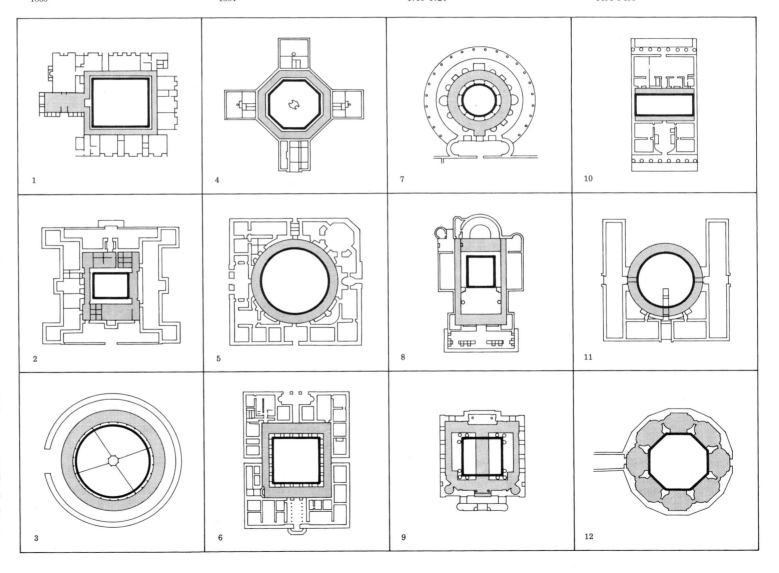

1. **LA ROTONDA**
ANDREA PALLADIO
1566–1571

2. **NORTH CAROLINA STATE CAPITOL**
TOWN AND DAVIS
1833–1840

3. **UNITED STATES CAPITOL**
THORNTON-LATROBE-BULFINCH
1793–1830

4. **HOUSE IN UR**
ARCHITECT UNKNOWN
2000 B.C.

5. **HOTEL DE BEAUVAIS**
ANTOINE LE PAUTRE
1656

6. **BLOEMENWERF HOUSE**
HENRY VAN DE VELDE
1895–1896

7. **BALTIMORE-OHIO RAILROAD DEPOT**
FRANK FURNESS
1886

8. **BURN HALL**
JOHN SOANE
c. 1785

9. **THE SALUTATION**
EDWIN LUTYENS
1911

10. **EXETER LIBRARY**
LOUIS I. KAHN
1967–1972

11. **CONVENT OF LA TOURETTE**
LE CORBUSIER
1957–1960

12. **STRATFORD HALL**
ARCHITECT UNKNOWN
1725

305　中心模式：交通性的

　　圆厅别墅（1）、北卡罗来纳州议会大楼（2）和美国国会大厦（3）都是古典式圆穹建筑的范例。在这些例子中，中央空间虽然在外形上起主要作用，它实际上是起着交通和组织其他空间的作用。乌尔的住宅（4）和博韦旅馆（5）中的庭院是古典式圆穹建筑的变形。在这两个建筑中，庭院是平面的主要部分，又是用来组织交通和小空间的，但在外表上并不显露。在凡·德·费尔德所做的布卢门韦夫住宅(6)、弗内斯所做的巴尔的摩—俄亥俄火车站（7）、索恩所做的伯恩府邸（8）和勒琴斯所做的萨吕泰兴府邸（9）中，中央空间是用作垂直交通和建筑的竖向组织。路易斯·康所做的埃克塞特图书馆（10），中央空间在主层上是个圆穹顶，而在上部各层则是由交通道环绕着这个空间。与此有些类似，在拉图雷特修道院（11）中的庭院具有中心模式的两种类型，有时候交通线路像回廊似地围绕着庭院，而有时又可穿过庭院。斯特拉特福大厅（12）的中央空间是主要的使用空间，并且像圆穹建筑那样，有交通流线穿过它通向较小的空间。

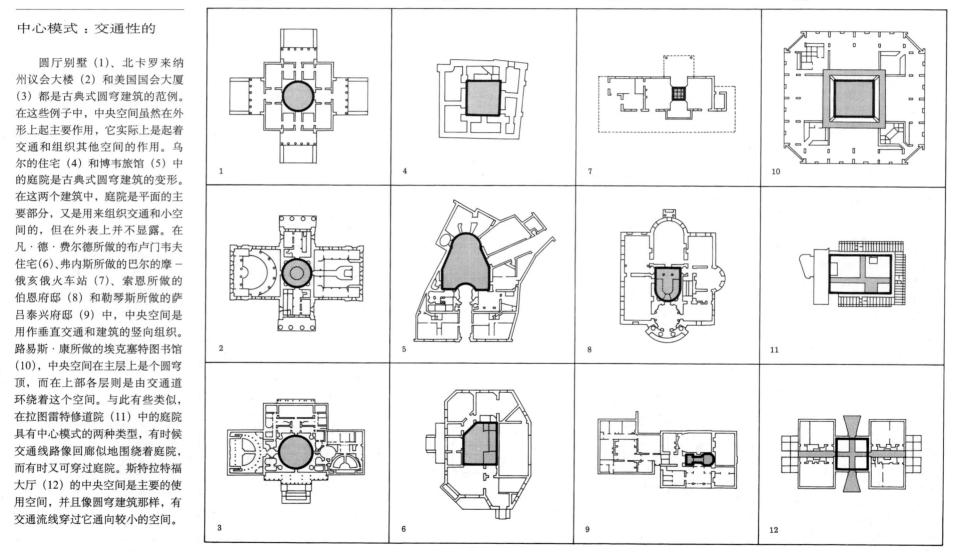

306　双中心模式

　　双中心式是指在同一个区域或场地内有两个同等重要的中心点。维纳斯与罗马神庙（1）有两个一样的、主要的房间，方向相反，位于神庙的其他部分的范围之内。每一个中心都是一个实体，放在形似空地的区域内。在法隆寺（2）和大莱普提斯市场（3），这种区域是个室外庭院，而在穆尔所做的奥林达的穆尔住宅（4）和议会大厦（5）中，这个区域是一间房间和室内空间。斯卡尔帕所做的布里永－维加公墓（6）有一个中心如同一个在室外场地中的实体，而另一个中心是建筑内部的一个房间。如果区域是实的，那么中心点可能是从"实"中刻出的"虚"的。在多佛城堡（7）中，虚的都是主要房间，在宾夕法尼亚美术学院（8），虚的都是特殊的地方，建筑的其余部分只是填充。这些虚的部分也可能作为双中心来组织周围的空间，并且将光线引入建筑内部，在耶鲁不列颠艺术中心（9）、孤儿院（10）、枢密院（11）和米拉公寓（12）都是这种情况。

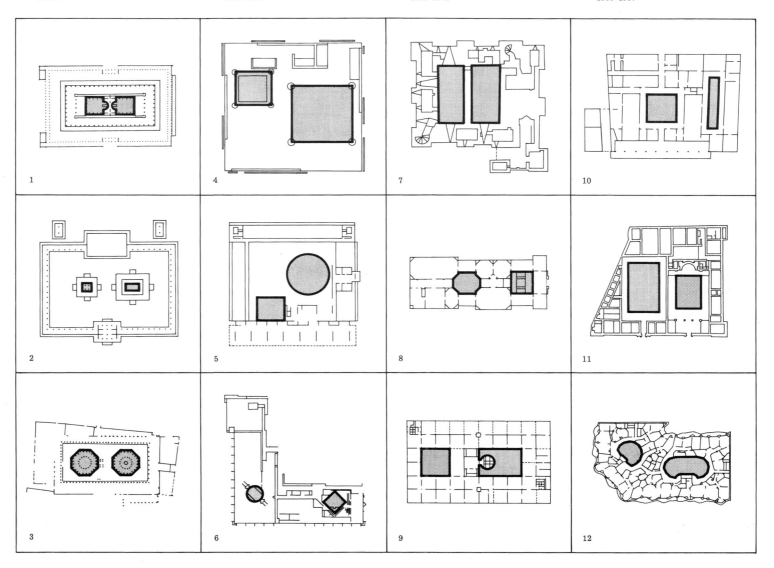

1. **TOWER OF LONDON**
   ARCHITECT UNKNOWN
   1070–1090
2. **FORTRESS NEAR RUDESHEIM**
   ARCHITECT UNKNOWN
   1000–1050
3. **HOUSE OF VIZIER NAKHT**
   ARCHITECT UNKNOWN
   1372 B.C.–1350 B.C.

4. **W. WATTS SHERMAN HOUSE**
   HENRY HOBSON RICHARDSON
   1874
5. **D. L. JAMES HOUSE**
   GREENE AND GREENE
   1918
6. **OLAVINLINNA CASTLE, FINLAND**
   ARCHITECT UNKNOWN
   1475

7. **CASTLE IN SOBORG, DENMARK**
   ARCHITECT UNKNOWN
   c. 1150
8. **OCCUPATIONAL HEALTH CENTER**
   HANDY-HOLZMAN-PFIEFFER
   1973
9. **CONVENT FOR DOMINICAN SISTERS**
   LOUIS I. KAHN
   1965–1958

10. **HOUSE IN TUCKER TOWN, BERMUDA**
    ROBERT VENTURI
    1975
11. **OLIVETTI TRAINING SCHOOL**
    JAMES STIRLING
    1969
12. **FONTHILL-MERCER CASTLE**
    HENRY MERCER
    1908–1910

307　聚集模式

空间或形体以不确定的模式集合起来，被称为聚集式。空间的聚集常常能够决定建筑的形体，至少对形体有很大影响，如在伦敦塔（1）和沃茨·合曼住宅（4）。然而，空间也可以聚集在一个事先已确定了外形的空间之内，在德国吕德斯海姆附近的要塞（2）和维齐耶·纳赫特住宅（3），是这一类型的空间聚集的实例。以上两种聚集式在詹姆斯住宅（5）中都很明显，而其中更主要的模式是空间聚集确定了形体变化。在芬兰的奥拉韦林纳城堡（6）和丹麦的索伯城堡（7）中，既有形体的也有空间的聚集体。聚集式的一个准则是聚集起的各个单体之间必须很靠近。在一定程度上，城堡的围墙造成了这种近距离的布局，而在职业保健中心（8）中，造成这种近距离布局的，是集合了各个形体的一个更大的房间。聚集起来的各个形体在其内部也可以再分割，只是这些分割的空间都不能太大。路易斯·康所做的多米尼加修女会修道院（9）、文丘里所做的百慕大塔克镇的住宅（10）、斯特林所做的奥利韦蒂专科学校（11）以及冯希尔－默瑟府邸（12）都是聚集式的例子。

1. **TEMPLE OF APOLLO**
ARCHITECT UNKNOWN
c. 400 B.C.

2. **TEMPLE OF KOM OMBO**
ARCHITECT UNKNOWN
181 B.C.–30 A.D.

3. **THE PALACE OF ASSEMBLY**
LE CORBUSIER
1953–1963

4. **MOORE HOUSE**
CHARLES MOORE
1962

5. **ENSO-GUTZEIT HEADQUARTERS**
ALVAR AALTO
1959–1962

6. **CAMBRIDGE HISTORY FACULTY**
JAMES STIRLING
1964

7. **J. J. GLESSNER HOUSE**
HENRY HOBSON RICHARDSON
1885–1887

8. **CHANDLER HOUSE**
BRUCE PRICE
1885–1886

9. **HOMEWOOD**
EDWIN LUTYENS
1901

308 筑巢模式

　　筑巢模式是指其中的每一个单元依次连续地套在下一个更大的单元中间，而每一个单元都有不同的中心。在意大利庞培的阿波罗神庙（1）和埃及的考姆翁布神庙（2）中，各个单元有共同的中心线。议会大厦（3）中的几何图形变化，说明筑巢式中的各个单元不一定是同样的形体语汇。查尔斯·穆尔所做的奥林达的穆尔住宅（4）含有两组筑巢的模式。由于巢中的各个单元没有共同的中心，它们可能在其形状中的某些部分是共同的。所有的单元可能共有一个边，例如阿尔托所做的恩索－古特蔡特公司总部（5）。虽然筑巢的各单元共有两条边和一个角的情况较为常见。然而更普遍的是单元以斜角线的方向筑巢而居。斯特林所做的历史系图书馆（6）、理查森所做的格莱斯纳住宅（7）、普赖斯所做的钱德勒住宅（8）和勒琴斯所做的霍姆伍德住宅（9）都是这种筑巢模式的实例。

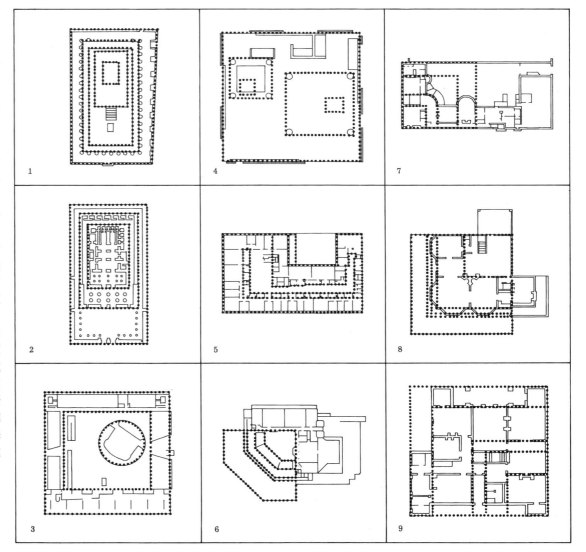

1. **EXETER LIBRARY**
   LOUIS I. KAHN
   1967–1972
2. **STOCKHOLM PUBLIC LIBRARY**
   ERIK GUNNAR ASPLUND
   1920–1928
3. **PANTHEON IN PARIS, FRANCE**
   JACQUES GERMAIN SOUFFLOT
   1756–1797
4. **SANTO STEFANO ROTONDO**
   ARCHITECT UNKNOWN
   468–483
5. **ALLEGHENY COUNTY COURTHOUSE**
   HENRY HOBSON RICHARDSON
   1883–1888
6. **UNITY TEMPLE**
   FRANK LLOYD WRIGHT
   1906
7. **FONTEVRAULT ABBEY**
   ARCHITECT UNKNOWN
   1115
8. **VILLA FARNESE**
   GIACOMO DA VIGNOLA
   1559–1564
9. **CHURCH OF SAN LORENZO**
   GUARINO GUARINI
   1666–1679
10. **ST. GEORGE-IN-THE-EAST**
    NICHOLAS HAWKSMOOR
    1714–1729
11. **THEATER IN BESANÇON, FRANCE**
    CLAUDE NICHOLAS LEDOUX
    1775
12. **PARTHENON**
    ICTINUS
    447–430 B.C.

309   同心模式

同心模式是指每个单元依次连续套在下一个更大的单元之中，而且各个单元共有一个中心。埃克塞特图书馆（1）是以简单几何形构成同心式的范例。在斯德哥尔摩公共图书馆（2）中，阿斯普隆德采用了不同语汇的简单形体。巴黎的万神庙（3）采取了略为复杂的单元，但基本上还是重复的。在圣斯提芬诺圆厅教堂（4），简单的圆几何形是重复的，但每个圆环的连接方式不同。阿勒格尼县法院（5）表示各个同心的形状的功能是不同的。在联合教堂（6）中只有主要空间中有同心式层次。丰特夫罗拉拜修道院（7）、法尔尼斯别墅（8）和圣洛伦佐教堂（9）等实例说明，由于同心单元中每个单元的几何图形都不一样，因此产生了复杂的形状。豪克斯穆尔在圣乔治教堂（10）中采用了筑巢和同心两种模式。勒杜所做的贝桑松剧院（11）的筑巢模式中隐含着平面的另一半，因此使整个平面仍然可以看成是同心式的。帕提衣神庙（12）的外层是同心式的，而到了内层就转变成了筑巢式。

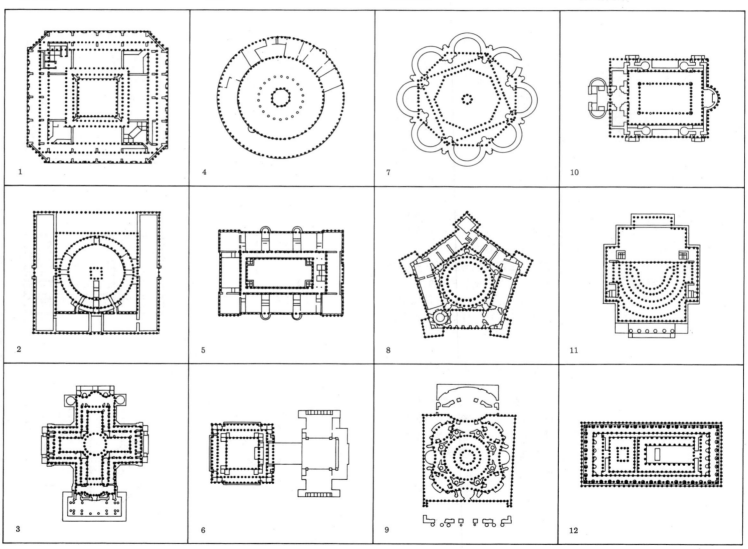

双核模式中包含两个同等重要的部分。两个核心部分之间的连接处可能是个作为入口的建筑形体，例如罗宾逊住宅（1）、威廉斯堡议会大厦（2）和联合教堂（4）。连接体也可能是主要的使用空间，如斯特拉特福大厅（3）；或者是一个桥，如在女王宫（5）。双核体之间也可以用一块空地或空间来连接，这个连接可以是实在的连接，如萨尔克生物研究所（9）；或者是暗示的连接，如邮政储蓄银行（6）、奥利韦蒂专科学校（8）；或纳希顿府邸（7）。大分医学楼（10）、赫尔辛基文化馆（11）和梅隆艺术中心（12），这些例子表现两个隔开的不同几何图形。在圣保罗教堂（13）和第波利会议中心（19）有两个不同几何形直接统一起来。柏林观象台（14）和里丹托教堂（15）把简单和复杂的形体放在一起。范斯沃斯住宅（16）、美国科学院（17）和鲍尔中心（18）中出现的双核式组成单体是一虚一实。相似的双核单体可以有不同的朝向，如在卡彭特中心（20）和费希尔住宅（21）。两个单体可以在形式上相似而功能上不同，如兰氏音乐中心（22）和罗比住宅（23）。双核式也能表现在立面上，例如勒·柯布西耶所做的苏黎世展览馆（24）。

1. **ROBINSON HOUSE**
MARCEL BREUER
1947

2. **THE CAPITOL AT WILLIAMSBURG**
ARCHITECT UNKNOWN
1701

3. **STRATFORD HALL**
ARCHITECT UNKNOWN
1725

4. **UNITY TEMPLE**
FRANK LLOYD WRIGHT
1906

5. **THE QUEEN'S HOUSE**
INIGO JONES
1629–1635

6. **POST OFFICE SAVINGS BANK**
OTTO WAGNER
1904–1906

7. **NASHDOM**
EDWIN LUTYENS
1905–1909

8. **OLIVETTI TRAINING SCHOOL**
JAMES STIRLING
1969

9. **SALK INSTITUTE**
LOUIS I. KAHN
1959–1965

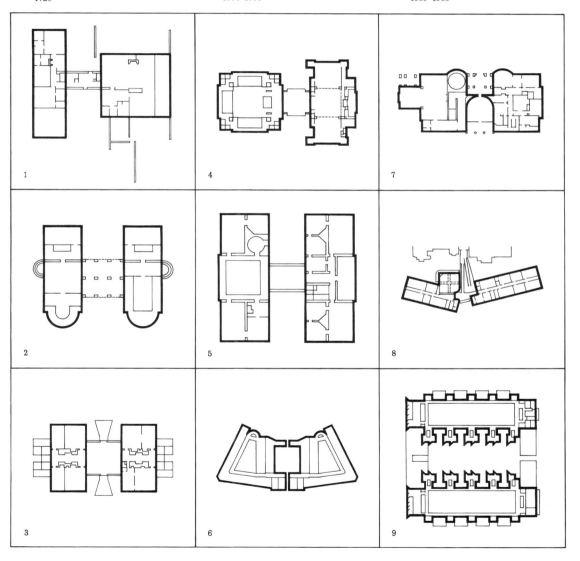

10. **ANNEX TO OITA MEDICAL HALL**
ARATA ISOZAKI
1970–1972

11. **HOUSE OF CULTURE IN HELSINKI**
ALVAR AALTO
1955–1958

12. **PAUL MELLON ARTS CENTER**
I. M. PEI
1970–1973

13. **ST. PAUL'S CHURCH**
LOUIS SULLIVAN
1910–1914

14. **OBSERVATORY IN BERLIN**
KARL FRIEDRICH SCHINKEL
1835

15. **REDENTORE CHURCH**
ANDREA PALLADIO
1576–1591

16. **FARNSWORTH HOUSE**
LUDWIG MIES VAN DER ROHE
1945–1950

17. **THE AMERICAN ACADEMY IN ROME**
McKIM, MEAD, AND WHITE
1913

18. **POWER CENTER**
ROCHE-DINKELOO
1965–1971

19. **DIPOLI CONFERENCE CENTER**
REIMA PIETILA
c. 1966

20. **CARPENTER CENTER**
LE CORBUSIER
1961–1963

21. **NORMAN FISHER HOUSE**
LOUIS I. KAHN
1960

22. **LANG MUSIC BUILDING**
ROMALDO GIURGOLA
1973

23. **FREDERICK G. ROBIE HOUSE**
FRANK LLOYD WRIGHT
1909

24. **ZURICH EXHIBITION PAVILION**
LE CORBUSIER
1964–1965

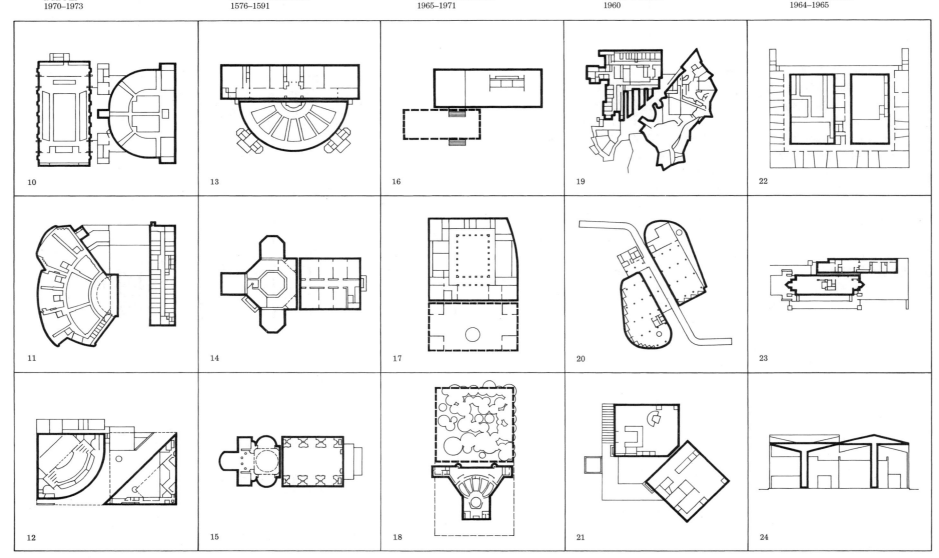

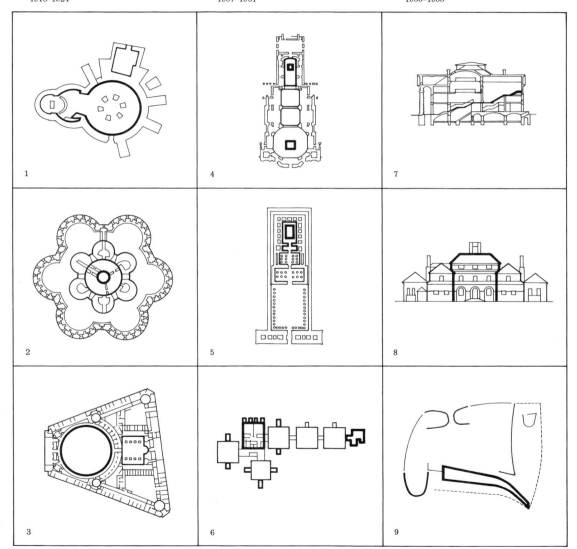

1. **OSTERLARS CHURCH**
   ARCHITECT UNKNOWN
   12th CENTURY
2. **DEAL CASTLE**
   ARCHITECT UNKNOWN
   c. 1540
3. **POLICE HEADQUARTERS**
   HACK KAMPMANN
   1918–1924

4. **EINSIEDELN ABBEY**
   KASPAR MOOSBRUGGER
   1719–1735
5. **TEMPLE OF HORUS**
   ARCHITECT UNKNOWN
   237 B.C.–57 B.C.
6. **RICHARDS RESEARCH BUILDING**
   LOUIS I. KAHN
   1957–1961

7. **DIRECTOR'S HOUSE**
   CLAUDE NICHOLAS LEDOUX
   1775–1779
8. **HEATHCOTE**
   EDWIN LUTYENS
   1906
9. **CHAPEL AT RONCHAMP**
   LE CORBUSIER
   1950–1955

312 **演　进**

演进是一种逐级变化。这种变化暗示着从一个状态或属性出发到另一种状态或属性的运动。变化的性质确定演进的类型。以下举例说明等级关系、转换、变形和中介等不同类型的演进。

等级关系

等级关系是根据各个单体的某个属性范畴做出的地位次序安排，所以每个单体的重要性和价值取决于这一属性存在或是不存在。在丹麦的厄斯特拉斯教堂（1）中的等级体系是以内部空间的大小来确定的。迪尔城堡（2）是一个同心式的范例，它表现了一种向心型的地位次序，越靠近中心的空间越重要。警察总署（3）的等级体系是由形体和空间的大小、完整性和给人印象深刻的程度来决定的，它的等级范围是从主要的形状到背景性或填充性的建筑。瑞士的艾恩西德伦修道院（4）、伊德富的何露斯神庙（5）和场长住宅（7）的建立，体现了这一从神圣到世俗的等级体系，这三者之间的差别是：艾恩西德伦修道院的神圣空间出现在两处，而何露斯神庙和场长住宅则出现在某一方向的尽端。其中最后一个例子还表现了剖面上的等级体系。在理查德医学研究中心（6），等级体系是从集中的设备空间到单个的设备空间再到非设备空间的进展过程。希思科特住宅（8）的立面的地位次序是根据对中心的距离远近来安排的，而在朗香教堂（9），等级体系是随着高度和门窗孔洞的复杂程度的变化而变化的。

1. **GUILD HOUSE**
   ROBERT VENTURI
   1961
2. **TEMPLE IN TARXIEN, MALTA**
   ARCHITECT UNKNOWN
   2100 B.C.–1900 B.C.
3. **BOYER HALL OF SCIENCE**
   GBQC
   1970–1972
4. **HOUSE OF THE FAUN**
   ARCHITECT UNKNOWN
   2nd CENTURY B.C.
5. **HOUSE IN CENTRAL PENNSYLVANIA**
   HUGH NEWELL JACOBSEN
   1980
6. **HOLY TRINITY UKRANIAN CHURCH**
   RADOSLAV ZUK
   1977
7. **SOUTH PLATFORM AT MONTE ALBAN**
   ARCHITECT UNKNOWN
   c.500
8. **MOORE HOUSE**
   CHARLES MOORE
   1962
9. **PAZZI CHAPEL**
   FILIPPO BRUNELLESCHI
   1430–1461
10. **WOODLAND CHAPEL**
    ERIK GUNNAR ASPLUND
    1918–1920
11. **THE PALACE OF ASSEMBLY**
    LE CORBUSIER
    1953–1963
12. **FALLINGWATER**
    FRANK LLOYD WRIGHT
    1935

转换是指一种表现特征在限定范围内的逐级变化。在老年人公寓（1）中，墙体在建筑的一条边上是简单的形状，逐级变换到另一边的复杂形式。马耳他的塔尔欣神庙（2）、博耶科学馆（3）、"农牧神"住宅（4）和中宾州住宅（5）都是大小尺寸转换的例子。在乌克兰圣三一教堂（6）、阿尔邦山南坛（7）和奥林达的穆尔住宅（8）也是如此。巴齐礼拜堂（9）、伍德兰礼拜堂（10）、议会大厦（11）、弗兰克·劳埃德·赖特的流水别墅（12）等实例中都表现了从敞开到封闭的演变。

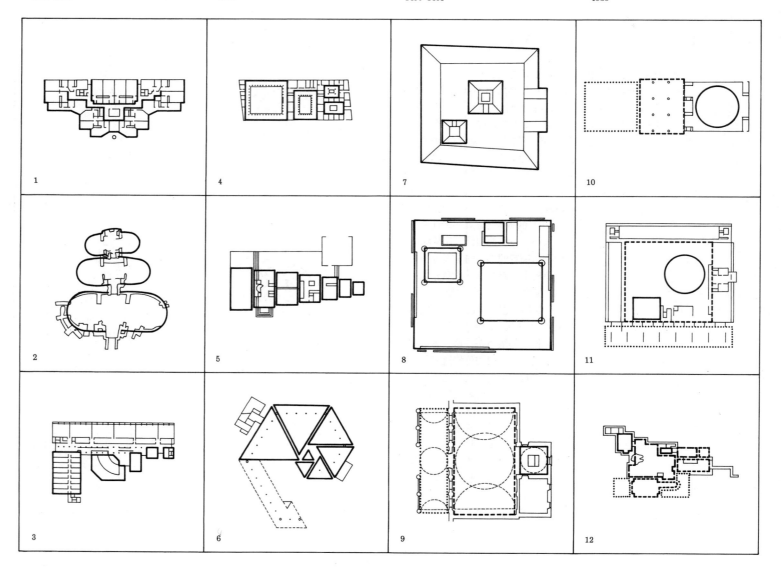

变形是从一种形体到另一种
不同形体的逐级变化。圣洛伦
佐教堂（1）、丰特夫拉罗拜修道院
（2）、哈德良离宫的水中央圆亭
（3）、孔索拉齐奥内圣玛丽亚教堂
（4）和达卡国民议会大厦（5）都
是同心式变形的范例。在这些建
筑中，中间的形体通过一连串的
变化，演变到周边的另一种形体。
在东京圣玛丽主教堂（6）和菲尔
米尼－韦尔教堂（7）中，从底
层到顶部垂直方向上出现了变形。
东京圣玛丽主教堂是从菱形变到
十字形，菲尔米尼－韦尔教堂是
从正方形到圆形。利斯特县法院
（8）和成人学习研究实验室（9）
的例子说明重要部位在建筑内部
由外到里的变化。方向和相邻的
形体的变化出现在卡尔斯基尔希
教堂（10）。在奥斯蒂亚浴室（11）
和蒙莫朗西旅馆（12）中可以看
到相邻单元的形状变化。

1. **CHURCH OF SAN LORENZO**
   GUARINO GUARINI
   1666–1679
2. **FONTEVRAULT ABBEY**
   ARCHITECT UNKNOWN
   1115
3. **HADRIAN'S MARITIME THEATER**
   ARCHITECT UNKNOWN
   125–135
4. **SAN MARIA DELLA CONSOLAZIONE**
   COLA DA CAPRAROLA
   1508
5. **NATIONAL ASSEMBLY IN DACCA**
   LOUIS I. KAHN
   1962–1974
6. **ST. MARY'S CATHEDRAL**
   KENZO TANGE
   1963
7. **CHURCH AT FIRMINY-VERT**
   LE CORBUSIER
   1963
8. **LISTER COUNTY COURTHOUSE**
   ERIK GUNNAR ASPLUND
   1917–1921
9. **ADULT LEARNING LABORATORY**
   ROMALDO GIURGOLA
   1972
10. **KARLSKIRCHE**
    JOHAN FISCHER VON ERLACH
    1715–1737
11. **BATHS AT OSTIA, ITALY**
    ARCHITECT UNKNOWN
    c.150
12. **HOTEL DE MONTMORENCY**
    CLAUDE NICHOLAS LEDOUX
    1769

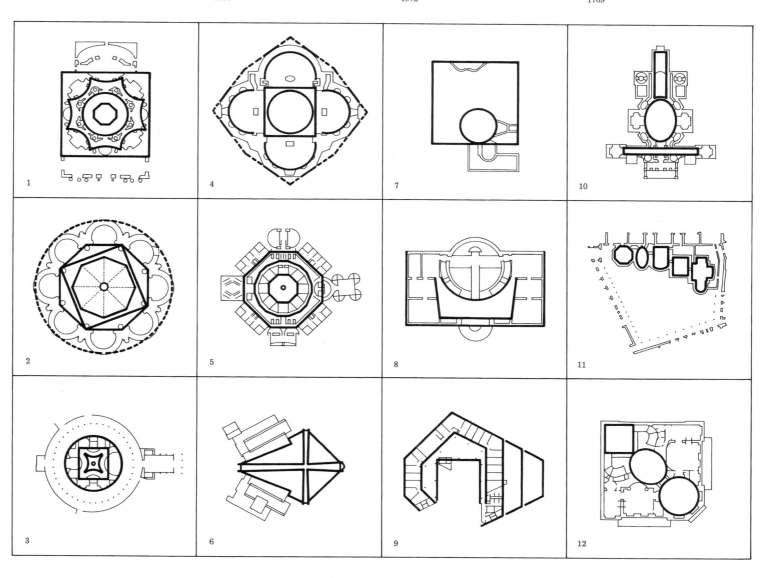

315　　中　介

　　中介，是指在建筑本身界线以外的两个环境之中插入的某种形体。这种演进方式一般出现在两个自然环境之间，或一个自然的和一个建筑形体之间，或两个建筑之间。王室官邸（1）、尤拉姆楼（2）、芬兰的阿拉耶尔维市政厅（3）、爱伦艺术中心（爱伦美术馆附属建筑）（4）和 AIA 总部（5）的中介体都是为了协调建筑环境中的现状而设计的。周末别墅（6）插入在两个自然环境中，一个是水平方向的水流、一个是垂直方向的树木。"雅典娜"游客中心（7）、特里迪弗林公共图书馆（8）和阿尔托所做在伊马特拉的伏克塞涅斯卡教堂（9）都是插入在一个自然形态和一个建筑形体中间。"雅典娜"处在弧形的河流和方格网形的城镇之间。在特里迪弗林图书馆，建筑是插入在一个以树为标志的点和正方格的建筑环境之间。伊马特拉的教堂的设计中，是把中介建筑插在另外的建筑和树木的自然延续之中。

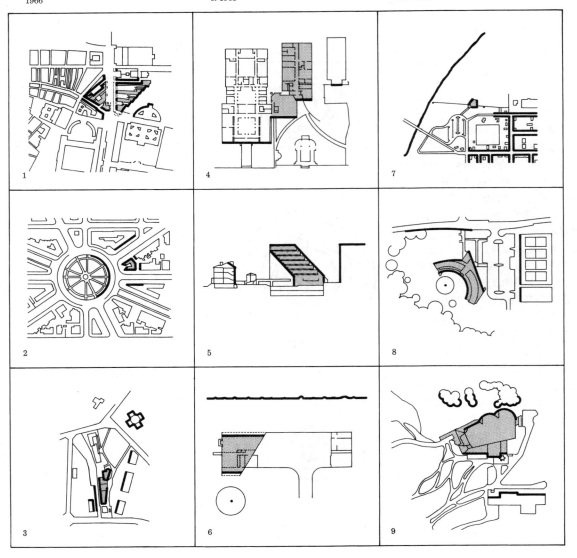

316 **缩　　小**

　　缩小是指建筑的整体或重要部分的缩小。尺度减小后的组成体可以包含在整体建筑内作为一个部分，也可以是另一个单体加到原来的形体一起。

大加小式的缩小

　　缩小的形体通常是作为辅助性用房的，例如萨吕泰兴府邸（1）、肖德汉别墅（2）、秀巧社大楼（5）、斯内尔曼住宅（6）、罗比住宅（7）和孔利住宅（8）。联合教堂（9）的缩小的形体也是辅助性的，但它的缩小形体出现在立面上。悉尼歌剧院（3）、歌德院 I（4）、穆默斯剧院（11）、伍德兰火葬场（12）、凡·布伦住宅（13）和沃尔夫斯堡教区中心大楼（14）等建筑中的缩小形体都有相近的用途。大加小的缩小型式中的每个尺度不限于仅有一个形体。蒙特城堡（10）是好多个小单元加在原来形体上的一个范例。运用缩小构思的很有趣的例子还包括在扩建设计中，在克拉格霍恩住宅（15）中的扩建部分是原有建筑的缩小，还有在阿尔托所做的赛于奈察洛市政中心（16）中会议厅（Council Chamber）的设计就是整个建筑的缩小。

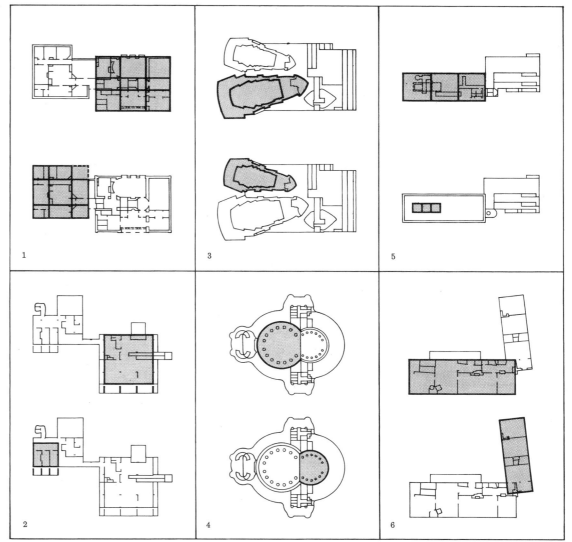

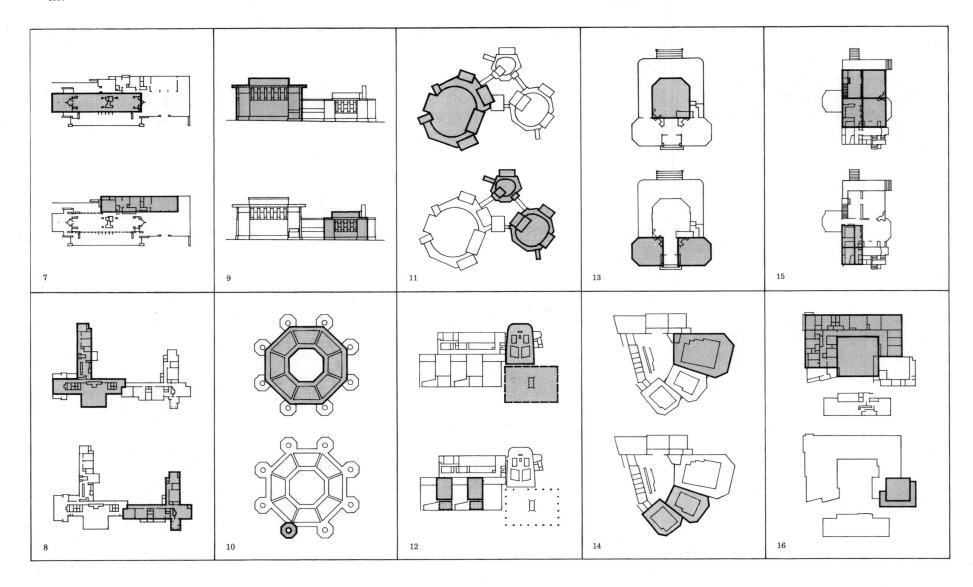

7

9

11

13

15

8

10

12

14

16

1. **EASTON NESTON**
NICHOLAS HAWKSMOOR
c. 1695–1710
2. **THE SALUTATION**
EDWIN LUTYENS
1911

3. **STRATFORD HALL**
ARCHITECT UNKNOWN
1725
4. **BANK OF PENNSYLVANIA**
BENJAMIN HENRY LATROBE
1798–1800

5. **ERDMAN HALL DORMITORIES**
LOUIS I. KAHN
1960–1965
6. **ALLEGHENY COUNTY COURTHOUSE**
HENRY HOBSON RICHARDSON
1883–1888

7. **OLD SACRISTY**
FILIPPO BRUNELLESCHI
1421–1440
8. **LANDERBANK**
OTTO WAGNER
1883–1884

318 整体缩小为局部

　　在伊斯顿·内斯顿府邸（1）、萨吕泰兴府邸（2）、斯特拉特福大厅（3）和宾夕法尼亚银行（4）中，主要房间、空间或空间组合体是整个建筑的缩小图形。在布林莫尔的厄德曼宿舍楼（5）、阿勒格尼县法院（6）和老年人公寓（14）也是同样情况。在老圣器堂（7）和兰德银行（8）中，其中的局部，一个祭坛和一个主要的楼梯，分别都是各自建筑的最重要空间或形体的缩小。基督教堂（9）和圣克莱门·达内斯教堂（10）与此类似，其中以柱子围成的相邻空间是建筑和塔楼的缩形。穆尔住宅（11）中两个神龛反映了整体，而在圣玛丽主教堂（12），中殿缩小成一个更小的穹顶和一个邻接的空间。在希思科特住宅（13），住所面向花园一侧的平面形状缩小而成为入口一侧的形状。在帕提农神庙（15）中，由墙体和柱子形成的空间互相交替反复地缩小（外圈柱子形成的空间缩小为中圈的墙，中圈的墙的空间又缩小成内圈柱子的空间——译者注）。汉泽尔曼住宅（16）和它的前院的虚实结合的外形被缩小后成为主要的使用空间。在矢野住宅（17）中的平面缩小而构成剖面的一部分。而在塔克住宅（18），外立面的缩小形成内部的壁炉。

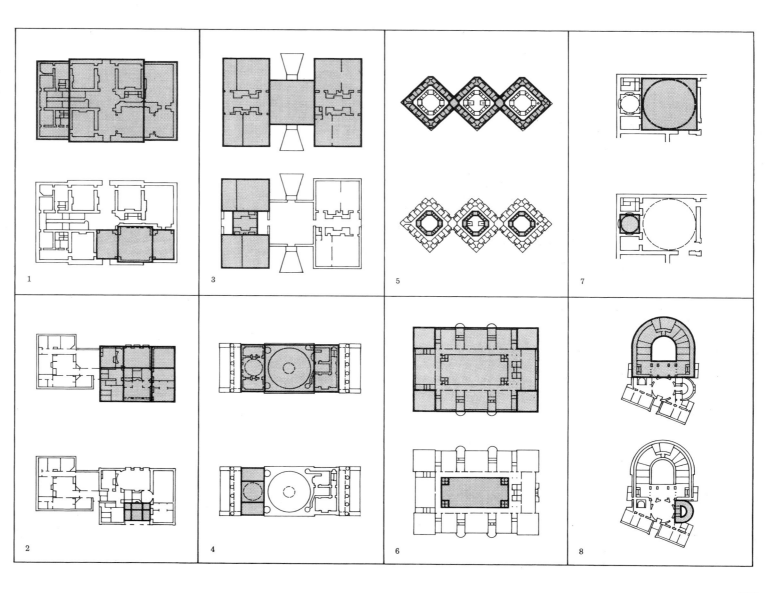

9. **CHRIST CHURCH**
NICHOLAS HAWKSMOOR
1715–1729

10. **ST. CLEMENT DANES**
CHRISTOPHER WREN
1680

11. **MOORE HOUSE**
CHARLES MOORE
1962

12. **ST. MARY'S CATHEDRAL**
BENJAMIN HENRY LATROBE
1814–1818

13. **HEATHCOTE**
EDWIN LUTYENS
1906

14. **GUILD HOUSE**
ROBERT VENTURI
1961

15. **PARTHENON**
ICTINUS
447 B.C.–430 B.C.

16. **HANSELMANN HOUSE**
MICHAEL GRAVES
1967

17. **YANO HOUSE**
ARATA ISOZAKI
1975

18. **CARLL TUCKER III HOUSE**
ROBERT VENTURI
1975

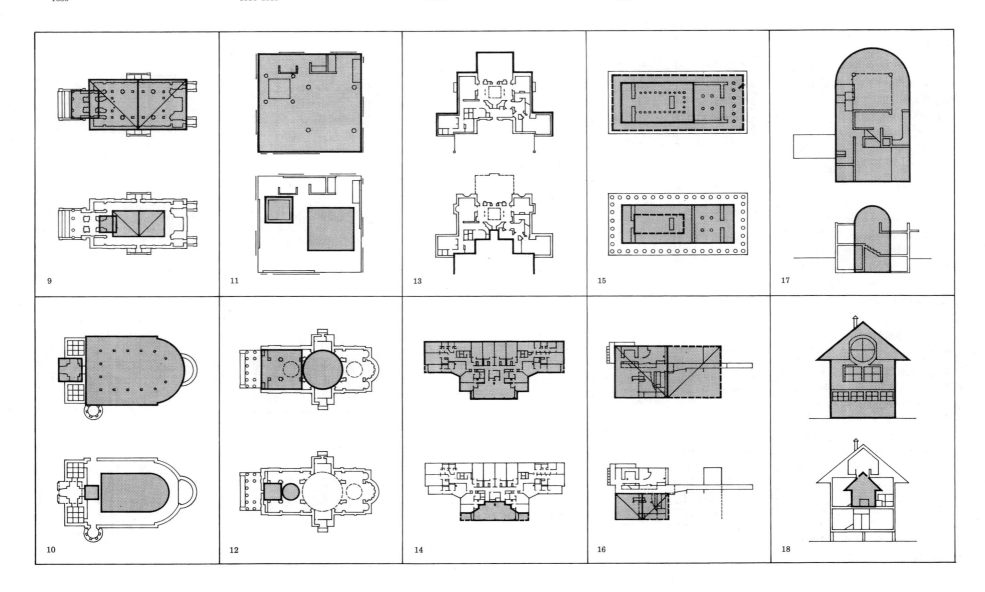

索 引 *

建 筑 师 索 引

本书中的资料重复编列在两种索引中：一种是以建筑师为条目的索引，另一种是以常用建筑物名称为条目的索引。建筑师索引中包括我们所知的每个人的生活年代、该建筑师所设计的建筑，在建筑名称之后是每个建筑的产生年代，最后是它在本书中的页码。

---

* 中译本只译出建筑师索引。索引中的页码均为英文版原书页码，我们将其作为边码标注在正文页边，便于读者查阅、检索。——编者注

Index

The information in this book has been indexed twice; by architect and by common building name. The index by architect includes the life dates of the person when known, the buildings by that architect that are included in this volume, and the dates of those buildings followed by the page number.

INDEX BY BUILDING
建 筑 物 索 引